AF443487

SURFACE DISORDERING:
GROWTH, ROUGHENING, AND PHASE TRANSITIONS

LES HOUCHES SERIES

Series Editor: N. Boccara

Hadronic Physics with Multi-GeV Electrons
B. Desplanques and D. Goutte (Eds.) (ISBN 1-56072-008-5).

**Coherent Detection Techniques at Millimeter Wavelengths
and their Applications**
*P. Encrenaz, C. Laurent, S. Gulkis, E. Kollberg and G. Winnesser (Eds.)
(ISBN 1-56072-012-3).*

The Physics of Granular Media
D. Bideau and J.A. Dodds (Eds.) (ISBN 1-56072-017-4).

Complex Dynamics
R. Livi, J.-P. Nadal and N. Packard (Eds.) (ISBN 1-56072-018-2).

Infrared Astronomy with ISO
Th. Encrenaz and M.F. Kessler (Eds.) (ISBN 1-56072-078-6).

Surface Disordering: Growth, Roughening, and Phase Transitions
R. Jullien, J. Kertész, P. Meakin and D.W. Wolf (Eds.) (ISBN 1-56072-098-0).

Centre de Physique

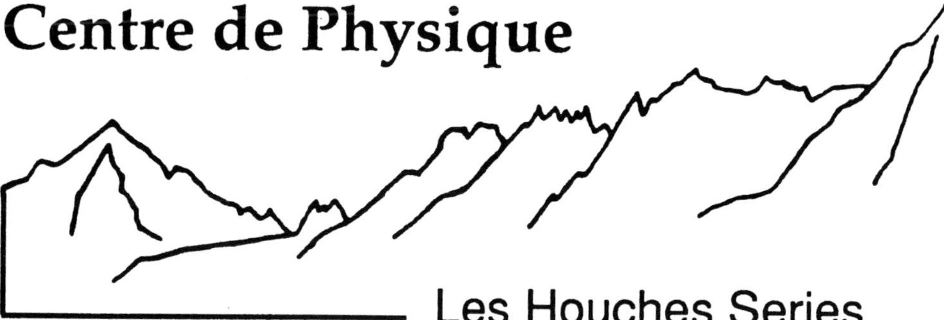

Les Houches Series

Proceedings of the Workshop on

SURFACE DISORDERING:
GROWTH, ROUGHENING, AND PHASE TRANSITIONS

Edited by
R. Jullien, J. Kertész,
P. Meakin, and D.E. Wolf

NOVA SCIENCE PUBLISHERS, INC.

Nova Science Publishers, Inc.
6080 Jericho Turnpike, Suite 207
Commack, New York 11725

Book Production Manager: Janet Glanzman White
Graphics: Elenor Kallberg, Christopher Concannon,
and Michael A. Masotti

Library of Congress Cataloging-in-Publication Data
available upon request

ISBN 1-56072-098-0

Printed in the United States of America

CONTENTS

PREFACE

Surfaces and interfaces play an essential role in many advanced technological processes and they are, at the same time, in the focus of basic research. After a coherent picture of *equilibrium roughening transitions* was reached, increasing activity has concentrated on the structure and properties of *growing surfaces*. This field was initiated by the application of the concepts of *dynamical scaling* and *fractal geometry* to irreversible surface growth and resulted in the discovery of new dynamical universality classes, where the surface roughness is described by exponents different from the equilibrium ones. Much of the progress has been due to extensive numerical studies of discrete models and their description in terms of continuum Langevin-type equations.

Until about two years ago there was little experimental activity on kinetic (or nonequilibrum) roughening of surfaces. Theoreticians could enjoy a comfortable dominance of the subject, and having reached a quite sophisticated level of theoretical understanding many of us did not expect dramatic new challenges in this field. When we set out organizing this workshop it was our primary aim to change this situation. At the time of the workshop the dramatic burst of vitality brought to this field of research by a variety of recent experiments ranging from molecular beam epitaxy (employing the latest techniques, surface science has to offer) to all kinds of "paper experiments" (wetting, burning, tearing, mostly performed by theoreticians) was obvious. The realization that surface roughening cannot be avoided during growth processes in general provides the common basis for a quite unique joint effort of theoretical, computational and experimental physicists to understand the emergence of nonequilibrium surface morphologies.

The matching between theory and experiment is not nearly as satisfactory as in the case of equilibrium surface phenomena. More experiments are needed which allow to extract by comparison whatever universal features there might be, and it is not clear to what extent the theoretical models can be applied to experimental findings. It became apparent at the meeting that more attention should be paid to other characteristics than just scaling exponents, too, such as *amplitudes and distributions* of statistical quantities, since quite distinct models and experimental systems lead to exponents that may be indistinguishable in practice. A quantitative understanding of *crossovers* and of possible surface *instabilities* is needed. The *nature of the disorder* (or "noise") is a crucial ingredient of any conclusive interpretation of experimental results. These keywords will be found in most of the talks throughout the chapters of these proceedings.

Let us mention just two episodes where the open and stimulating discussions have led to a new thinking about these aspects: J. Villain, summarizing some of the theoretical and experimental talks of the first week, pointed out that certain experiments compare favorably with the theory if one only considers the exponents characterizing the roughening of growing surfaces, but that in some of these cases the roughness amplitudes seem to be surprisingly large. A few days later J. Krim responded to this puzzle presenting her experimental point of view, which in turn was questioned e.g. by D. Eaglesham and J. Lapujoulade in the subsequent discussion. Similarly in his talk S. Balibar showed that some theoretical predictions are orders of magnitudes too weak to be measurable in his experiments on Helium crystals growing from the melt. This has challenged theoreticians like T. Nattermann or D. Wolf to estimate their phenomenological parameters in order to specify the material conditions where the different phenomena will show up.

Many of the arguments and ideas of these discussions will be found in the first three chapters of this book dealing with roughening of solid surfaces. The first one presents the state of the art from theoretical point of view. The second Chapter contains experimental results on kinetic roughening and their interpretations. The third one reviews what is known experimentally about the microscopic processes relevant for kinetic roughening.

In Chapter 4 the concept of kinetic roughening is shown to be useful for phenomena ranging from wetting to burning. Little is known about the microscopic background of such experiments in contrast to the crystal growth above. A phenomenological characterization of the noise is therefore at the focus of most of the contributions. The short abstract due to a multinational collaboration at the end of this Chapter deserves special attention!

Chapter 5 contains the latest results on directed polymers in a random medium, a problem intimately related to surface growth of the Kardar-Parisi-Zhang type. Here, one of the main aspects are universal quantities other than exponents.

The geometric scaling of random rough surfaces provides an important ingredient for the understanding of their physical and chemical behavior. Rough electrodes, surface catalysis or scattering by rough surfaces are examples. The present theoretical understanding is reviewed in the last Chapter, but high quality experimental results are, for the most part, lacking. The interdisciplinary aspect of surface growth phenomena is finally highlighted by beautiful experiments on the growth of bacterial cultures.

At the end of the workshop many of the participants felt that these two weeks had opened new perspectives for their research. What can be more satisfying for the organizers than this? We hope that these Proceedings will become a valuable reference and basis for further progress in the field.

It remains to thank the Commission of the European Communities, the CNRS, the DRED, and the DRET, without whose financial support the workshop would not have been possible. We are also grateful for the support provided by the Daimler-Benz AG and the Siemens AG in Germany. Thanks are due to Professor Nino Boccara, the Director of the Les Houches Organization at the Ecole de Physique and to the secretaries Birgitte Rousset and Daniele Choupin for their kind efficiency as well as to the restaurant manager Michelle Vittorelli.

Les Houches, April 1992

Rémi Jullien,	János Kertész,	Paul Meakin,	Dietrich E. Wolf
Université	Universität	Du Pont	IFF/KFA
Paris Sud	zu Köln and	Wilmington	Jülich and
	Inst. Techn. Phys.		Universität
	Budapest		Duisburg

KINETIC ROUGHENING OF SOLIDS I:

Theory and Computer Simulations

FROM MICROSCOPIC PARAMETERS TO MACROSCOPIC PROPERTIES AND CONVERSELY

Alberto Pimpinelli[*][†] **and Jacques Villain**
S.P.S.M.S./M.D.N., C.E.N.G., 85X, F-38041 Grenoble Cédex, France
**Laboratoire de Spectrométrie Physique, UJF, B.P. 87, F-38042 Saint-Martin-D'Hères Cedex, France*
†Dipartimento di Fisica, Università di Parma, I-43100 Parma, Italy

Abstract. We identify and describe in this note the microscopic parameters more relevant for surface growth and their relation to mesoscopic or macroscopic quantities, which are more easily accessible to measure. We explicitly show, for instance, how it is possible to determine the coefficient of diffusion of gradatoms along the step by investigating the shape of islands at low temperature, and how to know the rate of emission of gradatoms from kinks by observing the evaporation or the growth of stepped surfaces.

1. The microscopic parameters most relevant for surface physics

Experimental abilities in crystal growth (MBE, STM, ...) have by now left adolescence [1] and are steadily heading towards full maturity. The field is therefore reaching the stage where interactions between theory and applications are not only welcome, but indeed necessary.

These interactions involve different lenghtscales, ranging from first–principle quantum mechanical calculations to macroscopic considerations of non–equilibrium thermodynamics.

In the present note some of the information on microscopic features, which may be obtained from macroscopic or mesoscopic observations will be reviewed. We will consider the case of a high–symmetry surface (*i.e.* (001) or (111)) of a metal or a semiconductor at not–too–high temperatures, which means that one can find steps with well defined orientations, which change as little as possible in time. Our aim is to characterize the dynamics of atoms adsorbed on the surface (adatoms) in terms of a limited number of microscopic parameters, and to describe some of the experimental methods which allow a knowledge of these parameters or of combinations of them.

In the simplest cases, these microscopic quantities are (see Fig.1) (all lengths are in units of the interatomic spacing):

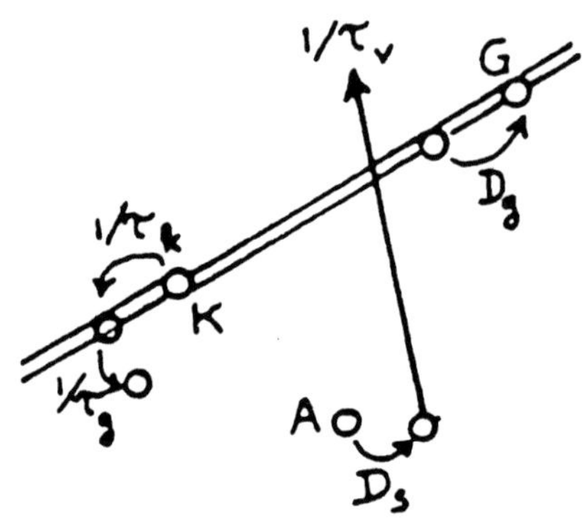

Fig.1 Artist's view of the microscopic parameters described in the text.

a) the surface diffusion constant D_s of an isolated adatom;
b) the line diffusion constant D_g along a step of a mobile atom adsorbed at the step (gradatom);
c) the rate of evaporation from the surface (desorption) $1/\tau_v$ of an isolated adatom;
d) the escape rate $1/\tau_g$ of a gradatom from a step;
e) the detachement rate $1/\tau_k$ of a bound gradatom from a kink, the gradatom becoming mobile along the step. We shall in general consider the steps with lowest energy per atom.

These quantities are all associated to activation energies. Arrhenius forms are commonly employed, although it is not clear that the prefactors should be independent of temperature (see for instance [2–4]). The following formulae define the activation energies:

$$D_s = D_s^o \exp(-\beta W_{sd}), \qquad D_g = D_g^o \exp(-\beta W_{gd})$$

$$\tau_v = \tau_v^o \exp\left(\beta \widetilde{W}_v\right), \qquad \tau_g = \tau_g^o \exp\left(\beta \widetilde{W}_g\right), \qquad \tau_k = \tau_k^o \exp\left(\beta \widetilde{W}_k\right)$$

Tildas indicate potential barriers. For instance, the energy difference between an adatom and a free atom in the vapor will be denoted as W_v. If a potential barrier has to be overcome, $\widetilde{W}_v > W_v$. $\beta = 1/k_B T$, as usual.

Escape rates are related by detailed balance to the reverse processes. For instance, an adatom near a step has a probability equal to $dt/\tilde{\tau}_g$ of being absorbed by the step during the time dt ; detailed balance implies

$$\tau_g = \tilde{\tau}_g \exp(\beta W_g).$$

Similarly, a gradatom next to a kink has a probability per unit time $1/\tilde{\tau}_k$ of being captured by the kink, with

$$\tau_k = \tilde{\tau}_k \exp(\beta W_k).$$

W_g <u>must</u> be different from $\widetilde{W}_g$. In fact, W_g is the difference in energy between an adatom and a gradatom, while $\widetilde{W}_g$ contains in addition a diffusional barrier. One may then expect

$$\widetilde{W}_g = W_g + W_{sd}.$$

In addition to these energies (or, rather, free energies) which describe the dynamics, there are energies associated with defects. Creating a kink on a step, for example, has a free energy cost W_0. The surface energy per atom W_s and the step energy per atom W_1 , will not play an important role in the following.

It is usual [5] and useful to know the values of the elementary energies obtained from simple geometrical models (broken bond models).

Consider a compact step on a compact face of an FCC Bravais lattice. If we assume that all broken interatomic links have an energy cost W_b , we can find estimates for various important energies, some of which we have previously introduced. These are:

the cohesion energy $W_{coh} = 6W_b$;
the surface energy $W_s = 3W_b/2 = W_{coh}/4$;
the energy of a step $W_1 = W_b$;
the energy of a kink $W_0 = W_b/2$.

The activation energy of a detachment process from a kink W_k , which requires the breaking of one bond, is equal to W_b ; the analogous quantities are $W_g = 2W_b$ and $W_v = 3W_b$.

Another important example is the (001) face of Si (diamond lattice); when only nearest–neighbour interaction are included,
$$W_{gd} = W_{sd} = \widetilde{W}_k = W_{coh}/2 = W_b ; \qquad W_k = 0.$$
Unfortunately, the latter estimates seem to be very far from reality.

2. Macroscopic properties

The cohesion energy of a crystal may be obtained considering that it is in principle possible to sublimate the whole crystal by sublimating kinkatoms. The normal procedure would be to convert the kinkatom into a gradatom, the gradatom into an adatom and finally to evaporate it. We thus find

$$W_{coh} = W_v + W_g + W_k \tag{1}.$$

Let now the macroscopic evaporation rate per site $1/\tau_v^{\mathrm{mac}}$ be evaluated. The evaporation rate per adatom is $1/\tau_v$, but the probability of finding an adatom at a given surface site cannot be larger than the equilibrium value ρ_1^s. Therefore, the maximum value of $1/\tau_v^{\mathrm{mac}}$ is

$$1/\tau_v^{\mathrm{mac}} = \rho_1^s/\tau_v \tag{2},$$

as we will show at the end of this Section.

The adatom concentration per adsorption site at equilibrium is $\rho_1^s = \exp(-\beta W_{ad})$, where W_{ad} is the energy needed to increase by one unity the number of adatoms, without any other modification of the surface. This operation may be performed by transforming a kinkatom into a gradatom and then into an adatom, since this procedure does not change the number of kinks. Therefore $W_{ad} = W_k + W_g$ and

$$\rho_1^s = \exp(-\beta W_k - \beta W_g) \tag{3}.$$

At equilibrium with the vapor, desorption must be compensated by particle adsorption. Let the sticking coefficient (ratio of adsorbed to incoming atoms) be equal to 1. If ρ is the number density of atoms in the vapor phase, the arrival rate per adsorption site is given by

$$1/\tau_{inc} = \rho \int_0^\infty dv_z p(v_z) v_z$$

where $p(v_z)dv_z$ is the probability that the z component of the velocity takes values between v_z and $v_z + dv_z$. If we allow for a potential barrier Δ at the surface, so that $\widetilde{W}_v \approx W_v + \Delta$, due to the conservation of energy and momentum we must require $\frac{1}{2}mv_z^2 \geq \Delta$. The particles of given velocity v_z hitting the surface in the (small) time interval δt, occupy a volume of thickness $v_z \delta t$; since the probability density p is the Maxwell distribution, one finds

$$1/\tau_{inc} = \rho\sqrt{\tfrac{\beta m}{2\pi}} \int_{\sqrt{2\Delta/m}}^\infty dv_z \exp\left(-\beta m v_z^2/2\right) v_z = \rho\sqrt{\tfrac{1}{2\beta m\pi}} e^{-\beta\Delta} \tag{4}.$$

Since the sticking coefficient has been assumed equal to 1, $1/\tau_{inc}$ must equal the evaporation rate $1/\tau_v^{\mathrm{mac}} = \rho_1^s/\tau_v$; from the definition of τ_v and from Eqs.(3) and (4) follows

$$\rho(2\beta m\pi)^{-\frac{1}{2}} e^{-\beta\Delta} = 1/\tau_{inc} = 1/\tau_v^{\mathrm{mac}} = 1/\tau_v^o e^{[-\beta(W_k + W_g + W_v + \Delta)]} \tag{5}.$$

The Δ's in both terms cancel, and the saturated vapor density is therefore

$$\rho_{sat} = \tfrac{1}{\tau_v^o}\sqrt{2\pi\beta m}\, \exp\left[-\beta\left(W_k + W_g + W_v\right)\right] \tag{6}.$$

Note that the energy between brackets in the exponential, is just the cohesion energy.

The microscopic evaporation rate of adatoms $1/\tau_v$ is not easily observed experimentally. Instead, the macroscopic evaporation rate is the generally measured quantity.

Similarly, the diffusion constant D_s of adatoms is difficult to measure. For instance, one could deposit a monolayer of radioactive tracers on a half–surface [6] and then observe how it spreads. One is thus measuring a coefficient $\tilde{D}_s$ defined by

$$\dot{\rho}_a = \tilde{D}_s \nabla^2 \rho_a$$

where ρ_a is the density of radioactive atoms. If penetration into the bulk is neglected, the tracers can only diffuse if they become adatoms, or if there is an "advacancy" nearby. Therefore

$$\tilde{D}_s = D_s \rho_1^s + D_{\mathrm{adv}} \rho_{\mathrm{adv}}$$

where ρ_{adv} is the surface density of advacancies and D_{adv} is the corresponding diffusion constant.

Essentially the same quantity is measured in other experiments such as those where the smoothening of an artificially corrugated profile is observed, at sufficiently low temperature. At higher temperature, evaporation will dominate. The crossover temperature depends on the wavelength λ of the profile. If advacancies are neglected, adatoms have to diffuse on a length of order λ, and this takes a time λ^2/D_s. During this time, the probabiity of evaporation is $\lambda^2/(D_s \tau_v)$. Therefore evaporation can be neglected if

$$\lambda^2 \ll D_s \tau_v.$$

Note that $D_s \tau_v = \tilde{D}_s \tau_v^{\mathrm{mac}}$. This is consistent with the fact that only $\tilde{D}_s$ and τ_v^{mac} are physically important in most of macroscopic experiments. The same conclusion also holds for the growth of stepped surfaces, as long as the interaction between adatoms is negligible [5]. In fact, the surface density $\rho_1^s(x, l)$ of adatoms between two parallel steps at distance ℓ satisfies the equation

$$\dot{\rho} = F + D_s \rho'' - \rho/\tau_v.$$

The motion of steps is slow in comparison with adatom diffusion, so that $\dot{\rho}$ may be replaced by 0. Choosing the origin midway between the steps and defining $\kappa = 1/\sqrt{D_s \tau_v}$, the solution is

$$\rho_1(x) = F\tau_v + (\rho_1^s - F\tau_v)\cosh(\kappa x)/\cos(\kappa \ell/2)$$

The quantity that can be actually measured is the step velocity, given by the sum of two terms, one for each adjacent terrace, of the form

$$D_s \rho_1'(x/\ell) = \kappa D_s (\rho_1^s - F\tau_v)\tanh(\kappa \ell/2) \tag{7}$$

Again this depends only on $D_s \rho_1^s$ and ρ_1^s/τ_v. Note that equation (7) also holds for evaporation, when $F = 0$. Then, if κ is small, as it is often the case, Eq.(7) reduces to Eq.(2). However, it will now be seen that also D_s can be measured when $\rho_1(x)$ becomes large enough, so that the growth equation is no longer linear.

3. Density of islands in MBE growth

D_s can in some cases be directly measured by field ion microscopy (FIM) [7], but microscopic methods are outside the scope of the present work. As remarked by Mo *et al* [8], the diffusion coefficient can be deduced, at low temperature, from macroscopic – or rather mesoscopic – observations. In fact, at temperatures where a pair of adatoms (dimer) is stable and immobile, and it is thus the critical nucleus for island growth, simple random–walk arguments show that the average distance ℓ between steps (related to the island density N_s through $N_s \approx 1/\ell^2$) is given in order of magnitude by [8,9]

$$1/N_s = \ell^2 \approx (D_s/F)^{\frac{1}{3}} \tag{8}$$

where F is the beam intensity in atoms per unit time and unit adsorption site. As always, lengths are given in unit of the atomic distance.

Since F is known, the measurement of the island density at constant beam intensity gives the order of magnitude of D_s. Numerical simulations are actually necessary for a quantitative determination of D_s in this range.

This program has been carried out by Mo *et al* [8] who have measured by STM the number density of Si islands grown on the (001) face of Si; in this case a complication is introduced by the important anisotropy in adatom surface diffusion.

On the other hand, Ernst *et al* [10] have performed similar measurements on Cu (001) by diffraction, finding $\ell \sim (D_s/F)^{1/4}$ in a situation where surface diffusion should be isotropic. This is in agreement with Irisawa *et al* [11], who derived the same formula both by numerical simulations and by an analytical argument which we believe to be incorrect.

All previous cited works always assume compact islands. But are they really so? To decide this point reciprocal–space techniques are powerless, and recourse to tunnel microscopy or similar methods is unavoidable.

4. Shape of islands and steps

Actually, fractal islands have been observed in heteroepitaxial growth of Au on Ru(0001) by Hwang *et al* [12] at room temperature. The aggregates look like clusters grown with DLA, with dendritic arms about ten nanometers thick. Dendritic–looking islands have also been observed in homoepitaxial Pt on Pt(111) [13].

More generally, one can address the question of the shape of islands and steps. It will be shown that it can give information on D_g, τ_k and τ_g.

At long lengthscales and for low fluxes, one can discuss the stability of compact island shapes in the following way, inspired by Mullins and Sekerka: there is a balance between a destabilizing effect and various stabilizing ones.

a) Destabilizing effect: at least for <u>small</u> islands, the growth is due mainly to atoms coming from outside. Then, as in DLA, if a tip forms, new atoms will go preferentially to the tip.

b) Stabilizing effects: the dominant one at low temperature is diffusion of gradatoms along the step which constitutes the edge of the island. The flow tends to decrease the free energy, and therefore to smoothen the step. This effect is appreciable if D_g and $1/\tau_k$ are not too large. A similar effect is due to the reemission of adatoms from the tip, followed by their diffusion on the surface near the step. This effect is important if τ_g is not too large, and presumably is negligible at low temperature.

The continuous, quasi–equilibrium theory outlined above is no longer applicable if the distance between kinks on a step is much shorter than the equilibrium distance

$$r_{eq} \simeq \exp(\beta W_0).$$

W_0 has been measured on Cu (001) as being 850 K [10] at about 200 K and 1140 K [14] at room temperature. Therefore, for instance, $r_{eq} \simeq e^4 \simeq 50$ (in units of the interatomic distance) at 200 K, the temperature of the experiments of Ernst $et\ al$ [10]. This is only four times smaller than the characteristic length ℓ measured in the same experiments.

Let us now estimate the distance r_c between kinks due to growth. A freshly arrived gradatom needs to explore a distance of the order of r_c along the step before reaching a kink; this requires a time $\tau \approx r^2/D_g$. The gradatom density on the sides of the island is $\rho_1^g = \widetilde{F}\tau$, where $\widetilde{F}$ is the flux of adatoms arriving to the island per unit time and length: $\widetilde{F} = F\ell^2/r$. Therefore $\rho_1^g = F\ell^2 r/D_g$. Since gradatoms perform a 1–dimensional random walk, the nucleation rate (meeting frequency of two gradatoms per unit length) of germs on the sides is given by

$$1/\tau_{gn} = \rho_1^g/\tau \rho_1^g \sqrt{D_g} = (\rho_1^g)^2 \sqrt{D_g/\tau} = (F\ell^2/D_g)^2 r.$$

Requiring one nucleation event in lenght r during the completion of one new layer, $1/\tau_{gn} = \widetilde{F}/r$, we find

$$\frac{\widetilde{F}}{r} = \left(\frac{F\ell^2}{D_g}\right)^2 r$$

and therefore

$$r_c^3 = \frac{D_g}{F\ell^2}$$

since adatoms arriving near the corners of the island would have then an higher probability of forming new pairs than to meet existing kinks (Fig.2).

In the previous formula F is known and both r_c and ℓ are directly measurable with STM, so that it allows a measure of the diffusion constant of gradatoms D_g ; in the case of Au on Ru [12] the pictures taken at 300 K suggest $r_c \approx 10$ nm, $\ell \approx 100$ nm, with $F^{-1} = 30$ s, which imply $D_g \approx 10^6$ s^{-1}.

The previous very simple estimate only gives an order of magnitude of the length r_c , beyond which the island is no longer able to maintain a polygonal shape. It does not necessarily imply that further growth will be fractal or dendritic, being possible that the shape only becomes more or less circular. Numerical simulations by Xiao $et\ al$ [15] suggest that the instability in the polygonal shape actually coincides with the

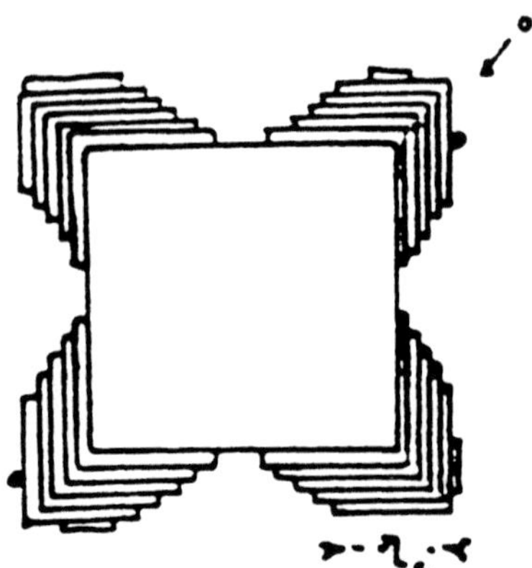

Fig.2 The shape of a low–temperature square island becoming unstable
as diffusing adatoms preferentially hit its corners.

appearance of dendrites. However, these authors made assumptions very different from ours; they introduced a sticking coefficient for adatoms on the step, which increases with temperature, and completely neglected line diffusion of gradatoms along the step itself. As a consequence, fractality increases on increasing the temperature! In our description, which we think is more appropriate to the present physical situation, it is just the opposite, since the ratio D_g/D_s , and therefore r_c , increases with temperature. Bales and Zangwill [16] in their study of step flow proposed a description very similar to that of Xiao *et al* , and they also did not take into account diffusion along the step edge.

5. Observation of a step by scanning tunneling microscopy

At room temperature, a step is not straight. This implies that between two scans the step has changed position. One can think of three cases:
  a) experimental artefact;
  b) presence of gradatoms;
  c) kink migration.

  Effect a) should be temperature independent. Effect b) should be weak at room temperature. We will examine effect c). Suppose that the step has a given fixed orientation, it is compelled to have some kink at a reciprocal distance ℓ_0 , probably shorter than the thermal distance. Each kink releases gradatoms at time intervals τ_k . Most gradatoms are reabsorbed by their parent kink. What matters is the probability $p(\ell_0)$ that a gradatom may diffuse to the next kink. It can be exactly evaluated, but we will give a simple argument. Ignoring the fact that each gradatom returning to its parent kink is reabsorbed, we can imagine the gradatom as diffusing back and forth, until it reaches the next kink in a time $t_1 \approx \ell_0^2/D_g$. During this time, the number of visits to the origin (the parent kink) is approximately equal to

$$\int_0^t \frac{D_g d\tau}{\sqrt{D_g \tau}} \approx \sqrt{D_g t} \approx \ell_0.$$

It means that the gradatom left its parent kink ℓ_0 times before reaching the next kink. Hence the success probability of each attempt is $1/\ell_0$. The time needed for a kink to displace of one atomic distance is therefore $\tau_k \ell_0$. We can define a kink diffusion coefficient as $D_k = 1/(\tau_k \ell_0)$. For it to cover a distance ℓ_0 , a kink needs a time $\ell_0^2/D_k = \tau_k \ell_0^3$. If such a time is shorter than the scanning time of a line (typically one second) the step will have a fuzzy appearance. Since ℓ_0 is known, we have thus a way to determine τ_k .

6. Conclusion

The connection between physical properties on different lengthscales is not only important in itself, but bears also a practical interest, since it allows the knowledge of microscopic parameters through the investigation of macroscopic, or, at least, mesoscopic quantities, which are in principle more directly accessible in experiments. We have reviewed some of these relations, and new ones we have explicited, which pertain to the dynamical properties of single atoms diffusing on high symmetry surfaces or along the steps on these surfaces.

REFERENCES

[1] G. Binnig and H. Rohrer, *Rev. Mod. Phys.* **59** 615 1987.
[2] H. A. Kramers,, *Physica* **7** 284 1940.
[3] G. Iche and P. Nozieres, *J. Physique* **37** 1313 1976.
[4] A. Zangwill *Physics at Surfaces* Cambridge University Press 1988.
[5] W. K. Burton, N. Cabreba and F. Frank, *Phil. Trans. Roy. Soc.* **243** 299 1951.
[6] J. Cousty, R. Peix and B. Perraillon, *Surface Sci.* **107** 586 1981.
[7] R. Gomer, *Rep. Prog. Phys.* **53** 917 1990.
[8] Y. W. Mo, J. Kleiner, M. B. Webb and M. G. Lagally, *Phys. Rev. Lett.* **66** 1998 1991.
[9] S. Stoyanov and D. Kaschiev *Current Topics in Material Science* Vol. 7 E. Kaldis ed., North Holland 1981.
[10] H. J. Ernst, F. Fabre and J. Lapujoulade,   preprint 1991.
[11] T. Irisawa, Y. Arima and T. Kuroda, *J. Crystal Growth* **99** 491 1990.
[12] R. Q.Hwang, J. Schröder, C. Günther and R. J. Behm, *Phys. Rev. Lett.* **67** 3279 1991.
[13] M. Bott, T. Michely and G. Comsa,   preprint 1991.
[14] J. Frohn, M. Giesen, M. Poensgen, J. F. Wolf and H. Ibach, *Phys. Rev. Lett.* **67** 3543 1991.
[15] R-F. Xiao, J. I. D. Alexander and F. Rosenberger, *Phys. Rev.* **A 38** 2447 1988.
[16] G. S. Bales and A. Zangwill, *Phys. Rev.* **B 41** 5500 1990.

Surface Disordering:
Growth, Roughening and
Phase Transitions

A CONTINUUM MODEL FOR KINETIC ROUGHENING WITH SURFACE DIFFUSION

Thomas Nattermann and Lei-Han Tang

*Institut für Theoretische Physik, Universität zu Köln,
Zülpicher Str. 77, D-5000 Köln 41, Germany*

Abstract. We extend a continuum model proposed by Villain [J. Phys. I (France) 1, 19 (1991)] to study equilibrium and nonequilibrium diffusion on a high symmetry surface under a fluctuating particle beam. Exponents characterizing dynamic scaling in various regimes are derived, as well as the relevant lengths which separate these regimes.

1. Introduction

In vacuum deposition experiments, epitaxial growth of a crystalline film is usually achieved through surface migration of newly arrived atoms to the energetically more favorable ledge and kink sites[1]. The morphology of the film surface is thus sensitive to the beam intensity which controls the nonequilibrium adatom population, and to the substrate temperature which influences the rate of surface diffusion. Even in the case of stable growth, one encounters different forms of kinetic roughness, ranging from mere terrace width fluctuations as in step-flow growth on a vicinal surface at elevated temperatures, to the nucleation of islands on islands which occurs at lower temperatures when the mobility of the surface atoms is much reduced[2,3].

Theoretical studies of driven interfaces have focused on the role of symmetry and conservation laws in determining dynamic universality classes[4-6]. These surfaces typically exhibit self-affine structures both in space and in time: the height fluctuation across a distance L in the surface scales as L^ζ, and has a lifetime proportional to L^z[7]. Here ζ and z are known as roughness and dynamic exponents, respectively. In molecular-beam epitaxy (MBE) the primary relaxation mechanism — surface diffusion — conserves the mass of the film. If the vacancy concentration in the film does not vary during growth, mass conservation translates to volume conservation described by a continuity equation[8-11]

$$\partial Z/\partial t + \Omega \nabla \bullet \mathbf{j} = F.$$

(1)

Here $Z(\mathbf{x}, t)$ is the film thickness above a substrate site $\mathbf{x}$ at time t, $\mathbf{j}$ is the particle number current density within the surface, Ω is the atomic volume, and F is the beam intensity which we take to be independent of Z. Equation (1) distinguishes surface diffusion from other relaxational processes such as desorption[12] or vacancy formation in the film[13]. The generic continuum description in the latter cases is given by the Kardar-Parisi-Zhang (KPZ) equation which is incompatible with the conservation law imposed by (1).

Villain[10] proposed a continuum model of the form (1) which includes nonequilibrium terms that are consistent with in-plane isotropy and a continuous translational symmetry in the growth direction. In this paper we consider an extension[14] of Villain's model which includes two additional terms: i) a lattice-pinning term which reduces the continuous translational symmetry to a discrete one; ii) a conserving component in the noise arising from the stochastic nature of surface diffusion on the atomic level. In the limit of vanishing beam intensity, our model corresponds to the equilibrium sine-Gordon model[15-18] with a conserving dynamics. As in the case of nonconserving dynamics[17], a finite growth velocity diminishes lattice-pinning on sufficiently large length scales, at which point the model proposed by Villain is recovered. On smaller length scales and below the thermal roughening temperature T_R of the equilibrium model, the lattice-pinning term may become dominant, leading to the formation of flat terraces on the surface.

2. The Model

For a sufficiently weak beam we write

$$\mathbf{j} = \mathbf{j}_{eq} + \mathbf{j}_{neq},$$

(2)

where $\mathbf{j}_{eq}$ is the (quasi)equilibrium part driven by the gradient of the chemical potential[19]

$$\mu = \Omega \delta \mathcal{F}/\delta Z,$$

(3)

and $\mathbf{j}_{neq}$ is the nonequilibrium correction to $\mathbf{j}_{eq}$. In the following we shall consider (3) on a semi-microscopic level by postulating a Ginzburg-Landau free energy of the form[15-17]

$$\mathcal{F} = \int d^d\mathbf{x}\left[\frac{\Gamma}{2}(\nabla Z)^2 + V(Z)\right],$$

(4)

where Γ is the surface stiffness constant, $V(Z + a) = V(Z)$ is a "lattice-pinning" potential which has its minima at integer multiples of the layer spacing a, and d is the *dimension of the surface*. Correspondingly, we take $\mathbf{j}_{eq}$ to be of the Langevin type and write, in accordance with the Nernst-Einstein relation,

$$\mathbf{j}_{eq} = -(\rho_s D_s/k_B T)\nabla \mu + \mathbf{j}_0,$$

(5)

where D_s is the surface diffusion coefficient, ρ_s is the surface density of the diffusing species, and $\mathbf{j}_0$ is the stochastic current due to thermal fluctuations, with

$$\langle j_{0i}(\mathbf{x}, t) j_{0k}(\mathbf{x}', t') \rangle = 2\rho_s D_s \delta_{ik} \delta^d(\mathbf{x} - \mathbf{x}') \delta(t - t').$$

The appropriate form for the nonequilibrium current $\mathbf{j}_{neq}$ in the context of MBE is less clear. $\mathbf{j}_{neq}$ may contain *new* terms which cannot be obtained from (3)-(5), and which govern dynamic scaling on sufficiently large length scales. For a high-symmetry surface, Villain[10] proposed the following expression

$$\mathbf{j}_{neq} = -(\nu/\Omega)\nabla Z - (\sigma/2\Omega)\nabla(\nabla Z)^2. \tag{6}$$

This choice is consistent with the translational symmetry $Z \to Z + a$ which excludes terms like ∇Z^2. Phenomenologically, the ν-term in (6) indicates a down (or up)-hill current for a surface tilted away from the high-symmetry direction. [1]Though absent in the equilibrium problem when gravity is negligible, such a current could be justified under the growth condition[10]. The σ-term cannot be obtained from a free energy and it also behaves differently from the ν-term under the transformation $Z \to -Z$[20]. Only stable growth ($\nu \geq 0$) will be considered here. [2]

To complete the definition of the model, we write $F = F_0 + f$, where f represents fluctuations away from an average beam intensity F_0. For simplicity, we shall also assume $V(Z) = -V_0 \cos(2\pi Z/a)$. The final equation of motion for Z takes the form

$$\frac{\partial Z}{\partial t} = \nu \nabla^2 Z - \gamma \nabla^4 Z + \frac{\sigma}{2}\nabla^2(\nabla Z)^2 + v\nabla^2 \sin(2\pi Z/a) + F_0 + \eta. \tag{7}$$

Here $\gamma = \rho_s D_s \Omega^2 \Gamma/(k_B T)$ and $v = 2\pi\rho_s D_s \Omega^2 V_0/(ak_B T)$. The noise $\eta = f - \Omega\nabla\mathbf{j}_0$ is assumed to be Gaussian distributed with zero mean and

$$\langle \eta(\mathbf{x}, t)\eta(\mathbf{x}', t') \rangle = 2\mathcal{D}\delta^d(\mathbf{x} - \mathbf{x}')\delta(t - t'), \tag{8a}$$
$$\mathcal{D} = D_0 - D_1 \nabla^2, \tag{8b}$$

where $D_0 \simeq \Omega F_0/4$ and $D_1 = \Omega^2 \rho_s D_s$.

3. The Linear Case

Equation (7) has sufficient complexity to allow for a number of different spatial scaling regimes which presumably also exist in experiments. To illustrate the point let us

---

[1] Villain pointed out to us that the coefficient ν may depend on $(\nabla Z)^2$. This dependence will be ignored in the present paper.

[2] Unstable (three-dimensional) growth takes place when ν becomes negative, see Ref. 10.

consider the solvable case $\sigma = v = 0$. Depending on the relative strength of ν and γ, two dynamical regimes are identified with

$$z = 4, \qquad L \ll L_\nu, \tag{9a}$$
$$z = 2, \qquad L \gg L_\nu, \tag{9b}$$

where $L_\nu = (\gamma/\nu)^{1/2}$ and L is the length of interest. A second length scale

$$L_f = (D_1/D_0)^{1/2} \sim (\Omega \rho_s D_s/F_0)^{1/2} \tag{10}$$

separates two scaling regimes of the noise such that

$$z = 2\zeta + d + 2, \qquad L \ll L_f, \tag{11a}$$
$$z = 2\zeta + d, \qquad\quad L \gg L_f. \tag{11b}$$

At low temperatures, where adatoms appear on the surface only due to the incoming flux, L_f has the meaning of the mean terrace size[21]. In this case L_f can be expressed as $L_f \simeq (\Omega/a)^{1/2}(D_s a^2/\Omega F_0)^{1/n}$. The value of n depends on details of the two-dimensional nucleation process on the terrace, and generally lies between 2 to 6[21].

The combination of (9) and (11) yields explicit expressions for ζ and z, applicable in respective regimes. For instance, if $L_f \ll L \ll L_\nu$, we have $\zeta = (4 - d)/2$, $z = 4$[9-11]. Asymptotically for $L \gg L_f, L_\nu$, we recover the well-known Edwards-Wilkinson (EW) result $\zeta = (2 - d)/2$, $z = 2$[22]. Other exponents are possible below L_f. In MBE, L_f may vary by many orders of magnitude due to the Arrhenius law $D_s \sim \omega_D \exp(-E_b/k_B T)$, where $\omega_D \sim 10^7$–10^{13}sec^{-1} is the phonon frequency and E_b is the activation energy for nearest-neighbor hopping.

4. Renormalization-Group Analysis

Simple power-counting shows that the ν-term also dominates the two nonlinear terms in (7) in determining the asymptotic scaling, so that the *asymptotic* result of the linear theory remains valid in general. In this section we consider the case $L \ll L_\nu = (\gamma/\nu)^{1/2}$, keeping in mind that γ is subject to renormalization by the nonlinear terms.

Due to the conserving form assumed by (7), the average growth velocity is not renormalized by surface fluctuations. By going to the comoving frame $Z \to Z + F_0 t$, the constant F_0 drops out, but a phase factor $2\pi F_0 t/a$ appears in the argument of the lattice potential. This suggests that growth averages out the effect of lattice-pinning on modes whose lifetime exceeds $\tau_0 = a/F_0$[17]. By invoking dynamic scaling, one may define a length scale $L_F \sim \tau_0^{1/z}$, such that only fluctuations with wavelengths smaller than L_F feel a nonzero v. The exponent z, of course, is determined by the dynamics below L_F. Equation (7) is now considered separately above and below L_F.

a) $L > L_F$. — In this regime the continuous translational symmetry in the Z-direction is restored. We analyzed Eq. (7) at $\nu = v = 0$ in a dynamic RG scheme similar to the one applied to the KPZ equation[4]. This calculation[14] as well as an independent investigation by Lai and Das Sarma[23] provides, *within the one-loop approximation*, a confirmation of the exponents proposed by Villain[10],

$$\zeta = (4 - d)/3, \qquad z = (8 + d)/3. \tag{12}$$

b) $L < L_F$. — Equation (7) is now considered within the time interval τ_0 over which the surface advances by one layer on average. The strength of the v-term on a given length scale is subject to renormalization due to fluctuations on smaller length scales. Here we discuss this effect in a perturbative RG scheme introduced by Nozières and Gallet[17], including only the γ-term, the v-term, and the noise on the right-hand-side of Eq. (7). This is justifiable since the σ-term in the equation could be expected to be less relevant in this regime if $a^2\sigma \ll \gamma$, which we assume here.

After integrating out fast modes in the momentum shell $e^{-l}\Lambda_0 \le |\mathbf{k}| \le \Lambda_0$ and performing the rescaling $\mathbf{x} \to e^l\mathbf{x}$, $t \to e^{zl}t$, $Z \to e^{\zeta l}Z$, the coefficients of various terms renormalize as, to the second order in v,

$$dv/dl = [z - \zeta - 2 - n]v + O(v^3), \tag{13a}$$
$$d\gamma/dl = [z - 4 + (\pi^2/2)A_d(n, \kappa)y^2]\gamma, \tag{13b}$$
$$dD_0/dl = [z - 2\zeta - d]D_0, \tag{13c}$$
$$dD_1/dl = [z - 2\zeta - d - 2]D_1, \tag{13d}$$

and $da/dl = -\zeta a$. Here $n = 2\pi^2 K_d(D_0 + D_1)/\gamma a^2$, $y = v/\gamma a$, $\kappa = D_0/D_1$, and $A_d(n, \kappa)$ is a positive function of n and κ in the range of interest. In terms of the dimensionless parameters y, n and κ, we have

$$dy/dl = y[2 - n - cy^2], \tag{14a}$$
$$dn/dl = n[2 - d_{\text{eff}} - (\pi^2/2)A_d y^2], \tag{14b}$$
$$d\kappa/dl = 2\kappa, \tag{14c}$$

where $d_{\text{eff}} = d - 2\kappa/(1 + \kappa)$. The exact form for c requires the knowledge of the third order term in (13a).

Equations (13) and (14) have the familiar look of the Kosterlitz-Thouless RG flow equations[18]. Indeed, by setting $\kappa = 0$ results for the equilibrium sine-Gordon model are correctly reproduced[15-18], though the dynamics here is purely conservative. Recall that in the equilibrium model, an arbitrarily small v grows under the RG transformation above $d = 2$, so that the surface is facetted at all temperatures. For $d = 2 - \epsilon$, $\epsilon \ge 0$, there is a roughening transition at a finite temperature T_R. As readily seen from Eqs. (14), the nonconserving noise can be accounted for qualitatively (but not quantitatively) by the varying effective dimension d_{eff}, such that $d_{\text{eff}} = d$ for $\kappa \ll 1$ ($L \ll L_f$) and $d_{\text{eff}} = d - 2$ for $\kappa \gg 1$ ($L \gg L_f$). As the length scale is varied, the RG flow samples the phase diagram of the equilibrium problem at various dimensions[18]. In particular, for $d \le 4$, there is the possibility that the RG trajectory starts out in the facetted phase, yielding flat terraces on small length scales, and ends up in the rough phase, meaning that terraces of linear size larger than L_f are destroyed by the shot-noise.

5. Summary

We presented a detailed analysis of a continuum model for kinetic roughening in MBE which includes all relevant terms compatible with the symmetries and the conservation law for the surface diffusion dynamics. The main results of this paper for the physically relevant case $d = 2$ are summarized in Fig. 1. On this schematic diagram, different regimes for kinetic roughening are presented as a function of the temperature T and length scale L but a fixed beam intensity F_0. For simplicity we assume $\nu = 0$. (For $\nu \neq 0$ there is a further crossover length $L_\nu = (\gamma/\nu)^{1/2}$ in the figure with $z = 2$ and $\zeta = 0$ (log) for $L \gg L_\nu$.) Then, for $L \gg L_f$ the shot noise (beam fluctuation) dominates the kinetic roughening, the lattice pinning potential turns out to be negligible (see below). In the asymptotic region $L \gg L_\sigma \simeq (\gamma^3/D_0\sigma^2)^{1/2}$ one expects the (Villain) roughness exponent $\zeta = 2/3$ and the dynamic exponent $z = 10/3$. For smaller σ there is an intermediate region $L_f < L < L_\sigma$ where the results of the linear theory, $\zeta = 1$ and $z = 4$, apply. On length scales *less than* L_f, the stochastic surface diffusion is responsible for roughening. For temperatures above the equilibrium thermal roughening transition temperature T_R, the pinning potential is wiped out by thermal fluctuations. The nonlinear σ-term is only marginally relevant and one obtains a logarithmically rough surface ($\zeta = 0$) and $z = 4$. A similar behavior is expected for T slightly less than T_R and $L < \xi_R$, where ξ_R is the correlation length of the thermal roughening transition[17,18]. For temperatures much lower than T_R the surface is facetted on small length scales. The advancing height of the surface, however, averages out the pinning potential on

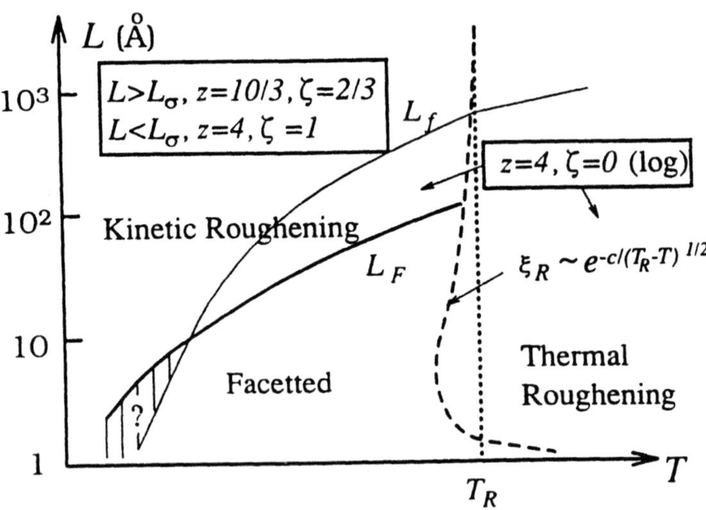

Figure 1. Schematic diagram of different regimes of kinetic roughening of the continuum model at $\nu = 0$ and a fixed F_0.

length scales $L > L_F$, where $L_F \sim F_0^{-1/z}$ is related to the *mean* beam intensity F_0. For $L_F < L < L_f$ we expect therefore a similar behavior as for $T > T_R$, i.e., $\zeta = 0$ (log) and $z = 4$. At very low temperatures L_f may become smaller than L_F. In this case, the pinning potential is renormalized to zero by the shot noise. However, it is not clear if such a regime exists in a more microscopic treatment of the problem[21].

REFERENCES

[1] For reviews of recent developments see, e.g., *Kinetics of Ordering and Growth at Surfaces*, edited by M. Lagally, Plenum, New York, 1990, and references therein.

[2] J. H. Neave, P. J. Dobson, B. A. Joyce and J. Zhang, *Appl. Phys. Lett.* **47** 100 1985.

[3] S. Clarke and D. D. Vvedensky, *Phys. Rev. Lett.* **58** 2235 1987.

[4] M. Kardar, G. Parisi and Y.-C. Zhang, *Phys. Rev. Lett.* **56** 889 1986.

[5] T. Sun, H. Guo and M. Grant, *Phys. Rev. A* **40** 6763 1989.

[6] T. Hwa and M. Kardar, *Phys. Rev. Lett.* **62** 1813 1989; G. Grinstein and D.-H. Lee, *ibid.* **66** 177 1991.

[7] F. Family and T. Vicsek, *J. Phys. A* **18** L75 1985.

[8] R. Kariotis, *J. Phys. A* **22** 2781 1989.

[9] R. Bruinsma, R. P. U. Karunasiri and J. Rudnick, in Ref. 1; L. Golubović and R. Bruinsma, *Phys. Rev. Lett.* **66** 321 1991. These authors considered F to depend on the instantaneous surface configuration.

[10] J. Villain, *J. Phys. I (France)* **1** 19 1991; D. E. Wolf and J. Villain, *Europhys. Lett.* **13** 389 1990.

[11] S. Das Sarma and P. Tamborenea, *Phys. Rev. Lett.* **66** 325 1991.

[12] T. Hwa, M. Kardar and M. Paczuski, *Phys. Rev. Lett.* **66** 441 1991.

[13] L. M. Sander, this volume.

[14] L.-H. Tang and T. Nattermann, *Phys. Rev. Lett.* **66** 2899 1991.

[15] S. T. Chui and J. D. Weeks, *Phys. Rev. Lett.* **40** 733 1978.

[16] D. J. Amit, Y. Y. Goldschmidt and G. Grinstein, *J. Phys. A* **13** 585 1980.

[17] P. Nozières and F. Gallet, *J. Phys. (Paris)* **48** 353 1987.

[18] J. M. Kosterlitz and D. J. Thouless, *J. Phys. C* **6** 1181 1973; J. M. Kosterlitz, *ibid.* **7** 1046 1974; *ibid.* **10** 3753 1977.

[19] W. W. Mullins, in *Metal Surfaces: Structure, Energetics and Kinetics* edited by W. D. Robertson and N. A. Gjostein Am. Soc. Metals, Metals Park, OH, 1963.

[20] Z. Rácz, M. Siegert, D. Liu and M. Plischke, *Phys. Rev. A* **43** 5275 1991.

[21] J. Villain, A. Pimpinelli and D. E. Wolf, submitted to *J. Phys. (France)*.

[22] S. F. Edwards and D. R. Wilkinson, *Proc. R. Soc. London, Ser. A* **381** 17 1982.

[23] Z.-W. Lai and S. Das Sarma, *Phys. Rev. Lett.* **66** 2348 1991.

Surface Disordering:
Growth, Roughening and
Phase Transitions

KINETIC ROUGHENING OF VICINAL SURFACES

Keye Moser* and Dietrich E. Wolf*†
**Inst. f. Theor. Physik, Univ. Köln, D-5000 Köln 41, Germany*
†IFF, Forschungszentrum Jülich, Pf. 1913, D-5170 Jülich, Germany

Abstract. Based on the Burton-Cabrera-Frank model of crystal growth at a vicinal surface the theoretical predictions for the surface morphologies in various growth modes of molecular beam epitaxy are reviewed. It is explained why in step flow growth with desorption the kinetic roughness is only logarithmic, whereas one expects algebraic roughness, if the formation of two dimensional islands on terraces cannot be neglected. Estimates of the parameters are given which specify the regimes in which the different surface morphologies should be observable.

1. Introduction

Although kinetic roughening i.e. the emergence of scale invariant surface fluctuations in growth processes is a more general phenomenon, we want to focus here on its manifestation when a crystal film is being deposited onto a flat substrate from an atomic beam. The substrate is generally close to, but not exactly at a high symmetry crystallographic orientation, so that it may be regarded as an array of high symmetry terraces separated by parallel, monoatomic steps (vicinal surface). Depending on the temperature and the deposition rate various growth modes have been observed, notably the step flow growth at high temperatures (still below the roughening temperature so that the equilibrium surface would be flat). At these temperatures it is possible to grow crystalline films with very few defects (molecular beam epitaxy, MBE). Nonetheless the fluctuating beam intensity renders the surface rough. The morphology develops coarser and coarser features during the growth, characterized by a length ξ parallel to the substrate, and by the root-mean-square surface height fluctuation,

$$w = (\overline{h^2} - \overline{h}^2)^{1/2} \sim \xi^\zeta \quad \text{and} \quad \xi \sim t^{1/z}. \tag{1}$$

These asymptotic power laws define the roughness exponent ζ and the dynamic exponent z [1]. It is the aim of this paper to predict the large scale surface morphology for various growth modes on vicinal surfaces.

2. Microscopic processes and growth modes

The following model is considered [2]: Atoms are deposited at a rate F onto the terraces of a vicinal surface with slope s. There they diffuse until they reach a step or desorb. They may also cluster together with other adatoms and form two dimensional islands on the terraces.

The typical distance l between island edges for a surface without tilt ($s = 0$) is

$$l \approx c(D/Fa^4)^\gamma. \tag{2}$$

D is the surface diffusion coefficient and a the lattice constant. The exponent γ depends on the minimum size of stable islands, varying between 1/6 if dimers are stable and immobile and 1/2 if islands of any size are unstable [3-5]. c depends on temperature T, but not on F.

Stability of islands has a kinetic meaning in this context: A surface step propagates by one lattice constant a within a time $\sim 1/Fla$ so that after the time $1/Fa^2$ the whole surface has been shifted by one lattice constant. Therefore, if an island emits atoms onto the terrace with frequency

$$1/\tau_I < Fla \tag{3}$$

the island can be regarded as stable. In the same sense we assume that adatoms which reach a kink site at a step are permanently bound. Similarly desorption only has to be taken into account if it happens with a frequency

$$1/\tau_0 \approx (k_B T/\hbar)\exp(-\beta E) > Fa^2, \tag{4}$$

(E = adsorption energy).

Surface diffusion has to be fast enough for epitaxial growth. a^2/D is the typical time an adatom remains on a site. Epitaxial growth requires

$$Fa^4/D < 1. \tag{5}$$

Otherwise atoms will be deposited on top of earlier ones, before they can diffuse to favorable sites. Then the film is expected to develop a lot of defects as illustrated by the model of ballistic deposition [6,7].

When (5) is fulfilled, several epitaxial growth modes can be observed. Here we give a (incomplete) list and mention the theoretical predictions for the roughness and dynamical exponents. At sufficiently high temperatures, such that (4) is fulfilled and $l > a/s$ one has step flow growth with desorption. Island formation on the terraces can be neglected, the number of steps on the vicinal surface remains basically constant. Roughening happens through fluctuations of the terrace widths and is predicted [8] to be only logarithmic: $\zeta = 0(log)$, $z = 2$. For lower temperatures diffusion will be slower so that l decreases, cf.(2). If $l < a/s$, island formation on the terraces can no longer be ignored. If (4) still holds, an algebraic roughness according to the Kardar-Parisi-Zhang (KPZ) equation [9,10] is expected [8] with $\zeta + z = 2$ and the numerically determined roughness exponent $\zeta \approx 0.387$ [11]. If desorption is absent, surface diffusion is the only mechanism left for smoothing the surface. Hence one expects a stronger roughness than obtained from the KPZ-equation, which no longer applies. In fact, Villain [2] predicts $\zeta = 2/3$ and $z - 2\zeta = 2$ [12]. For very low temperatures where (5) is no longer valid, diffusion also becomes ineffective. As mentioned above, the film will then have many pores: Large surface fluctuations become "sealed off" by overhangs so that the roughness of the active zone, where the growth proceeds, becomes smaller. For reasons quite different than above the KPZ equation should again apply [13]. Other growth morphologies are less well characterized, in particular those due to the destabilizing influence of the Schwoebel effect [2,14,15]. In the following we restrict ourselves to the first two growth modes.

3. Equation of motion for the surface in MBE with desorption

In this section we show that the time evolution of the epitaxial film thickness h is given by the anisotropic generalization of the KPZ equation [2,8]:

$$\partial_t h = \frac{\lambda_\parallel}{2}(\partial_\parallel h)^2 + \frac{\lambda_\perp}{2}(\partial_\perp h)^2 + \nu_\parallel \partial_\parallel^2 h + \nu_\perp \partial_\perp^2 h + \text{noise}, \tag{6}$$

where ∂_t, $\partial_\parallel$ and $\partial_\perp$ denote the partial derivatives with respect to deposition time and parallel, resp. perpendicular to the tilt direction. This continuum description is valid on length scales much larger than the average spacing l or a/s between surface steps.

Due to desorption the adatom density ρ and hence also the growth rate

$$\kappa = F - \rho/\tau_0 = F(1 - \tau/\tau_0) \tag{7}$$

depends on the slope s as in Fig.1. For large tilt angles the lifetime τ of the adatoms corresponds to the time needed to diffuse across a terrace of width a/s:

$$\tau \approx a^2/Ds^2. \tag{8}$$

This s-dependence is characteristic of step flow growth and extends approximately down to $s \approx a/l$. For smaller slopes more and more adatoms are caught by island edges. At $s = 0$,

$$\tau \approx l^2/D. \tag{9}$$

Expanding κ around s for small deviations of the local tilt, $[(s+\partial_\parallel h)^2+(\partial_\perp h)^2]^{1/2}$, one obtains the first two terms of (6), with

$$\lambda_\parallel = \kappa''(s), \qquad \lambda_\perp = \kappa'(s)/s, \tag{10}$$

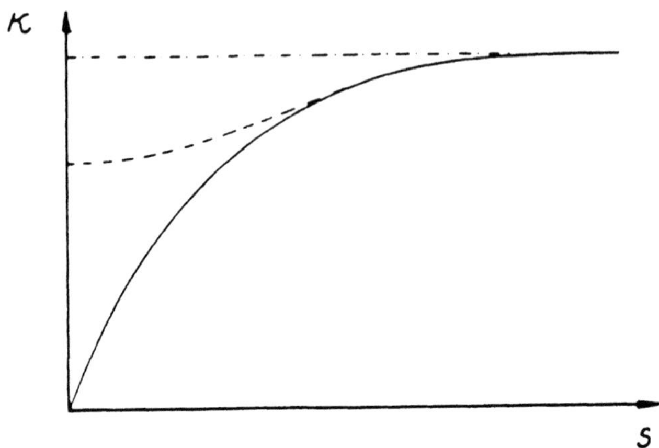

Figure 1. Schematic dependence of the growth velocity κ on the tilt s. Solid line: Only attachment to the steps taken into account. Dashed line: Island formation on sufficiently large terraces included. Without desorption the growth velocity becomes independent of the tilt in this model (dash-dotted line). For simplicity we assume that the terraces have at least a fourfold in-plane symmetry so that κ depends only on the absolute value of s.

where κ' and κ'' denote the first and second derivatives. Two additional terms (constant and linear in $\partial_{\parallel} h$) can easily be transformed away (comoving frame and Galileo transformation). From Fig.1 it is immediately clear that $\lambda_{\perp}$ is positive, whereas $\lambda_{\parallel}$ changes sign as one goes from step flow growth to a regime where island formation dominates [2,8]. In this regime, not too close to the transition, the anisotropy due to vicinality should vanish, as the few extra steps should not play any role compared to the island edges. Hence, $\lambda_{\parallel} \approx \lambda_{\perp}$, implying $\kappa(s) \propto s^2$ for small tilts. The λ's are of the order of magnitude $Fa^5/D\tau_0 s^4$ or $Fal^4/D\tau_0$ for step flow and island growth, respectively.

The preference of the surface for having a uniform tilt is expressed by the curvature terms with positive coefficients $\nu_{\parallel}$ and $\nu_{\perp}$. In step flow growth these coefficients will in general be different. Consider for example a surface curved in the tilt direction, $\partial_{\parallel}^2 h \neq 0$. The steps are not uniformly spaced on average. Step-step interactions will then modify the sticking coefficients so as to enhance the desorption probability, where $\partial_{\parallel}^2 h < 0$, or to reduce it, where $\partial_{\parallel}^2 h > 0$. The relaxation back to a planar surface is governed by the coefficient $\nu_{\parallel}$. Curvature $\partial_{\perp}^2 h \neq 0$ means actually curved steps. Kink-antikink annihilation then is the main mechanism for reducing the curvature, leading to an enhanced (reduced) desorption at convex (concave) parts of the surface. This determines the coefficient $\nu_{\perp}$. As a rough estimate one can argue that $\nu_{\parallel}$ and $\nu_{\perp}$ should be of order $\rho\beta\sigma a^6/\tau_0$, where σ denotes the components of the surface stiffness tensor [16]. For simplicity it has been assumed that the tilt direction is a principal direction of the stiffness tensor so that no mixed derivatives occur in (6).

Desorption and redeposition are not the only ways, by which the surface can reduce its curvature. In fact, surface diffusion is often the more important mechanism. It would be described by additional terms

$$-K\nabla^4 h, \qquad K \approx \rho D \beta \sigma a^6, \tag{11}$$

in (6) [17]. However, for the surface smoothening on lengthscales larger than $\sqrt{K/\nu} \approx \sqrt{D\tau_0}$ desorption is more efficient than diffusion. For the asymptotic scaling behavior the terms (11) are not important [2, 18].

The last term in (6) is the shot noise of the beam: The number of particles deposited on a site per unit time is $F \pm \sqrt{F}$. Hence the fluctuations of the growth velocity can be represented by white Gaussian noise η with second moment

$$\langle \eta(\mathbf{x}, t)\eta(\mathbf{x}', t') \rangle = a^6 F \delta(\mathbf{x} - \mathbf{x}')\delta(t - t'). \tag{12}$$

Other sources of nonconserved noise are conceivable, but are neglected here. The thermal fluctuations of the surface currents give rise to conserved noise, for which F on the right hand side of (12) is replaced by $\rho D \nabla^2$ [18]. Inserting $\rho = F\tau$ and (9) one sees that the conserved noise dominates on length scales up to the terrace width l, while on much larger distances the shot noise is more important.

4. Results

Rather than summarizing the results of the dynamical renormalization group calculation [8] we now explain the physical behavior expected from (6) giving intuitive arguments and supporting this picture by computer simulations [19].

If both λ's are zero, (6) is equivalent to the Edwards-Wilkinson equation [20] which gives logarithmic roughness, $w^2 \sim \log \xi$, and a diffusive spreading of fluctuations along the surface, $\xi \sim t^{1/2}$. The isotropic case, $\lambda_\parallel = \lambda_\perp = \lambda$ and $\nu_\parallel = \nu_\perp = \nu$, is the well known Kardar-Parisi-Zhang (KPZ) equation [9], which leads to a much stronger algebraic roughness [11], $w \sim \xi^{0.387}$, and a faster spreading of fluctuations along the surface, $\xi \sim t^{0.620}$, than in the linear theory. On first sight this might seem odd as in the absence of noise the KPZ-nonlinearity provides an additional healing mechanism (lateral growth) for deviations from a flat surface. How does the same mechanism make the surface more susceptible to noise and hence enhance the kinetic roughening? The reason is that the lateral growth transports small wavelength perturbations towards larger wavelengths (mode coupling): For $\lambda > 0$ bumps grow in diameter (for $\lambda < 0$ the same is true for holes). Of course there is competition between bumps of different size so that the fluctuations do not spread with a finite velocity, but faster than diffusively. Like in the linear theory the lifetime of perturbations increases with their wavelength. Hence the KPZ-nonlinearity enhances the randomness.

What do we expect for the anisotropic case (6)? If the two λ's have equal sign the mechanism of lateral growth is not changed much, apart from an anisotropic rescaling of the lengths parallel to the surface. Therefore one should get the same algebraic roughness as for the isotropic theory. By contrast, if $\lambda_\perp > 0$ and $\lambda_\parallel < 0$, bumps try to spread along $x_\perp$ but not along $x_\parallel$, giving rise to "ridges". Similarly holes should

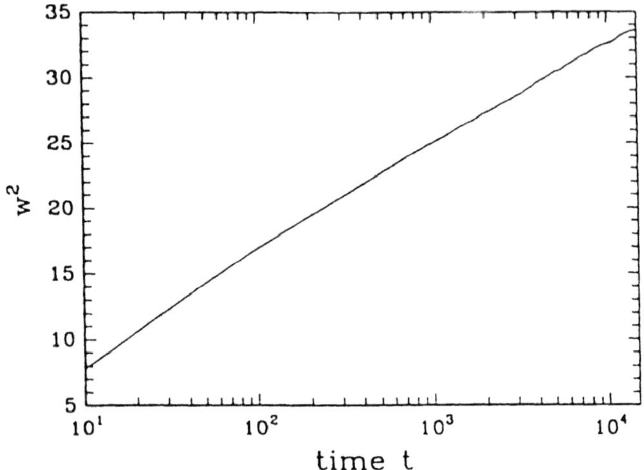

Figure 2.a) Logarithmic roughening for the anisotropic KPZ equation in 2+1 dimensions for $\lambda_\parallel = -\lambda_\perp/2$ and coupling constant $g_\perp = 40$.

spread into "canyons" perpendicular to them. One can imagine that the destructive interference at crossings between "ridges" and "canyons" effectively stops the further spreading of the fluctuations. This explains why the lateral spreading mechanism is no longer active, and the nonlinearity becomes irrelevant. Hence one expects logarithmic roughness as in the linear theory. This is in agreement with what is found in a dynamical renormalization group calculation (one loop approximation) [8].

We used the Euler-algorithm [21] to integrate equation (6) numerically. The method has already been tested and described in detail for the isotropic case in an earlier publication [22], so that we show here only the dimensionless discretized version of (6):

$$
\begin{aligned}
h_{\underline{n}}(t+\Delta t) = h_{\underline{n}}(t) + \frac{\Delta t}{\Delta x^2} &\left[\left((h_{\underline{n}+\underline{e}_\perp}(t) + h_{\underline{n}-\underline{e}_\perp}(t) - 2h_{\underline{n}}(t)) \right. \right. \\
&+ \frac{\nu_\parallel}{\nu_\perp}\left(h_{\underline{n}+\underline{e}_\parallel}(t) + h_{\underline{n}-\underline{e}_\parallel}(t) - 2h_{\underline{n}}(t) \right) \Big) \\
&+ \frac{1}{8}\left((h_{\underline{n}+\underline{e}_\perp}(t) - h_{\underline{n}-\underline{e}_\perp}(t))^2 + \frac{\lambda_\parallel}{\lambda_\perp}\left(h_{\underline{n}+\underline{e}_\parallel}(t) - h_{\underline{n}-\underline{e}_\parallel}(t) \right)^2 \right) \Big] \\
&+ \sqrt{12\Delta t}\, R_{\underline{n}}(t),
\end{aligned}
\tag{13}
$$

where R are random numbers taken from a uniform distribution between $-1/2$ and $1/2$. All simulations start at $t = 0$ with a flat surface (i.e. $h_{\underline{n}} = 0$). For all data presented here we chose a system length of 280 sites and $\Delta t = 2.0$ (guaranteeing numerical stability) and averaged over 100 realisations of the noise. We used helical boundary conditions to achieve optimal vectorization. On a Cray Y-MP we reached about 195 MFlops + 80 MIps, on a NEC SX-3 even 570 MFlops + 1180 MIps, which is close to the maximum performance of the machines.

As $\nu_\parallel/\nu_\perp$ is irrelevant under renormalization it was set equal to 1. For the coupling constant $g_\perp = 2\Delta x^2$ the two values 12.6 and 40 were considered. The results are qualitatively the same for both cases.

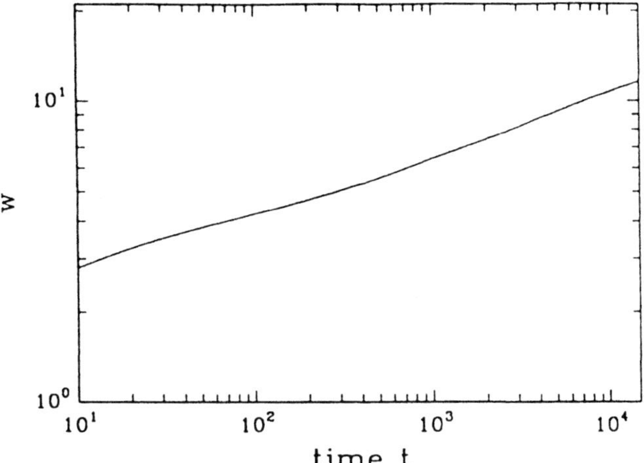

Figure 2.b) Algebraic roughening for $\lambda_\parallel = \lambda_\perp/2$.

Figure 2a shows our results for $\lambda_\parallel = -\lambda_\perp/2$. We find a slight downward curvature on a w^2 vs. $\log t$ plot, which indicates roughening that is not stronger than logarithmic. This is in agreement with the theoretical prediction. Figure 2b shows $\log w$ vs. $\log t$ for $\lambda_\parallel = \lambda_\perp/2$. As in the isotropic case [22] we find algebraic roughening after some initial transients. Between $t = 10^3$ and $t = 10^4$ the effective exponent is $\beta = \zeta/z = 0.23\pm0.01$. This agrees with the best numerical value known today: $\beta = 0.240 \pm 0.001$ [11,22].

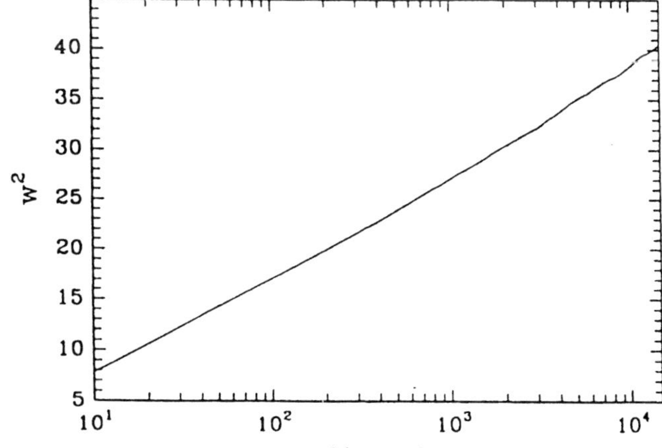

Figure 2.c) Behavior at the transition, $\lambda_\parallel = 0$.

Figure 2c shows the roughening at the transition point $\lambda_\parallel = 0$. The theory predicts a straight line on a w^2 vs. $\log t$ plot. We find a slight upward curvature, but this may be due to initial transients as in figure 2b. Longer runs for larger systems are needed to check if the asymptotic behavior is really logarithmic. After the last time step of the numerical simulations we also evaluated the height-height correlation function depending on the distance in both the $\perp$ and $\parallel$ directions. For $\lambda_\parallel \neq 0$ we find isotropic scaling, whereas for $\lambda_\parallel = 0$ the data are difficult to interpret. [1]

In summary we have shown that in the step flow growth mode with desorption the kinetic roughness is only logarithmic as the nonlinear terms in the appropriate KPZ equation have opposite sign. If two dimensional island formation becomes important, one gets a morphological transition to algebraic roughness when one of the λ's changes sign. It has been shown recently [23, 24] that it may be difficult to observe the algebraic roughness, because for small coupling constants $g_\perp = \lambda_\perp^2 F a^6 / \nu_\perp^3$ one gets extremely slow crossovers. In the present case the situation is more promising. Inserting the estimates for the parameters given in section 3 one obtains

$$g_\perp \approx (D\tau_0/a^2)(l/a)^2(\beta\sigma a^2)^{-3}, \tag{14}$$

which needs not be small.

Acknowledgements

We thank S. Balibar, V. K. Horváth, J. Kertész, L.-H. Tang, and J. Villain for stimulating discussions. This work was supported by the Deutsche Forschungsgemeinschaft within SFB 341. We thank the supercomputer center HLRZ/Jülich and the RRZ Köln for providing us with time on the Cray Y-MP and the NEC SX-3, respectively.

REFERENCES

[1] F. Family and T. Vicsek, *J. Phys.* **A 18** L75 1985.

[2] J. Villain, *J. Physique I* **1** 19 1991.

[3] S. Stoyanov and D. Kashchiev *Current Topics in Materials Science* 7 (E. Kaldis ed.) North Holland 1981.

[4] Y. W. Mo, J. Kleiner, M. B. Webb and M. G. Lagally, *Phys. Rev. Lett.* **66** 1998 1991.

[5] J. Villain, A. Pimpinelli and D. E. Wolf, *Comments on Condensed Matter Physics* to be published 1992.

[6] M. J. Vold, *J. Colloid Sci.* **14** 168 1959.

[7] P. Meakin, *J. Phys.* **A 20** L1113 1987.

[8] D. E. Wolf, *Phys. Rev. Lett.* **67** 1783 1991.

---

[1] At the workshop we learnt about independent numerical work on the anisotropic KPZ equation by A. Assdah and T. Halpin-Healy also confirming the theoretical predictions.

[9] M. Kardar, G. Parisi and Y.-C. Zhang, *Phys. Rev. Lett.* **56** 889 1986.

[10] E. Medina, T. Hwa, M. Kardar and Y.-C. Zhang, *Phys. Rev. A* **39** 3059 1989.

[11] B. Forrest and L.-H. Tang, *Phys. Rev. Lett.* **64** 1405 1990.

[12] D. E. Wolf and J. Villain, *Europhys. Lett.* **13** 389 1990.

[13] J. Krug, *J. Phys. A* **22** L769 1989.

[14] R. L. Schwoebel and E. J. Shipsey, *J.Appl.Phys.* **37** 3682 1966.

[15] R. Kunkel, B. Poelsema, L. K. Verheij and G. Comsa, *Phys. Rev. Lett.* **65** 733 1990.

[16] P. Nozières *Solids Far From Equilibrium: Growth, Morphology and Defects* (C. Godrèche ed.) Cambridge University Press 1992.

[17] W. W. Mullins *Metal Surfaces: Structure, Energetics and Kinetics,17* Am. Soc. Metal, Metals Park, Ohio 1963.

[18] L.-H. Tang and T. Nattermann, *Phys. Rev. Lett.* **66** 2899 1991.

[19] K. Moser, D. E. Wolf, V. K. Horváth and J. Kertész, to be published 1992.

[20] S. F. Edwards and D. R. Wilkinson, *Proc. R. Soc. A* **381** 17 1982.

[21] A. Greiner, W. Strittmatter and J. Honerkamp, *J. Stat. Phys.* **51** 95 1988.

[22] K. Moser, J. Kertész and D. E. Wolf, *Physica A* **178** 215 1991.

[23] L.-H. Tang, T. Nattermann and B. M. Forrest, *Phys. Rev. Lett.* **65** 2422 1990.

[24] S. Balibar, contribution to these proceedings .

REALISTIC MODELS FOR MBE GROWTH

David A. Kessler*, **Herbert Levine†**, **and Leonard M. Sander***
**Physics Department, The University of Michigan, Ann Arbor, MI 48109-1120
USA*
*†Department of Physics and Institute for Nonlinear Science, University of
California, San Diego, La Jolla, CA 92093 USA*

Abstract. We show that recent models for MBE growth which emphasize the effects
of surface diffusion neglect physically important effects such as the fact that diffusion
always takes place along the surface, independent of its inclination, and that voids
can be formed. When such effects are taken into account growth followed by surface
diffusion gives, in the long time limit, the conventional KPZ form. The crossover has
several unexpected features.

1. Introduction

Much of the work on surface roughening [1] which is the subject of this conference
takes its inspiration from simulations of the Eden model and the ballistic aggregation
model. These were originally thought to describe only the growth of disordered rough
surfaces and amorphous films. It is natural for a materials scientist to try to describe
refined growth techniques such as molecular beam epitaxy (MBE) in a different way.
This has led to the notion that the KPZ equation (2), is not the proper description of
MBE growth [3-6]. In this paper we will criticize this point of view, and try to show
that not only is the KPZ equation the correct asymptotic description, but that the
crossover to this behavior is quite complex.

The papers which purport to describe MBE growth [3-6] are solid-on-solid models
(no voids) in accord with the belief that MBE films are very nearly perfect. Surface
diffusion is the restructuring mechanism considered. The driving force for the diffusion
is supposed to be the gradient of a chemical potential proportional to curvature. The
growth equation for the height, h, is of fourth order:

$$\partial h/\partial t = -D_4 \nabla^4 h + \lambda_4 |\nabla h|^2 + \eta \tag{1}$$

Here η is a white noise term and the non-linear term is the lowest order one allowed by the assumed symmetry. Equation (1) is different, and has different scaling, from KPZ-type models (or their linear version, the Edwards-Wilkinson (EW) model [7]):

$$\partial h/\partial t = D_2 \nabla^2 h + \lambda_2 |\nabla h|^2 + \eta \tag{2}$$

Terms such as $\nabla^2 h$ can arise from a chemical potential which depends on height, i.e. in a model driven by gravity [7]. Gravity is, however, negligible compared to the chemical forces that actually cause film growth.

However the $\nabla^2 h$ term in Eq. (2) does not necessarily arise from a tendency for aggregated particles to move downhill. It can result from the lateral spreading of columns (which leads to an effective diffusion of h). This is most easily seen in the ballistic aggregation model [1], where a particle sticks as soon as it encounters any particle of the aggregate thus producing voids and overhangs; adjacent columns are coupled. We emphasize that this 'sticking on the side' is physically correct and would be present in any experiment or realistic molecular dynamics simulation [1] with finite-range forces. It leads to an $\partial h/\partial t$ which depends on the local slope, since the larger the slope, the higher up the next particle will stick. This produces the nonlinear term in Eq.(2). The $\nabla^2 h$ arises in the same way from lateral spreading. The magnitude of D_2 is inevitably of order 1, in units of (lattice spacing)^2 per monolayer deposition time. Ballistic aggregation certainly obeys KPZ scaling [1].

We also recover KPZ scaling in the single step model [8] of Meakin et al. even though it produces no voids or overhangs, because the sticking rule (sticking allowed only in highly coordinated sites) couples the columns. Since the terms of Eq. (2) are relevant perturbations even a slight amount of partial sticking will invalidate Eq. (1).

In the physical situation of MBE growth it may be reasonable to ignore desorption and sticking coefficients less than unity, and we will do so. However, it is our contention that voids and overhangs cannot be ignored. In any case, we should look carefully at the crossover. To this end we present a simple model which captures the essence of the physics of deposition and true surface diffusion so that the crossover can be elucidated.

2. Surface diffusion model with overhangs

Let us return to Eq. (1) and the mechanism of surface diffusion. The growth of the surface should be described by [9]:

$$v_n = D_4 \nabla^2_{LB}(\kappa) + \mathcal{F} \tag{3}$$

where v_n is the normal velocity of the interface, κ the curvature, ∇^2_{LB} the Laplace-Beltrami (intrinsic Laplacian) operator on the surface, and $\mathcal{F}$ the (fluctuating) flux. The use of intrinsic variables (measured in the local frame of the surface) is justified because gravity is physically irrelevant to MBE growth. If we assume that the flux is incident so that the surface moves primarily in one growth direction we can expand about a flat interface:

$$v_n = -D_4 \nabla^4 h + \mathcal{F} + ... \tag{4}$$

This is similar to Eq. (1) with some important differences. First, a nonlinear term of the form $\nabla^2 |\nabla h|^2$ *cannot occur* because the chemical potential is independent of the orientation of the surface. This symmetry argument would seem to imply that such a term cannot be generated by coarse-graining, and so λ_4 should be zero to all orders. This is consistent with the fact that λ_4 is not renormalized [6] to one loop order, so that if the bare theory has λ_4 equal to zero, it cannot be induced by other nonlinear terms. Furthermore, with $\lambda_4 = 0$ the nonlinearities implicit in Eq.(4) are all relevant in d=1 and marginal in d=2 so that the scaling behavior of surface diffusion dominated regime is in fact not obvious.

The second important difference between Eq. (4) and Eq. (1) is that the flux $\mathcal{F}$ has not been specified. The crossover to KPZ behavior is, as we shall see, due to the fact that the flux $\mathcal{F}$ *cannot* be modelled by a white noise η as in Eq. (1), as a result of the process of 'sticking on the side' discussed above.

Now consider the linear model of Refs. [4] and [5]. In this model particles are dropped at a random value of $\mathbf{x}$ onto the top of that column so that the $h(\mathbf{x})$ is increased by 1. Then the just-dropped particle is relaxed, moving to the top of a neighboring column if it increases the particle's coordination. This model appeared to give a surface scaling according to Eq. (1) with $\lambda_4 = 0$. In $d = 1$ the width of the surface, w, grew as t^β, $\beta = 3/8$ for $t \ll L^z$, $z = 4$. As a model for MBE growth this model suffers from two defects which render it inappropriate for our purposes. First, the particles drop directly onto the top of a column, bypassing an arbitrary number of neighboring taller columns on the way down. This 'drop-through' neglects 'sticking on the side'. Second, relaxation is allowed to occur between nearest-neighbor columns regardless of the difference in heights involved: this is height diffusion, not surface diffusion. This would be extremely anisotropic diffusion, finite parallel to the substrate but infinite in the growth direction. Note that both these defects arise from the solid-on-solid constraint which arises naturally in some contexts, like the single-step model [8]. Here it is not appropriate since it enforces a symmetry (particle number conservation, on average) not present in the physical situation.

Our model of MBE growth treats surface diffusion and deposition physically. The 'drop-through' rule is replaced by ballistic aggregation sticking. For surface diffusion, a randomly chosen surface (not fully coordinated) particle moves to a new position chosen from among the best-coordinated sites in a hypercubical box with sides of length $2L_d + 1$ centered on the particle's current position. Each particle moves, on average, D times. The distance moved up and down is counted this way: there is no physical distinction since gravity plays no role. The diffusivity is governed by L_d and D, and scales for small D as $L_d D^{1/2}$. Note that we let all the surface particles diffuse, not just the last dropped.

This new model is related to our previous work [10] which had height diffusion plus ballistic aggregation sticking rules. Our new model, ballistic aggregation with surface diffusion (BASD), is probably the physically correct analogue of ideal MBE growth and should describe the statistical properties of surfaces formed in this manner. Even more realistic models involving activated hopping [11] should behave similarly, we think, but their complexity makes them unsuited for scaling studies.

The essential question we now address is whether this new model makes a physical difference for fairly flat films, or whether it is simply identical in this regime to the work of Refs. [3-6]. We will look at the long-time crossover to ballistic aggregation behavior.

3. Simulation results: exponents and crossovers

We have simulated the BASD Model in 1+1 dimension for varying L_d. After a short transient, there is a regime, lengthening with L_d, where the surface width grows with time, apparently as t^β, with $\beta \approx 0.25$. Then follows a period of rapid growth of the width, following which a power law again sets in. This latter regime then terminates in a saturation regime at very long times. In the first power-law regime the deposit has initially essentially no defects or overhangs. In the second power-law regime we find a coarsened version of a standard ballistic aggregate. We thus interpret the first power-law regime as representing surface diffusion scaling, whereas the second power-law regime and subsequent saturation are described by the KPZ equation, with the generation of overhangs and voids.

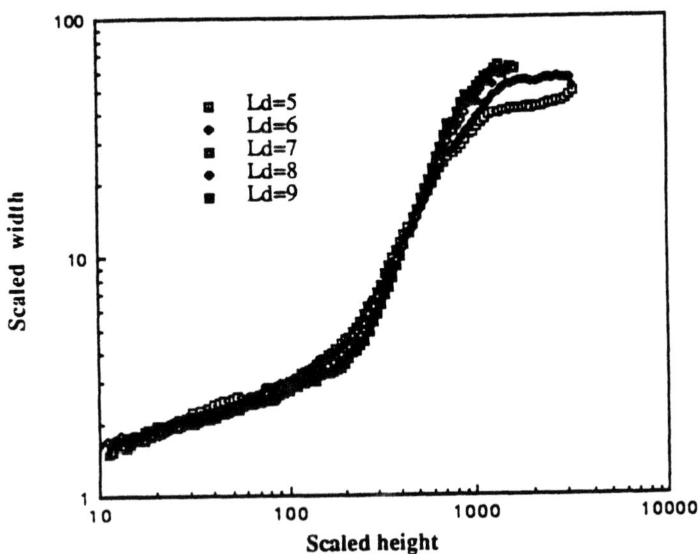

Figure 1. Scaled width versus scaled height for the BASD model for $L_d = 5 - 9$.

One interesting and unexpected feature to note is the sharp rise in width separating the two regimes. If we had the simplest model we could expect (with both $\nabla^2 h$ and

$\nabla^4 h$ terms) the crossover would be much smoother. This crossover is shown in data-collapsed form in Fig. 1 where we have scaled the height (i.e. the time) by the crossover time in order to find the best fit. We find that the crossover time goes approximately as L_d^2. The good collapse as overhangs set in is evidence that this is a real effect, and could have observable consequences.

Even more surprising is the value of β in the surface-diffusion regime. While we do not expect the KPZ value of $\beta = 1/3$, since λ_4 is zero, we do not see the $\beta = 3/8$ characteristic of Refs. [3-5] either but something very close to the EW value of 1/4! We have checked this point by returning to the original model of Ref. [4]. For the conditions of that paper ($L_d = 1$) we recover the quoted value of $\beta = 0.365$. However, if we increase L_d to larger values we find a decrease of the exponent, apparently towards 1/4. This is shown in Fig. 2. We have no explanation for this either, except to speculate that that the dynamics of Eq. (1) is so 'fragile' that even lattice effects, e.g., pinning of the surface are sufficient to induce terms like those of Eq. (2).

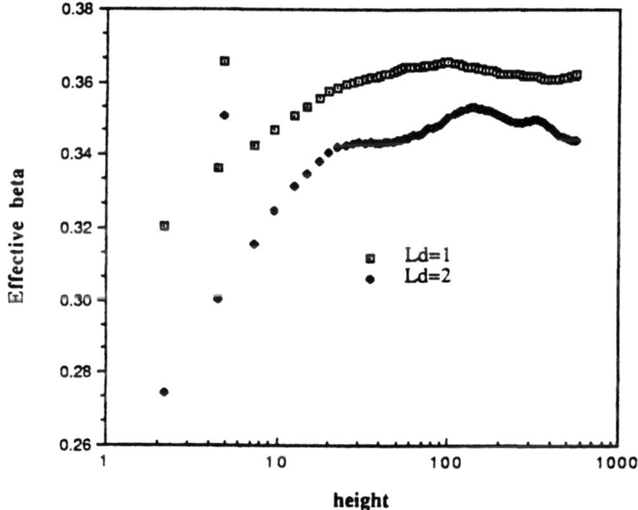

Figure 2. The effective β for the model of Ref. 4 for $L_d = 1$ and $L_d = 2$.

The short distance structure in the KPZ regime reflects the smoothing due to surface diffusion. This smoothing leads to the surprising result that the saturated width of our BASD model increases with L_d. Essentially, the BASD Model produces ballistic aggregates made with blocks, L_B, whose size increases with L_d. The effective length of the system is thus L/L_B so that w seems to be reduced. However, excursions in width are now measured in units of L_B so that the measured width in the original scale has increased, by a factor $L_B^{1-\alpha}$, where a is the roughness exponent. In this regime we find porous films whose scaling appears to be close to $\beta = 1/3$. Their density approaches 0.83 as L_d increases. This is to be expected for a structure made of blocks of size L_B.

4. Discussion and possible relationship to experiments

The crossover we observe is striking and rather mysterious. Its scaling with L_d seems to indicate the onset of a non-linear term as the surface features develop with time scale $\sim L_d^2$, and it does correspond to the development of overhangs. However, since we do not understand why we are seeing EW scaling at early times we cannot further interpret the crossover.

We can speculate about what might happen if these results hold up for real systems (in 2+1 dimensions). The rapid crossover from very flat to very rough might correspond to the experimenters' distinction between 'good' and 'bad' films. For small L_d roughness could develop very quickly. Also, we should note that porous films like those produced by the BASD model are not unknown. Growth conditions in the laboratory are usually chosen to produce 'good' films by growing at high temperatures where diffusion lengths are of order 0.1 microns. Low temperature films do become porous if grown too thick [12]. It is an important challenge to identify the operative mechanism responsible for 'good' films in experimental conditions.

We have seen how a physically reasonable model for ideal MBE growth leads to a crossover to KPZ scaling on a time scale which grows with diffusion length. We maintain that similar results would obtain from a study of any physically correct model, including more realistic Monte-Carlo or molecular-dynamics simulations. We believe further study of our BASD Model, especially of the 2+1 dimensional case, will prove fruitful.

Acknowledgements

We acknowledge useful conversations with Y. Tu, B. Orr, and H. Yan, and the insightful comments of J. Villain. DAK is supported by U.S. DOE grant DE-FG-0285ER54189, HL by U.S. NSF grant DMR91-15413, and LMS by U.S. NSF grant DMR91-17249.

REFERENCES

[1] M. J. Vold, *J. Colloid Interface Sci.* **14** 168 1959.

J. Leamy, G. H. Gilmer, and A. G. Dirks, in *Current Topics in Materials Science* (Vol. 6) E. Kaldis ed., North-Holland, Amsterdam 1980; for a recent review, see L. Sander, in *Solids Far From Equilibrium: Growth, Morphology and Defects* C. Godreche ed., Cambridge, 1991.

[2] M. Kardar, G. Parisi, and Y.-C. Zhang, *Phys. Rev. Lett.* **56** 889 1986.

[3] J. Villain, *J. Physique I* 1 19 1991.

[4] D. E. Wolf and J. Villain, *Europhysics Lett.* **13** 389 1990.

[5] S. Das Sarma and P. Tamborenea, *Phys. Rev. Lett.* **66** 325 1991.

[6] Z. W. Lai and S. Das Sarma, *Phys. Rev. Lett.* **66** 2348 1991.

[7] S. F. Edwards and D. R. Wilkinson, *Proc. Roy. Soc. Lond.* **A381** 17 1982.

[8] P. Meakin, P. Ramanlal, L. M. Sander, and R. C. Ball, *Phys. Rev* **A34** 5091 1986.

[9] W. W. Mullins, *J. Appl. Phys.* **28** 333 1957.

[10] H. Yan, D. A. Kessler, and L. M. Sander, *Phys. Rev. Lett.* **64** 926 1990.

[11] S. Das Sarma, *J. Vacuum Sci. Tech.* **A8** 2714 1990.

[12] D. D. Perovic, G. C. Weatherly, J.-P. Noel, and D. C. Houghton, *J. Vac. Sci. Technol.* **B9** 2034 1991.

Surface Disordering:
Growth, Roughening and
Phase Transitions

SURFACE-DIFFUSION INDUCED INSTABILITIES

Zoltán Rácz[t*], Martin Siegert[‡] and Michael Plischke[‡]

*Department of Physics, Clarkson University, Potsdam, N.Y. 13676 USA
‡Department of Physics, Simon Fraser University, Burnaby, B.C. V5A 1S6,
Canada

Abstract.   The effects of surface diffusion on both surface relaxation and surface growth are investigated. We show that the scaling properties of surface fluctuations depend essentially on whether the diffusion satisfies detailed balance and on whether particle deposition is present. Morphological instabilities are found to exist if detailed balance is violated or if a detailed-balance diffusion is coupled to particle deposition.

1. Introduction

The dominant mechanisms of surface relaxation are desorption and surface diffusion. At low temperatures, however, desorption is absent and the relaxation is governed by surface diffusion [1,2]. Thus, in order to understand the surface morphologies observed in experiments [3], we have to understand in detail the effects of surface diffusion. When trying to do this, one faces the problem that a growing surface may be in a far from equilibrium state either because of the high particle deposition rates or because of the temperature gradients produced by the impact of the deposited particles (this effect may be present at low deposition rates as well). The dynamics then may violate detailed balance, and once detailed balance is lost, we have lost one of the restrictive elements in building a model of the process. The arbitrariness of model construction can be reduced somewhat by introducing a 'minimal' model [4] that neglects particle deposition and contains only surface diffusion that violates detailed balance. In the first part of this paper we summarize what we learned from such a model, the main conclusions being that i) one can observe flat and rough steady-state surfaces as well as an instability towards the formation of grooves with surface width becoming pro-

---

* Permanent Address: Institute for Theoretical Physics, Eötvös University, 1088 Budapest, Hungary

39

portional to the length of the substrate; ii) the scaling properties of the surfaces are always different from those produced by 'equilibrium' diffusion satisfying detailed balance; and iii) the much discussed [2,5-8] surface relaxation term $\nabla^2(\nabla h)^2$ that appears e.g. in the conserved Kardar-Parisi-Zhang equation [5] can arise naturally as a result of diffusion that violates detailed balance.

In the second part of the paper we study a one-dimensional model in which surface diffusion that satisfies detailed balance is combined with random deposition of particles [9]. The dynamics in this case can be considered as a mixture of a finite temperature process (diffusion) and an infinite temperature process (random deposition). The resulting non-equilibrium dynamics is remarkable in the sense that i) the surfaces produced in steady states are unstable (the surface slopes are proportional to a positive power of the size of the system); ii) the surface morphologies depend on details of the Hamiltonian that would be irrelevant in the equilibrium case; and iii) a phase transition between a rough and a grooved surface is observed.

2. Diffusion that violates detailed balance

The model of surface diffusion we discuss is a dynamical, restricted solid-on-solid (RSOS) model [4]. A one-dimensional periodic lattice of L adsorption sites is considered (the lattice constant is ϵ). The surface is given by a single-valued function, h_i, of the height of the deposit at site i and the moving particles are blocks of height Δ and width ϵ. The dynamics consists of randomly moving particles to nearest-neighbor or next-nearest-neighbor sites with the condition that a move is permitted only if the constraint $|h_i - h_{i+1}| \leq H$ is satisfied for every i at the end of the process. More precisely, the above dynamics was implemented in two different ways.

(I) First a particle is randomly selected for the trial move and then the direction of the move is chosen randomly. If the move would violate the constraint $|h_i - h_{i+1}| \leq H$, the attempted move is abandoned and a new particle is selected.

(II) The procedure is the same as in (I) but if the move is rejected in the chosen direction then a move in the opposite direction is attempted. If this too is rejected, only then a new particle is selected for the next attempted move.

Although the above two procedures do not seem to be significantly different, a detailed analysis [4] reveals that the diffusion described by Procedure (I) obeys detailed balance, whereas the diffusion corresponding to Procedure (II) does not. Furthermore, it follows from the violation of the detailed balance that Procedure (II) does not respect the reflection symmetry $h_i \to -h_i$. The consequences of the violation of the detailed balance are highly nontrivial as will be explained below first by considering the continuum limit of Procedures (I) and (II), and then by describing our Monte Carlo simulations of the above diffusion processes.

2.1. Continuum limit

The Langevin equation describing the continuum limit of Procedures (I) and (II) can be derived by following the steps of the method we used [10] for deriving the Kardar-Parisi-Zhang (KPZ) equation [11]. First the Master equation of the process is written down. Next one derives an equation for the time derivative of $< h_i >$

$$\frac{\partial < h_i >}{\partial t} = < G(\{h_i\}) > \quad , \tag{1}$$

where the average $<>$ is over the probability distribution $P(\{h_i\}, t)$ which is the solution of the Master equation. Then $G(\{h_i\})$ that is a complicated function of h_i, $h_{i\pm1}$, and $h_{i\pm2}$ is expanded in height differences $h_i - h_{i-1}$. Keeping only the lowest order terms, the differences are replaced by spatial derivatives $\partial h(x,t)/\partial x$. Finally, one assumes that equation (1) has been obtained from a Langevin equation of the form $\partial h(x,t)/\partial t = F(h, \partial h/\partial x, ...) + \eta$ that has been averaged over the noise η. Then the functional F is given by the expression under $<>$ on right hand side of eq. (1) and the only problem that remains is to choose the noise η in accord with the conservation laws in the system. In our case the result of the above steps lead to the following equation [4]:

$$\frac{\partial h}{\partial t} = \nu_2 \frac{\partial^2 h}{\partial x^2} - \nu_4 \frac{\partial^4 h}{\partial x^4} + \frac{\lambda}{2}\frac{\partial^2}{\partial x^2}\left[\frac{\partial h}{\partial x}\right]^2 + \eta \quad . \tag{2}$$

where $\nu_2, \nu_4, and \lambda$ are constants and η represents a conservative Gaussian white noise. An infinite sequence of other terms are, in principle, present in eq.(2). They can be shown, however, to be irrelevant in the renormalization-group sense provided the noise $\eta(x,t)$ is conservative.

In case of the detailed-balance diffusion, Procedure (I), one can show that $\lambda = 0$ since the reflection symmetry $(h_i \rightarrow -h_i)$ is not broken. Furthermore, one can argue [4] that the Laplacian term is also absent $(\nu_2 = 0)$. The resulting equation is exactly solvable and its solution describes a rough surface with the width of the surface w scaling as $w \sim L^{1/2}$ in $d = 1$ dimension. The dynamic critical exponent z is also known $(z = 4)$ in this case.

Procedure (II) violates both detailed balance and the reflection symmetry. Thus, in general, $\lambda \neq 0$ and the lowest order nonlinear term is of the form $\nabla^2(\nabla h)^2$. A term of this form has been used in many studies of surface evolution [2,5-8] but its origin was discussed only in [2] where it was suggested that this type of term may appear if diffusion occurs on a high-symmetry plane of a crystal. As we see, the term $\nabla^2(\nabla h)^2$ arises naturally in case of diffusion that does not obey detailed balance.

Another difference between the continuum limits of Procedures (I) and (II) is that one does not have an argument that would exclude the Laplacian term $(\nu_2 \neq 0)$ in case of violation of detailed balance. Below we shall analyse the results of our Monte Carlo simulations in terms of equation (2) with nonzero ν_2, ν_4, and λ.

2.2. Monte Carlo simulations

We carried out simulations of both Procedures (I) and (II) for one-dimensional systems of sizes $L = 10, 20, ..., 160$, and for various values of the maximum hight-difference parameter H/Δ. Details of the finite-size scaling analysis of the steady-state and of the relaxational properties of the surfaces can be found in [4]. Here we summarize the results in the following observations.

i) In case of detailed-balance diffusion, the usual equilibrium roughness in the steady state ($w \sim L^{1/2}$) is observed, and furthermore one finds $z \approx 4$ for the dynamical critical exponent. These results are independent of H/Δ and they are in agreement with the notion that Procedure (I) is described by the detailed-balance version ($\nu_2 = 0$ and $\lambda = 0$) of equation (2).

ii) In case of violation of detailed balance (Procedure (II)), the results depend on H/Δ and three cases should be distinguished.

ii.a) If $H/\Delta < 3$ then the surface is flat ($w \sim$ constant) and $z \approx 2$. This result can be understood in terms of eq.(2) if one assumes that $\nu_2 > 0$.

ii.b) If $H/\Delta = 3$ then the surface is dynamically rough ($w \sim L^{0.32}$) and $z \approx 3.67$. The exponents found in this case are very close to those calculated [5] for the conserved KPZ equation $w \sim L^{1/3}$ and $z = 11/3$. The conserved KPZ equation, however, is equation (2) with $\nu_2 = 0$. Thus we can interpret our results in terms of equation (2) by assuming that ν_2 is a function of H/Δ and $\nu_2(H/\Delta = 3) = 0$.

ii.c) The entire regime of $H/\Delta > 3$ can be characterized as a grooved phase in which the width of the surface is proportional to L. The relaxation towards the steady state seems to be an activated process that does not show obvious scaling behavior, and we did not characterize the relaxation by a dynamical exponent. The steady-state configuration of the surface in the grooved phase (for all $H/\Delta > 3$ we considered) appears to be the deterministic (noise-free), time-independent solution of the conserved KPZ equation,

$$h(x) = h_0 - a \ln\left[\cosh(x/b)\right] \quad,$$

where h_0, a and b are fitting parameters. The fact that the surface developes slopes which are close to the maximum allowed in the RSOS model indicates that the system is unstable. This behavior can also be explained by equation (2). If the trend $\nu_2(H/\Delta < 3) > 0$ and $\nu_2(H/\Delta = 3) = 0$ continues then it is expected that $\nu_2(H/\Delta > 3) < 0$ and the changing of the sign of ν_2 produceses the instability observed in the simulations of the discrete model.

In summary, quite a variety of surface behavior can be observed when diffusion violates detailed balance. The scaling of these surfaces is essentially different from the scaling found in the equilibrium case.

3. Detailed-balance diffusion coupled to particle deposition

This model was designed [9] to describe conditions that apply to molecular-beam epitaxy when the deposition of particles comes from a particle beam perpendicular to the substrate. Evaporation is neglected and the surface is assumed to relax exclusively by diffusion. The deposition rate is assumed to be so small that the effect of particle deposition on diffusion can be neglected so that the diffusion satisfies detailed balance at the substrate temperature T. The Hamiltonian in the detailed balance condition is assumed to be that of an unrestricted SOS model

$$H = \frac{J}{2} \sum_i [\ (h_{i+1} - h_i)^2 + g_4(h_{i+1} - h_i)^4\ ]\ , \tag{3}$$

where h_i is a single-valued function of i, and periodic boundary conditions are used. The probability of deposition of a particle at a given site is τ while the probability that a surface particle attempts a diffusive move to one of the neighboring site is $1 - \tau$. The Monte Carlo simulations of this model are described in detail in [9]. Here we summarize the results.

i) In the zero deposition limit ($\tau \to 0$) one recovers the equilibrium diffusion results $w \sim L^{1/2}$ and $z = 4$.

ii) The scaling properties are different for $\tau \neq 0$ but the exponents depend neither on the actual value of τ nor on the temperature of the substrate ($\tau = 0.01$ and $\tau = 0.1$ as well as $J/T = 0.01$ and $J/T = 1$ were investigated). The surface morphology, however, depends crutially on the value of g_4.

ii.a) In case of the Gaussian model ($g_4 = 0$), one finds a strongly fluctuating surface ($w \sim L^{1.3}$) but no 'shape' emerges (translational invariance is obeyed). The dynamics is slow, it is characterized by an exponent $z = 3.6$.

ii.b) For $g_4 > 0$, a shape instability occurs. Translational invariance is broken and the height, $< h(x) >$, becomes a nontrivial function of x. The shape one observes [9] is a slightly rounded letter V with the maximum height-difference proportional to $L^{3.6}$. Preliminary results indicate that the same shape instability occurs in two dimension as well with the surface having a pyramidal shape. The occurance of an instability when g_4 becomes positive is a rather puzzling. Since the $(h_{i+1} - h_i)^2$ term in the Hamiltonian has a positive coefficient, one expects that the quartic term with a positive coefficient is irrelevant. It is an interesting and unresolved question how the interplay between deposition and diffusion makes this term relevant. A clue to what is happening may be in the time-development of the instability that can be monitored by the time-dependent height-height correlation function

$$g(x,t) = \frac{1}{L} \sum_r < h(x + r, t)h(r, t) >\ .$$

The amplitude and the characteristic wavelength obtained from $g(x, t)$ seem to grow as power laws of time. This is similar to spinodal decomposition and, indeed, it has been argued [12] that the slope $< \partial h(x,t)/\partial x >$ behaves like the order parameter of Ising systems in spinodal decomposition. It should be noted, however, that our results for the exponents describing the time evolution of the characteristic wavelength and amplitude are different from those predicted by the Lifshitz-Slyozov theory. This indicates that actually a different growth mechanism is at work.

In summary, we can see that instabilities develop in rather generic circumstances ($g_4 \neq 0$) when diffusion is coupled to particle deposition, i.e. when diffusion takes place in far-from-equilibrium circumstances.

4. Final remarks

SOS type models become unphysical if the slopes of the height profile are larger than 1 (overhangs must be included). The onset of instabilities caused by far-from-equilibrium diffusion, however, occurs when the roughness is still small, and therefore the instability itself is physically relevant and might be observable. Whether it is observable depends largely on the answer to the question: What is the mechanism that prevents the divergence of the surface width in real deposition experiments?

Acknowledgement

We would like to thank M. den Nijs, H.-W. Diehl, M. Grant, H. Guo, B. Schmittmann, H. Spohn, and R.K.P. Zia for helpful conversations. This research was supported by the Deutsche Forschungsgemeinschaft, by NSERC of Canada, and by US NSF grant CHE-90 08033.

REFERENCES

[1] W.W. Mullins, *Metal Surfaces: Structure, Energetics, and Kinetics*, Am. Soc. Metals, Metals Park, 1963.

[2] J. Villain, *J.Phys.I (France)* **1** 19 1991.

[3] See experimental papers in this Volume.

[4] Z. Rácz, M. Siegert, D. Liu and M. Plischke, *Phys.Rev.A* **43** 5275 1991.

[5] T. Sun, H. Guo and M. Grant, *Phys.Rev.A* **40** 6763 1989.

[6] D.E. Wolf and J. Villain, *Europhys.Lett.* **13** 389 1990.

[7] Z.-W. Lai and S. Das Sarma, *Phys.Rev.Lett.* **66** 2348 1991.

[8] L.-H. Tang and T. Nattermann, *Phys.Rev.Lett.* **66** 2899 1991.

[9] M. Siegert and M. Plischke, *Phys.Rev.Lett.* **68** 2035 1992.

[10] M. Plischke, Z. Rácz and D. Liu, *Phys.Rev.B* **35** 3485 1987.

[11] M. Kardar, G. Parisi and Y.-C. Zhang, *Phys.Rev.Lett.* **56** 889 1986.

[12] L. Golubović and R.P.U. Karunasiri, *Phys.Rev.Lett.* **66** 3156 1991.

ROUGHENING TRANSITION OF A DRIVEN INTERFACE WITH A CONSERVATION LAW

Tao Sun, Bertrand Morin, Hong Guo, and Martin Grant

Department of Physics, and Centre for the Physics of Materials, Rutherford Building, McGill University, 3600 University Street, Montréal, Québec, Canada H3A 2T8

Abstract. The dynamic roughening transition of a driven interface, with conservation of total volume under the interface, has been studied using a dynamic renormalization group technique. This conserved system exhibits a true phase transition, rather than crossover behavior, as has been found to occur for nonconserved driven interfaces. Indeed, the conserved nonlinear driving force is favorable for smoothing the interface, implying that the roughening transition shifts to higher temperatures as that driving force is increased. The nature of the phase transition remains the same as the equilibrium transition: The renormalization-group transformation is controlled by a Kosterlitz-Thouless fixed point.

1. Introduction

The roughening transition of an interface separating two phases has been of long-standing interest in surface science, both because of its practical importance in understanding the process of crystal growth and because of the fundamental interest in the nature of the phase transition. The first suggestion of the existence of a roughening transition in equilibrium interfaces was made by Burton, Cabrera, and Frank [1] more than forty years ago. Following their study, much theoretical and experimental work has been done [2–8] and the main features of the equilibrium roughening transition are now well understood. A flat solid facet will become unstable to the formation of ledges as temperature is increased. Beyond the roughening transition temperature T_R, these ledges appear and disappear with no cost of free energy, giving rise to capillary-wave excitations, and a rough interface. At the roughening transition, it has been shown that the free energy and all its derivatives are continuous, and that the transition is a Kosterlitz-Thouless transition [9–16].

The dynamics of the equilibrium roughening transition was first studied within linear response by Chui and Weeks [17], and later by Nozières and Gallet [18] using dynamical renormalization group techniques. Both groups used linearized models and obtained height fluctuations characterized by logarithmic growth. However, these results are believed to capture the essential physics of systems near equilibrium only.

The morphology of a growing interface which is *far from equilibrium* has attracted increasing attention during the last decade. Of particular interest is the problem of kinetic roughening of a growing surface. The understanding of this nonequilibrium process was improved substantially by Kardar, Parisi, and Zhang (KPZ) [19], who first realized that the essential physics of this dynamic process can be described by the non-linearity due to lateral growth. Recently, a number of investigations have been made of the possibility of a nonequilibrium roughening transition by extensive numerical simulations of driven interface growth models [20–27]. These apparent transitions were thought to be possibly the far-from-equilibrium analog of the equilibrium roughening transition.

Motivated by this work Hwa, Kardar, and Paczuski (HKP) recently generalized the linear theories to include the nonlinearity due to lateral growth [28]. They found that the dynamics was no longer diffusive, and an anomalous decrease in mobility was predicted. Furthermore, they showed that the apparent transition seen numerically was actually only crossover behavior. At sufficiently large length scales, the apparent roughening transition in these far-from-equilibrium systems disappeared. This was due to a non-zero velocity of the interface, which was magnified under rescaling by the renormalization group. Physically, the driving force causes a constant velocity by the nucleation of ledges on ledges; however, these ledges destroy the flat facet, regardless of how low the temperature may be.

2. Model

In this work, we consider a closely related problem, where a conservation law is present [29], so that the above-mentioned crossover effects play no role. It is well known that conservation laws often determine the long-length-scale, late-time behavior of dynamic systems. That is, conservation laws are relevant for determining a dynamical universality class. In fact, a conservation law leads to a new universality class from that of the non-conserved KPZ model [29–31]. That conserved model is:

$$\frac{\partial h}{\partial t} = -\nabla^2[\nu\nabla^2 h + \frac{\lambda}{2}(\nabla h)^2] + \eta(\mathbf{x},t),$$

where $h(\mathbf{x},t)$ is the interface height variable at space-time point $(\mathbf{x},t)$, and the noise $\eta(\mathbf{x},t)$ satisfies Gaussian statistics with

$$\langle\eta(\mathbf{x},t)\rangle = 0,$$

$$\langle\eta(\mathbf{x},t)\eta(\mathbf{x}',t')\rangle = -2D\nabla^2\delta\,(\mathbf{x}-\mathbf{x}')\delta(t-t').$$

Here ν, λ, and D are constants.

The conservation of total volume below the interface is evident because $\partial h/\partial t$, including the noise, can be written as the divergence of a current. Aside from the Laplacians acting on the deterministic part of the equation of motion, and in the noise correlation, the model is the same as the KPZ equation. The nonlinear driving force here, proportional to λ, arises from breaking detailed balance of the dynamics. A thorough discussion of this has been given by Rácz et $al.$ [31].

The usefulness of the model is in its description of, for example, the surface diffusion of atoms rearranging on a surface due to a driven flux, where detailed balance does not apply. It should be noted, though, that the model does not describe the potentially important effects of conserved long-range diffusion [32], or nonconserved stress-strain relaxation at a surface [33]. Furthermore, the numerical study of the lattice-gas model of Rácz et $al.$ [31] implies that one may require considerable tuning to find a system which has neither a finite correlation length, nor an instability (these correspond to an additional term on the right-hand side of the equation above of the form $m\nabla^2 h$, where the constant m would be positive or negative, respectively). Nevertheless, we think it plausible that the model would describe some aspects of driven growth when a conservation law is important. We also note that this and related conserved models are being studied to further understand epitaxial growth by molecular beams [34,35].

The model has interesting properties [29]. First, if the width of a rough interface is given by $w \sim L^\chi f(t/L^z)$, for a system of linear size L at time t after it has been prepared with $w = 0$, the exponents χ and z are nontrivial below the critical dimension $d_c = 3$ (where, in our notation, $2 + 1 = 3$). The exponents can be calculated by the $\epsilon = 3 - d$ expansion, since the fixed point is stable, giving $\chi = (3 - d)/3$ while $z = (9 + d)/3$, enforcing $\chi + z = 4$. These correspond to a significantly less rough interface than the trivial case where $\lambda \equiv 0$; There, $\chi_0 = (3 - d)/2$ and $z_0 = 4$, so that $\chi \leq \chi_0$. This feature arises from the conservation law, which smooths out a rough interface by rearranging matter by surface diffusion. In contrast, the nonconserved KPZ equation has an unstable fixed point, and $\chi \geq \chi_0$.

To investigate the possibility of a roughening transition, we add a piece describing a lattice pinning potential [35] of constant strength y:

$$\mu^{-1}\frac{\partial h}{\partial t} = -\nabla^2[\nu\nabla^2 h - y\sin(2\pi h) + \frac{\lambda}{2}(\nabla h)^2] + \eta(\mathbf{x}, t),$$

where μ is a mobility. Again, aside from the Laplacians in the noise and in the deterministic part of the right-hand side of the equation, this model is the same as the nonconserved HKP equation. However, the conservation law explicitly forbids a constant velocity of the interface. Thus, unlike the nonconserved case, our model has, as we show below, a true roughening transition in the critical dimension $d_c = 3$.

3. Method and results

We have used the dynamic renormalization group to analyze this model. Unfortunately, the nonlinear coupling terms rule out Galilean invariance, and there is no obvious fluctuation-dissipation theorem for this model. Therefore we have to calculate simultaneously the two-point correlation functions, response functions and vertex functions in order to obtain the long-time, large-distance properties of the dynamic model. This is an involved calculation, and we will present the details elsewhere [36]. Here, we give the results. Because of the existence of the pinning potential, it is convenient to use Nozières and Gallet's dynamic renormalization technique [18]. Treating y and λ perturbatively, and including all terms to second order, we obtain the following flow equations

$$\frac{d\ln\nu(l)}{dl} = \frac{1}{2}\pi^2\zeta(n)\left(\frac{y}{\nu}\right)^2 + \frac{1}{8\pi^2}n\left(\frac{\lambda}{\nu}\right)^2,$$

$$\frac{d\ln y}{dl} = 2 - n - \pi^2\xi(n)\left(\frac{y}{\nu}\right)^2,$$

$$\frac{d\ln\lambda(l)}{dl} = -\frac{1}{2}\pi^2\zeta(n)\left(\frac{y}{\nu}\right)^2,$$

$$\frac{d\ln D(l)}{dl} = 0,$$

$$\frac{d\ln\mu(l)}{dl} = 0,$$

where $n = \pi D\mu/\nu$, and the values of $\zeta(n)$ and $\xi(n)$ around the fixed point $n^* = 2$ are: $\zeta(n) = 0.22 - 0.30(n-2) + O((n-2)^2)$ and $\xi(n) = 0.14 + O(n-2)$. In deriving these equations, a standard rescaling of $x' = e^{-\delta l}x$, $t' = e^{-4\delta l}t$, and $h' = h$ have been used. The new feature here is that the conservation law makes the pure dynamical parameters μ and D unrenormalized, while the lattice effect is relevant to the renormalization of λ. Note that the renormalized mobility μ stays constant at the phase transition. This is a novel difference from the non-conserved system where the mobility decreases anomalously.

Following the method of HKP, we examined the renormalization group flows near the equilibrium roughening transition fixed point by setting $t = \pi D\mu/\nu - 2$, $\bar{y} = \pi\zeta(2)^{1/2}y/\nu$, and $\bar{\lambda} = \lambda/(2^{1/2}\pi\nu)$. Then to the next-leading order in $(t,\bar{y},\bar{\lambda})$, we have the following reduced dynamic Kosterlitz-Thouless flow equations

$$\frac{dt}{dl} = -\bar{y}^2 - \bar{\lambda}^2 - \alpha t\bar{y}^2 - t\bar{\lambda}^2,$$

$$\frac{d\bar{y}}{dl} = -t\bar{y} - \beta\bar{y}^3 - \gamma\bar{y}\bar{\lambda}^2,$$

$$\frac{d\bar{\lambda}}{dl} = -\bar{\lambda}\bar{y}^2 - \gamma\bar{\lambda}^3,$$

with $\alpha = -0.86$, $\beta = 1.14$, and $\gamma = 0.50$. Note that with $\bar{\lambda} = 0$, this gives the equilibrium Kosterlitz-Thouless flow equations with the universal number [16] $\alpha + 2\beta \approx 1.4$, which is close to the correct number 1.5, the difference arising from our approximate evaluation of integrals.

For finite $\bar{\lambda}$, the flow equations have a fixed line ($\bar{y} = \bar{\lambda} = 0$, $0 \leq t < -1/\alpha$), analogous to the usual flow close to a Kosterlitz-Thouless fixed point, which is stable both to $\bar{\lambda}$ and to $\bar{y}$. Away from the fixed line, the leading-order terms show that, on the $t > 0$ side, $\bar{\lambda}$ tends to drive t to zero and then helps $\bar{y}$ to persist. Indeed, the flow of $\bar{y}$ to zero follows the slow $\sqrt{1/l}$, rather than the usual $1/l$, as $l \to \infty$. This indicates that the lateral growth effect is favorable for smoothing the interface when the conservation law is present, as mentioned above. This feature is essentially different from that of the non-conserved model where $\bar{\lambda}$ is favorable to roughening. In some sense, the origin of this effect is because the conserved equation has a stable Gaussian fixed point at its critical dimension $d_c = 3$.

It is easy to show that the invariant of the flow equations to third order is

$$F(t, \bar{y}, \bar{\lambda}) = t^2 - \bar{y}^2 - \bar{\lambda}^2 + 2\beta t \bar{y}^2 + (4 - \alpha - 2\beta)t\bar{\lambda}^2 - \frac{2}{3}(\alpha + 2\beta)t^3;$$

$$G(t, \bar{y}, \bar{\lambda}) = -\bar{y}^2 - \bar{\lambda}^2 + (\alpha + 2\beta - 1)t\bar{y}^2 + t\bar{\lambda}^2.$$

The RG flow lines are then given by $F = F_0$ and $G = G_0$ for different values of constants F_0 and G_0 (that is, F gives the separatrix, while G gives the flow on that separatrix). When $F_0 = 0$, one obtains a two-dimensional separatrix in the three dimensional parameter space $(t, \bar{y}, \bar{\lambda})$

$$\bar{y}^2 = t^2 - \bar{\lambda}^2 + \frac{2}{3}(\beta - \alpha)t^3 + (4 - \alpha - 4\beta)t\bar{\lambda}^2,$$

The volume of phase space roughening the interface decreases as $\bar{\lambda}$ increases. Indeed, the roughening transition shifts to *higher* temperatures when the driving force is present. On the separatrix, asymptotically $t \sim 1/l$, $\bar{y}^2 \sim 1/l$, and $\bar{\lambda}^2 \sim 1/l$, which is the same as the nonconserved HKP model. This implies that, on small length scales, the transition could be similar to a second-order transition, although asymptotically it is a Kosterlitz-Thouless transition, since the lattice pinning potential y iterates more slowly to zero than the equilibrium case.

At the transition, the height-height correlation function is

$$C(\rho, \tau) \equiv \langle [h(\mathbf{x}, t) - h(\mathbf{x}', t')] \rangle = \frac{2}{\pi^2} \ln \left\{ \rho g \left[\frac{\tau}{\rho^4 \exp(\text{const.}/\ln \rho)} \right] \right\},$$

where $\rho = |\mathbf{x} - \mathbf{x}'|$, $\tau = |t - t'|$. The scaling function $g(x) \sim x^{\frac{1}{4}}$ for $x >> 1$ but becomes constant for $x << 1$. The static logarithmic growth is recovered when $\tau \sim 0$. However, the new scaling regime characterized by τ/ρ^4 for large ρ is due to the conservation law. The correlation length can be identified as the value of e^l for which the deviations of the reduced parameters from the fixed points become significant. Indeed, it is easy to show that the correlation length to leading order

$$\xi(T) \sim \exp \left\{ \frac{\text{const.}}{(T_R - T)^{1/2}} \right\}, \text{ as } T \to T_R^{(-)},$$

which signals that the transition is a dynamic analog of the equilibrium roughening transition.

In summary, our dynamical renormalization group analysis of the conserved equation shows that it has a true roughening transition of the Kosterlitz-Thouless type. Furthermore, the transition temperature shifts to higher temperatures as the nonlinear driving force is increased. We are presently studying the feasibility of testing these predictions experimentally.

4. Acknowledgements

This work was supported by the Natural Sciences and Engineering Research Council of Canada, and *les Fonds pour la Formation de Chercheurs et l'Aide à la Recherche de la Province de Québec*.

REFERENCES

[1] W. K. Burton, N. Cabrera, and F. C. Frank, Philos. Trans. R. Soc. **243A**, 299 (1951).

[2] J. D. Weeks, G. H. Gilmer, and H. J. Leamy, Phys. Rev. Lett. **31**, 549(1973).

[3] G. H. Gilmer, K. A. Jackson, H. J. Leamy, and J. D. Weeks, J. Phys. C7, L123 (1974).

[4] H. J. Leamy, G. H. Gilmer, and K. A. Jackson, in *Surface Physics of Materials* (Academic, New York, 1975), Vol. 1, p. 121.

[5] S. T. Chui and J. D. Weeks, Phys. Rev. B **14**, 4978 (1976).

[6] J. D. Weeks, in *Ordering in Strongly Fluctuating Condensed Matter Systems*, edited by T. Riste (Plenum, New York, 1980), p. 293.

[7] D. Jasnow, in *Phase Transitions and Critical Phenomena*, edited by C. Domb and J. L. Lebowitz (Academic, London, 1986), Vol. 10, p, 269.

[8] H. van Beijeren and I. Nolden, in *Structure and Dynamics of Surfaces*, edited by W. Schommers and P. von Blackenhagen, Vol. 43 of *Topics in Current Physics* (Springer, Berlin, 1987).

[9] J. M. Kosterlitz and D. J. Thouless, J. Phys. C **6**, 1181(1973); J. M. Kosterlitz, *ibid*, **7**, 1046 (1974); **10** 3753 (1977).

[10] D. R. Nelson, J. M. Kosterlitz, Phys. Rev. Lett. **39**, 1201 (1973).

[11] J. V. Jose, L. P. Kadanoff, S. Kirkpatrick, and D. R. Nelson, Phys. Rev. B **16**, 1217 (1977).

[12] P. B. Wiegmann, J. Phys. C **11**, 1583 (1978).

[13] T. Ohta and D. Jasnow, Phys. Rev. B **20**, 139 (1979).

[14] L. P. Kadanoff, Ann. Phys. **120**, 39 (1979).

[15] J. B. Kogut, Rev. Mod. Phys. **51**, 696 (1979).

[16] D. J. Amit, Y. Y. Goldschmidt, and G. Grinstein, J. Phys. A **13**, 585 (1980).

[17] S. T. Chui and J. D. Weeks, Phys. Rev. Lett. **40**, 733 (1978).

[18] P. Nozières and F. Gallet, J. Phys. (Paris) **48**, 353 (1987); F. Gallet, S. Balibar, and E. Rolley, *ibid* **48**, 369 (1987).

[19] M. Kardar, G. Parisi, and Y. Zhang, Phys. Rev. Lett. **56**, 889 (1986); E. Medina, T. Hwa, M. Kardar, and Y. Zhang, Phys. Rev. A **39**, 3053 (1989).

[20] J. Kertesz and D. E. Wolf, Phys. Rev. Lett. **62**, 2571 (1989).

[21] J. G. Amar and F. Family, Phys. Rev. Lett. **64**, 543 (1990).

[22] H. Yan, D. Kessler, and L. M. Sander, Phys. Rev. Lett. **64**, 926 (1990).

[23] H. Guo, B. Grossmann, and M. Grant, Phys. Rev. Lett. **64**, 1262 (1990); B. Grossmann, H. Guo, and M. Grant, Phys. Rev. A **43**, 1727 (1990).

[24] B. M. Forrest and L. Tang, Phys. Rev. Lett. **64**, 1405 (1990).

[25] Y. P. Pellegrini and R. Jullien, Phys. Rev. Lett. **64**, 1745 (1990).

[26] J. Krug, Phys. Rev. Lett. **64**, 2332 (1990).

[27] J. M. Kim, T. Ala-Nissila and J. M. Kosterlitz, Phys. Rev. Lett. **64**, 2333 (1990).

[28] T. Hwa, M. Kardar, and M. Paczuski, Phys. Rev. Lett. **66**, 441 (1991). M. Paczuski, Phys. Rev. Lett. **66**, 1545 ((1991).

[29] T. Sun, H. Guo, and M. Grant, Phys. Rev. A **40**, 6763 (1989).

[30] A. Chakrabarti, J. Phys. A **23**, L919 (1990).

[31] Z. Rácz, M. Siegert, D. Liu and M. Plischke, Phys. Rev. A **43**, 5275 (1991).

[32] I. M. Lifshitz and V. V. Slyozov, J. Phys. Chem. Solids **19**, 35 (1961).

[33] W. W. Mullins and J. Viñals, Acta Metall. **37**, 991 (1989).

[34] J. Villain, J. Phys. (France) **1**, 19 (1991). J. Villain, preprint (1991). D. E. Wolf, Phys. Rev. Lett. **67**, 1783.

[35] L.-H. Tang, and T. Nattermann, Phys. Rev. Lett. **66**, 2899 (1991).

[36] T. Sun, B. Morin, H. Guo, and M. Grant, in preparation.

RESTRICTED STEP HEIGHT BALLISTIC DEPOSITION WITH STICKY AND NON-STICKY PARTICLES

Rémi Jullien, Yves-Patrick Pellegrini and Paul Meakin

Laboratoire de Physique des Solides, Bât. 510, Université Paris-Sud, Centre d'Orsay 91405, Orsay, France

Abstract. A model for the deposition of a mixture of sticky and non sticky particles with restricted height steps was numerically investigated in dimensions $d=2$, 3 and 4. In contrast with a previous numerical analysis of a similar model with unrestricted height steps, our results are not consistent with the existence of a morphological phase transition when the concentration of sticky particles is varied in $d = 3$.

The ballistic deposition model [1] is one of the simplest model which has been introduced to simulate the growth of rough surfaces from a flat substrate. In the on-lattice two-dimensional version of this model, particles are deposited one after another along randomly positionned vertical trajectories onto a basal horizontal line of L occupied sites. In the generalized d-dimensional model particles (represented by filled lattice sites) are deposited onto a $(d-1)$-dimensional hypercubic substrate with sides of length L. Particles become part of the deposit as soon as they find a nearest-neighbor occupied site on the lattice.

In this model, the surface of the deposit can be described by as a statistically self-affine fractal [2] and the following scaling form has been proposed [3] to describe the growth of the surface width (σ):

$$\sigma \simeq L^\alpha f(\frac{h}{L^{\alpha/\beta}}) \tag{1}$$

where h is the mean height of the deposit. The scaling function $f(x)$ satisfies:

$$f(x) \longrightarrow \text{const} \quad \text{when} \quad x \longrightarrow \infty \tag{2a}$$

$$f(x) \sim x^\beta \quad \text{when} \quad x \longrightarrow 0 \tag{2b}$$

Hence the dependence of σ on L for large h ($h >> L^{\alpha/\beta}$):

$$\sigma \sim L^{\alpha} \tag{3a}$$

is different than the dependence of σ on h for small h:

$$\sigma \sim h^{\beta} \tag{3b}$$

In both $d = 2$ and higher dimensions, all numerical results available up to now are very close to the following conjecture, due to Kim and Kosterlitz (KK) [4]:

$$\alpha = \frac{2}{d+2}; \quad \beta = \frac{1}{d+1} \tag{4}$$

based on numerical results from a modified version of the model that includes restricted step heights. Some extensions of the on lattice ballistic model were considered in which surface diffusion, or particle restructuring, was included. If a particle, after landing on the top of a randomly chosen column, is allowed to "slide" to the nearest neighboring column if it is of smaller height ("one-step restructuring") [5] or if it can slide further until it reaches the nearest local minimum ("multi-restructuring") [6], it is found that the scaling form (1) is still valid with a different β exponent, $\beta = \frac{1}{4}$, while α remains equal to $\frac{1}{2}$ in two dimensions [5-6]. For $d = 3$, $\sigma \sim (\ln L)^{\frac{1}{2}}$ for large h and $\sigma \sim (\ln h)^{\frac{1}{2}}$ for small h and for $d \geq 4$, σ remains small for all values of h and L.

It is generally believed that such models are discrete versions of the continuous model introduced by Edwards and Wilkinson (EW) [7] and subsequently modified by Kardar, Parisi and Zhang (KPZ) [8]. These authors describe the growth of a surface using the following Langevin equation:

$$\frac{\partial h}{\partial t} = v + \nu \Delta h + D\eta + \lambda |\nabla h|^{2} \tag{5}$$

In this expression, $h(\vec{r},t)$ is the vertical coordinate of a surface point whose other $(d - 1)$ coordinates are described by $\vec{r}$, v is a constant equal to the mean velocity of the surface, the Laplacian term describes surface tension effects and favours surface smoothing, $\eta(\vec{r},t)$ is a Gaussian random variable with zero mean and unit mean square which describes the noise and the non-linear term, which has been added by KPZ to EW, describes the lateral growth of the surface. In terms of the discrete ballistic models, the full KPZ equation corresponds to the regular ballistic deposition model while the EW equation (KPZ without the non linear term) corresponds to the ballistic models with restructuring.

The KPZ equation has been extensively studied by means of renormalization group techniques in d-dimensions [9-11]. The behavior of the solution $h(\vec{r},t)$ is governed by the strength of the reduced coupling constant $\overline{\lambda} = \frac{\lambda D}{\nu^{3/2}}$. For $d < 3$, it is found that strong coupling prevails for all $\overline{\lambda} \neq 0$ while for $d > 3$ there should be a phase transition between a weak-coupling and a strong-coupling phase as $\overline{\lambda}$ is increased. Dimension $d = 3$ is a marginal dimension where nothing rigorous can be deduced from

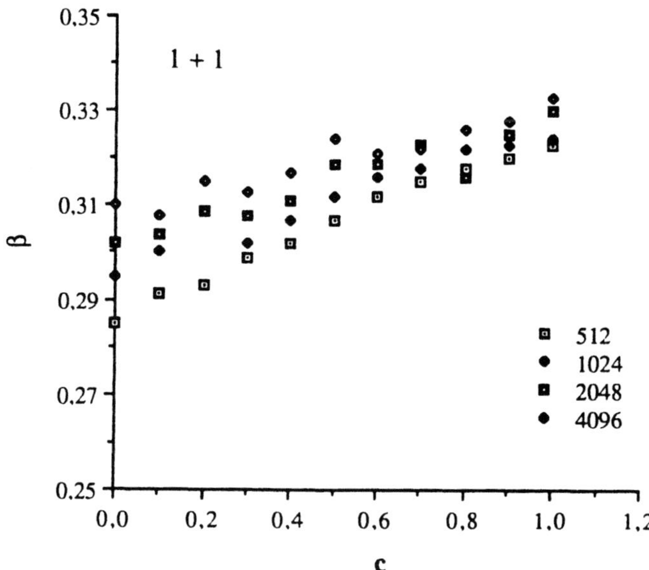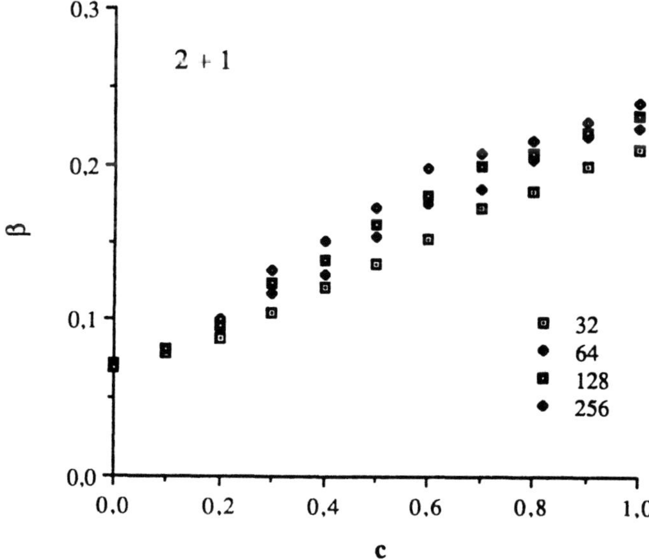

Figure 1. The β exponent as a function of c for $d = 2$ with $L = 512, 1024, 2048$ and 4096.

Figure 2. The β exponent as a function of c for $d = 3$ with $L = 32, 64, 128$ and 256.

renormalization group analysis. Numerical simulations on discrete models in which the magnitude of the reduced coupling constant λ can be varied give controversial results [11-13]. Two of us (YPP and RJ) have already introduced a simple model which considers the deposition of a mixture of sticky and non-sticky particles [13]. This model is a discrete version of the KPZ model which interpolates between $\lambda = 0$ and $\lambda \neq 0$ situations. In the previous publications [13-14], the results in $d = 3$ where interpreted in favour of the existence of a morphological phase transition. Here we would like to introduce a modified version of this model which includes restricted height steps conditions in a manner very similar to KK [4].

In our model, we start from a fully occupied $(d - 1)$-dimensional hypercubic base. To "deposit" a particle, we first choose a column at random in this base. Then if a random variable (uniformly distributed between 0 and 1) is smaller than c, the particle is a "sticky" particle otherwise it is a "non-sticky" particle. A sticky particle is deposited using the rule of the standard ballistic model [1] while a non-sticky particle is deposited using the rule of the one-step restructuring ballistic model [5]. As soon as the final position is reached according to these rules, we make a check of the step height conditions. If the absolute value of the height difference between the final column and any of the neighboring columns is larger than unity, the trial is discarded and another column is chosen at random (keeping the sticky or non sticky character of the particle) and this is repeated as long as the height steps conditions are not satisfied. The main difference with the previous model is that now there are neither holes in the bulk nor overhangs in the surface. Thus we can expect that there will be substantially reduced finite size corrections to scaling.

Here we report on numerical estimates of the exponent β for various L-values up to a maximum value of 4096, 256, 64 in dimensions 2, 3 and 4, respectively. This exponent has been estimated from the slope of the best linear region in the dependence of $\ln \sigma$ on $\ln h$ well below the saturation regime, for h-values smaller than $AL^{1.5}$ with $A = = \frac{1}{1024}, \frac{1}{64}, \frac{1}{4}$, in $d = 2$, 3 and 4, respectively. The numerical results are reported as a function of c in figures 1, 2 and 3. In $d = 3$ the effective value of the exponent β increases with increasing L except for $c = 0$ where it stays at a very small value (it even decreases very slightly with increasing L). Although there might be strong corrections to scaling, these results are consistent with an unstable fixed point at $c = 0$ (β is equal to the KK value, $\frac{1}{4}$, in the infinite-L asymptotic limit, for all c-values different from $c = 0$). In $d = 4$, the effective β decreases for c-values smaller than 0.6 while it increases for larger c-values. This is consistent with existence of a "phase transition" when c is changed in the thermodynamic limit.

We obtain a surprising result in two dimensions. Here, the effective β-exponent tends to the KK value 1/3 for all c-values, even for $c = 0$. It appears that our model, which includes both restricted step heights and restricted restructuring, does not reproduce the EW fixed point for $c = 0$. We have checked that, when multi-restructuring is used instead of one-step restructuring, the EW behavior is recovered. The two-dimensional simulation results with multi-restructuring are reported in figure 4 where they are compared with the one-step restructuring case. While the effective β exponent tends to the EW value $\frac{1}{4}$ when L tend to infinity for one step restructuring, it tends to the KPZ value, $\frac{1}{3}$, for multi-restructuring. As a further check, we have verified

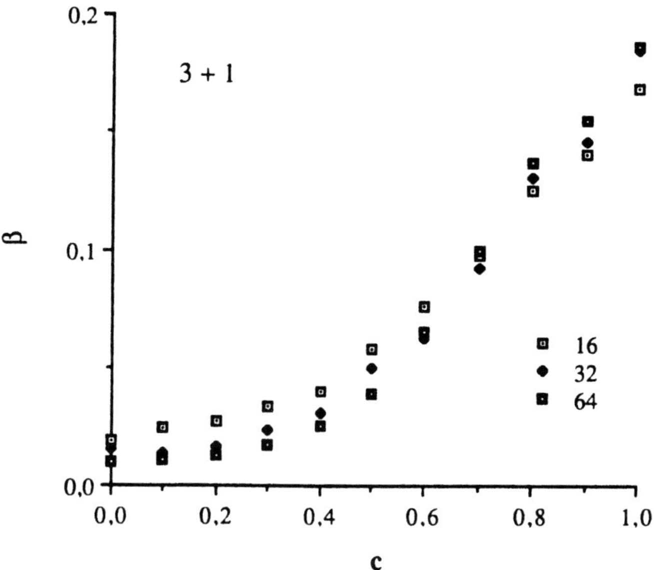

Figure 3. The β exponent as a function of c for $d = 4$ with $L = 16, 32,$ and 64.

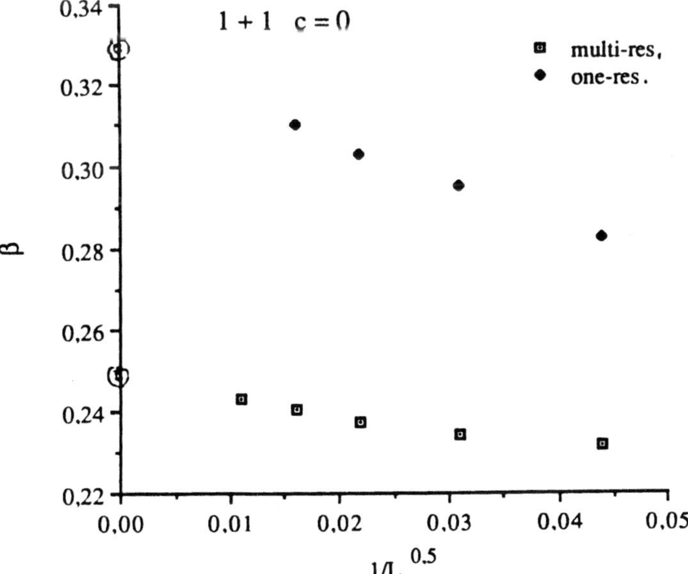

Figure 4. The β exponent as a function of $L^{-1/2}$ for $c = 0$ and $d = 2$ in the two cases of one-step restructuring and multi-restructuring.

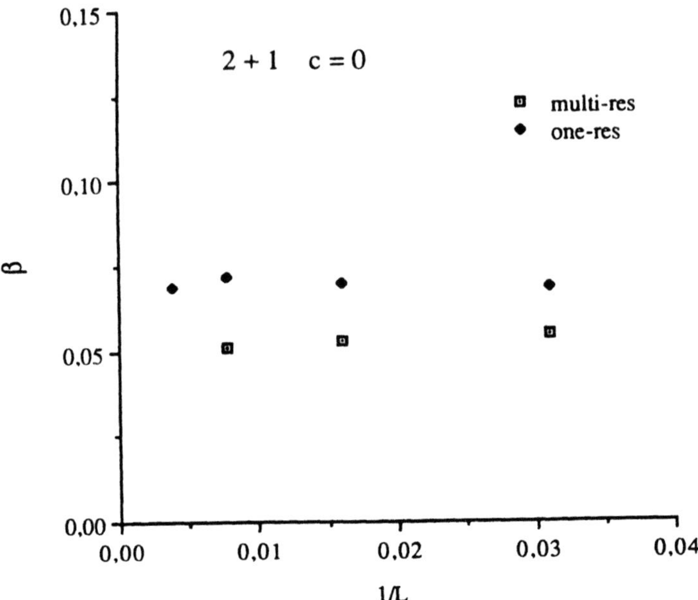

Figure 5. The β exponent as a function of L^{-1} for $c = 0$ and $d = 3$ in the two cases of one-step restructuring and multi-restructuring.

that such a difference occurs only in two dimensions. In three dimensions we have found that the value of the effective β exponent is not sensitive to the type of restructuring that we can consider. In figure 5, we have compared the L-dependence of the effective β exponent for $c = 0$ for the one-step restructuring and multi-restructuring models. In both cases β seems to converge towards a very small value perfectly consistent with a logarithmic growth of the surface width as expected for the EW fixed point.

In conclusion, the numerical results that we have reported here are now in good agreement with the results obtained from a direct numerical investigation of the KPZ equation [12]. The earlier evidence for the existence of a phase transition in three dimensions in a similar model with unrestricted height steps [13] might have been biased by strong finite size crossover effects. The existence of a strong maximum in the derivative of the compactness versus concentration curve and its coincidence with the percolation of sticky particles in $(d - 1)$ dimensions might be due to other effects such as a non linear variation of the λ parameter with c, as suggested to us by J. Krug. This will be checked by a direct estimation of λ from simulations on the unrestricted model with a tilted interface.

REFERENCES

[1] P. Meakin, P. Ramanlal, L. M. Sander and R. C. Ball, *Phys. Rev. A* **34** 5081 1986.

[2] B. B. Mandelbrot, *The Fractal Geometry of Nature*, Freeman 1982.

[3] F. Family and T. Vicsek, *J. Phys. A* **18** L75 1985.

[4] J. M. Kim and J. M. Kosterlitz, *Phys. Rev. Letters* **62** 2289 1989.

[5] F. Family, *J. Phys. A* **19** L441 1986.

[6] P. Meakin and R. Jullien, *J. Physique (France)* **48** 1651 1987.

[7] S. F. Edwards and D. R. Wilkinson, *Proc. Roy. Soc. London* **A 381** 17 1982.

[8] M. Kardar, G. Parisi and Y. C. Zhang, *Phys. Rev. Letters* **56** 889 1986.

[9] E. Medina, T. Hwa, M. Kardar and Y. C. Zhang, *Phys. Rev. A* **39** 3053 1989.

[10] T. Halpin-Healy, *Phys. Rev. Letters* **64** 926 1990.

[11] H. Yan, D. Kessler and L. M. Sander, *Phys. Rev. Letters* **64** 926 1990.

[12] J. G. Amar and F. Family, *Phys. Rev. A* **41** 3399 1990.

[13] Y. P. Pellegrini and R. Jullien, *Phys. Rev. Letters* **64** 1746 1990.

[14] Y. P. Pellegrini and R. Jullien, *Phys. Rev. A* **43** 920 1991.

KINETIC ROUGHENING OF SOLIDS II:

Experiments

ABOUT TWO DIFFERENT TYPES OF DYNAMIC ROUGHENING

S. Balibar, J.P. Bouchaud, F. Gallet, C. Guthmann and E. Rolley
Laboratoire de Physique Statistique de l'E.N.S. 24 rue Lhomond, 75231 Paris Cedex 05, France

Abstract. There are at least two different types of dynamic roughening which need to be clearly distinguished. Let us call them type I and type II. Type I only concerns crystal surfaces and type II is more general.

1. Introduction

A usual definition of the roughness of a crystal surface is the height-height correlation function $C(r) = < (h(r) - h(0))^2 >$, where h is the height at the horizontal position r (the surface is supposed to be horizontal). At equilibrium, and below the roughening temperature T_R, the roughness $C(r)$ saturates at large distance r with a finite value $C(\infty) \approx a^2 \ln(\xi/a)$ (a is the lattice spacing and x is the correlation length of the surface, which is smooth [1]). Above T_R, the roughness $C(r) \approx a^2 \ln(r/a)$ diverges as in the case of the free surface of a liquid. However, the change from finite to logarithmic roughness not only occurs as a consequence of increasing temperature, it can also result from the growth at a finite velocity v, under a certain departure from equilibrium dm or driving force $F = \rho_c \delta\mu$ (where ρ_c is the crystal density, and $\delta\mu$ is the difference in chemical potential per unit mass between the adjacent fluid and the crystal). This is the dynamic roughening of type I.

By type II, we mean a change from logarithmic roughness to algebraic roughness ($C(r) \approx r^{\chi}$); it is the main topic of this conference and seems to concern many different sorts of moving surfaces, not only crystal surfaces.

The type I was first introduced by Chui and Weeks in 1978 [2] and extensively studied theoretically by Nozières and Gallet [3] as well as experimentally by Gallet, Balibar and Rolley [4] in the case of helium crystals. Type II was predicted by Kardar, Parisi and Zhang [5] as a consequence of a particular non-linear coupling between fluctuations in their model. Eventually, in a recent letter [6], Hwa, Kardar and Paczuski presented an interesting theory which predicts mixed effects of the two types of dynamic roughening. It is possible to illustrate this problem by considering helium 4 crystals where the surface properties are particularly well known. However, as explained in a comment by S. Balibar and J.P. Bouchaud [7], the type II roughening is negligible in experiments such as those performed by Gallet et al. [4]. This is what we would like to explain in some more detail here. We need first to recall the experimental results of Gallet et al., as well as their interpretation in terms of type I roughening. We then estimate the various orders of magnitude which are necessary to compare the relevance of type I and type II. In a short conclusion, we also briefly try to imagine how the two types could perhaps occur simultaneously.

2. The experiments in helium 4 and type I dynamic roughening

Helium 4 can be made extremely pure, and both superfluid helium 4 and good quality single crystals of helium 4 are extremely good thermal conductors. As a consequence, thermodynamic equilibrium can be obtained in short times and very small departures from this equilibrium can be accurately controlled. The first measurements of the surface stiffness g were obtained by Balibar, Wolf et al. [8] from an analysis of equilibrium crystal shapes. More complete measurements were obtained by Babkin et al. [9] and by Andreeva et al. [10] who studied the propagation of crystallization waves. The critical behaviour of the step energy β was obtained by Gallet et al. [4] who measured the growth velocity v as a function of the driving force F in the temperature interval $1.15 < T < 1.21$K, below $T_R = 1.30$K. This measurement was made possible from the experimental evidence of the 2D nucleation of terraces in this temperature interval. Eventually, the critical behaviour of the growth rate $k = v/\delta\mu$ was studied by Gallet et al. in the critical region $T = T_R \pm 0.08$K and for various applied forces F (typically $10^{-7} k_B T/a^3$). All these data are very well described by Nozières'theory [3]. This theory is based on the following Langevin equation:

$$\rho_c k^{-1} \partial h/\partial t = F + \gamma\Delta h - (2\pi/a)V\sin(2\pi/a) + \eta(x,t) \tag{1}$$

where ρ_c, h, F and k have already been defined, k can also be considered as the interface mobility, γ is the surface stiffness, V the amplitude of a lattice potential which favors integer values of h/a and η a thermal noise.

From a renormalization calculation, Nozières and Gallet obtain the variation of γ, k and $U = VL^2$ as a function of the scale λ at which fluctuations are averaged. They define $n = \pi k_B T/(\gamma a^2)$ and $\varepsilon = \ln(L/L_0)$ and write

$$\partial U/\partial\varepsilon = (2 - n)U \tag{2}$$

$$\partial\gamma/\partial\varepsilon = 2A(n)\pi^4 U^2/(\gamma a^4) \tag{3}$$

$$\partial k/\partial\varepsilon = 8B(n)[\pi^4 U^2/(\gamma a^4)]/(k/\gamma) \tag{4}$$

The type I dynamic roughening is a finite size effect. As explained in refs.[3] and [4], a finite velocity v introduces a finite time $\tau = a/v$. If there is a time scale in the problem, there is also a length scale $L^* = (\gamma a/F)^{1/2}$. Indeed, if one neglects F, V and h, eq.(1) becomes a diffusion equation for height fluctuations in which the diffusion coefficient is $k\gamma/\rho_c$. L^* is nothing but the diffusion length of fluctuations on a time $\tau = a/v$. On a scale larger than about L^*, fluctuations see a potential which oscillates in time and is thus averaged to zero. Since it is the non linear sin term which is responsible for the coupling between fluctuations, consequently for the renormalization of γ and k, one sees that a finite growth velocity liberates the large wavelength fluctuations and changes the roughness from finite to logarithmic.

Some theorists write that any finite F is relevant and destroys the smooth facetted state: this is true in principle, for an infinite system at infinite time, but it may be of no practical meaning since the growth velocity may be exponentially small and the logarithmic roughness negligible. However, close to T_R, where the step energy is small, Gallet et al. observed on helium crystals that a very small force F produces a substantial growth velocity and a definite dynamic roughening.

In order to compare with experiments, a modification was introduced by Nozières and Gallet[3] in eq.(1): h was replaced by $h - vt$ in the lattice term. The same renormalization equations were obtained, but with modified coefficients $A(n, L/L^*)$ and $B(n, L/L^*)$ which now depend on scale. These two coefficients were found to vanish at a scale L of order L^*, as expected [3][11]. From the integration of the new renormalization equations, a good description of the experiments is obtained [4][11]. The measurements of β and γ were also well described and the three parameters in the theory [4][11]were adjusted to

$$T_R = 1.30\text{K} \tag{5}$$

$$t_c = 1.25\pi^2 U_0/(\gamma_0 a^2) - 0.58D \tag{6}$$

$$\Lambda_0 = 1/L_0 = 0.3/a. \tag{7}$$

The same theory, without any further adjustment, also describes very well the data from refs. [9] and [10] on the angular variation of γ, and this has important consequences on the step degeneracy transition and the measurement of step-step interactions[11]. One can thus describe Gallet's work as a complete check that Nozières' theory accurately describes the equilibrium roughening transition as well as the type I dynamic roughening of helium crystals.

3. The algebraic roughening

Let us now consider the type II dynamic roughening and restrict ourselves to the one described by KPZ [5]. This restriction is justified by the fact that helium 4 atoms are directly incorporated in the crystal from the superfluid above, in the absence of any bulk or surface diffusion field. KPZ introduced another non-linear term which couples surface fluctuations. It is usually written as $(\lambda/2)(\nabla h)^2$ and was added by Hwa et al. [6] in eq.(1) in order to improve Nozières' description of dynamic roughening. As a consequence, they predicted new renormalization equations, a displacement of the roughening temperature and new critical behaviours. However, it seems to us that their corrections are negligible in the case of experiments close to T_R and with small forces, such as Gallet's experiments on helium 4. We first need to understand the physical origin of the KPZ term, then to calculate its magnitude and finally to compare it with the other terms, namely the capillary term $\gamma \Delta h$ and the lattice term V.

The KPZ term has a double origin. First there is a geometrical effect: the crystal grows with a certain shape, with a local slope ∇h, and the vertical component $\partial h/\partial t$ must be related to the local velocity v, which is normal to interface, by the relation

$$\partial h/\partial t = v(1 + 1/2(\nabla h)^2 + ...) \tag{8}$$

where we assume that the slope is small. Secondly, there is no reason why the mobility k should be isotropic for a crystal and we can describe this anisotropy by

$$k = k_0(1 + \varepsilon_k (\nabla h)^2 + ...) \tag{9}$$

(More general and complex anisotropies are described by D. Wolf in ref. [12] and in this conference). The KPZ equation is simply obtained by equating $\rho_c k^{-1} v$ to the sum of applied forces, and by keeping only those terms which respect the symmetry under the change ($h \Rightarrow -h$, $F \Rightarrow -F$). This leads to the very simple equation

$$\lambda = F(1 + \varepsilon_k) \tag{10}$$

which shows that λ vanishes for $F = 0$, as explained by Medina et al.[13]. For rough surfaces, ε_k is of order 1 or less. Below T_R, the occurence of steps may create large anisotropies since in the limit $F = 0$, the mobility k has a cusp $k = k_s|\phi|$ in the angular range $|\phi| < a/\xi$. However, this cusp is rounded by the 2D nucleation of terraces or from the existence of screw dislocations and Frank-Read sources under a finite F. Close to T_R, the step width ξ is very large and the narrow cusp is washed out by a small F since the 2D nucleation of terraces is very easy. In Gallet's experiments, ek is of order 1 and λ of order F.

Let us now compare the KPZ and capillary terms. The width of the interface is at most logarithmic (we anticipate on our conclusion),

$$(\nabla h)^2/\Delta h < a\ln(x/a) \approx a \tag{11}$$

and the ratio of the two terms is less than aF/γ. This is the dimensionless version of λ in ref. [6] and it is less than 10^{-7} in Gallet's experiments.

Let us now compare the two non-linear terms. The exact quantities to be compared are λ and $2\pi U/a^3$. As explained by Hwa et al. [6], λ is not renormalized. At the microscopic scale, U_0 was found to be equal to $5.10^{-2} k_B T_R$ since $t_c = 0.58$ and $\gamma_0 = 0.245$ erg/cm^2. At the microscopic scale, the KPZ term is thus negligible again. At larger scale we need to distinguish between the two sides of T_R. According to eq.(2), one has

$$U \approx U_0(L^*/L_0)^{2-n} \tag{12}$$

Below T_R, n is smaller than 2 and U increases with scale. Above T_R, U decreases and one could imagine that the lattice effect becomes comparable to the KPZ non-linearity. However, in Gallet's experiments, the renormalization stops at $L^* \approx 10^4 a$ so that even up to 1.40K where $2 - n = 0.78$, one has

$$U > U_0(L^*/L_0)^{-0.78} \approx 5.10^{-5} k_B T_R \qquad (13)$$

and the KPZ term is once more negligible.

The above analysis illustrates why, close to T_R, a small driving force F is enough to produce the type I dynamic roughening and why it does not mix with the type II. Equivalently, one could say that type I occurs on length scales of order $L^* = 1\mu$ for driving forces which are so small that type II would occur for unrealistic scales. Indeed, as explained by Tang et al. [13], the crossover length to algebraic roughness is

$$L^{**} \approx a \exp(\gamma^3/\lambda^2 D) \qquad (14)$$

where D is the noise amplitude. In our case, since D is $k_B T$, and the universal relation of roughening says $\gamma a^2 \approx k_B T_R$, we get

$$L^{**} \approx a \exp(\gamma^2/a^2 F^2) \qquad (15)$$

and since $\gamma/aF > 10^7 7$, L^{**} is unrealistically large.

An interesting question remains open. How could one imagine a situation in which the dynamic roughening of types I and II could mix? L^{**} varies so rapidly that it seems hard to escape from the condition $F \approx k_B T_R/a^3 \approx \gamma/a$. One probably needs to work at low enough temperature and with non-thermal noise so that the lattice potential is non-zero and the anisotropy is large, and then use a moderate driving force to increase λ without destroying the large possible anisotropy of k. Such simultaneous conditions seem hard to achieve with helium crystals. They may be met in MBE experiments and a precise prediction about such a regime would be very intersting. One should check that the two lengths L^* and L^{**} can be brought to the same order of magnitude.

Acknowledgement

Sebastien Balibar and Jean Philippe Bouchaud wish to thank Terrence Hwa, Mehran Kardar, Maya Paczuski and Dietrich Wolf for very stimulating discussions during the conference.

REFERENCES

[1] For a review, see P. Nozières, in *Solids Far from Equilibrium*, (C. Godreche ed.) Cambridge University Press 1992.
[2] S.T. Chui and J.D. Weeks, *Phys. Rev. Lett.* **40** 733 1978.
[3] P. Nozières and F. Gallet, *J. Physique* **48** 353 1987.

[4] F. Gallet, S. Balibar and E. Rolley, *J. Physique* **48** 369 1987.

[5] M. Kardar, G. Parisi and Y.C. Zhang, *Phys. Rev. Lett.* **56** 889 1986.

[6] T. Hwa, M. Kardar and M. Paczuski, *Phys. Rev. Lett.* **66** 441 1991.

[7] S. Balibar and J.P. Bouchaud, Comment submitted to *Phys. Rev. Lett.* 1992.

[8] S. Balibar, D.O. Edwards and C. Laroche, *Phys. Rev. Lett.* **42** 782 1979.

P.E. Wolf, F. Gallet, S. Balibar, E. Rolley, and P. Nozières, *J. Physique* **46** 1987 1985.

[9] A.V. Babkin, D.B. Kopeliovitch and A. Ya. Parshin, *Zh. Eksp. Teor. Fiz.* **89** 2288 1985 (*Sov. Phys. JETP* **62** 1322 1985).

[10] O.A. Andreeva and K.O. Keshishev, *Pis'ma Zh. Eksp. Teor. Fiz.* **46** 160 1987 (*Sov. Phys. JETP Lett* **46** 200 1987).

O.A. Andreeva, K.O. Keshishev and S. Yu. Osip'yan *JETP Lett.* **49** 756 1989.

[11] S. Balibar, F. Gallet, F. Graner, C. Guthmann and E. Rolley, *Physica B* **169** 209 1991.

S. Balibar, C. Guthmann and E. Rolley, to be published

[12] D. Wolf, *Phys. Rev. Lett.* **67** 1783 1991.

[13] E.Medina, T.Hwa, M.Kardar and Y.C.Zhang, *Phys. Rev. A* **39** 3053 1989.

[14] L.H. Tang, T. Nattermann and B.M. Forrest, *Phys. Rev. Lett.* **65** 2422 1990.

ROUGHENING DURING SI DEPOSITION AT LOW TEMPERATURES

D.J. Eaglesham and G.H. Gilmer
AT&T Bell Laboratories, Murray Hill, NJ 07974 USA

Abstract. We briefly review the experimental grounds for invoking rapid surface roughening in Si epitaxy as a mechanism for nucleation of the amorphous phase in low-temperature deposition. The basic crystal growth models show a very slow build-up of roughness during crystal growth, and alternative models are discussed. We then describe experiments on measuring the surface roughness build-up in Si MBE using marker layer width measurements. The interface width (Δ) is shown to increase rapidly during growth of the film, approximately linearly with layer thickness (h): $\Delta \propto \Delta_0 \cdot h$, even when diffusion is rapid. Monte Carlo simulations show that roughness increasing with exponents $\beta > 0.5$ (where $\Delta \propto \Delta_0 \cdot h^\beta$) can arise through asymmetric sticking at steps. We further sugg est that nucleation of the amorphous phase at the epitaxial thickness occurs near a critical roughness of $\Delta \approx 15\text{Å}$.

1. Introduction

The low-temperature limit to epitaxial growth processes in MBE is currently a subject of renewed interest. Traditionally this limit has been viewed as a simple competition between deposition and surface diffusion: at temperatures sufficiently high for s urface diffusion to be activated (at a rate that exceeds the [weighted] rate of deposition) the film is crystalline, while at lower temperatures it is amorphous [1]. In practice, several groups have now suggested that at low temperatures there is an epitaxial thickness, and deposition always initially proceeds epitaxially. While the initial studies [2] viewed this simply as the continuous nucleation of the amorphous phase below an epitaxial temperature, later work [3] implies that there is a true thickness-dependent crystallinity, with formation of the amorphous phase occuring abruptly once some critical thickness (the epitaxial thickness h_{epi}) is exceeded. This latter conclusion is forcing us to rethink models for epitaxy in a radical way.

The mechanism for the breakdown of epitaxy remains unclear at this time, but one obvious candidate is the increase in surface roughness that occurs during growth. Damped RHEED oscillations are observed during low-T Si MBE, and it is possible that the in crease in roughness during deposition that this implies can cause the formation of the amorphous phase once some critical surface roughness has built up. Specifically, one can argue that sufficient surface roughness could in principle raise the free ener gy of the crystalline surface to exceed that of the amorphous phase (assuming that in a-Si the additional degrees of structural freedom require a lower density of steps). A quantitative estimate of the degree of roughness required for such a transformation may be obtained by comparison of the free energy difference between c-Si and a-Si at these temperatures (120meV/atom) with the step energies (30meV/atom for the "good" S_A step, 90meV/atom for the "bad" S_B step); these numbers suggest that once roughness on a crystalline surface builds up to the extent that step densities approach one-per-atom row (and locally exceed this), the surface may become unstable with respect to the amorphous phase. The remainder of this paper will be devoted to addressing two questions with regard to this possible scenario: first, can roughness build up sufficiently fast as a function of thickness to explain the apparently abrupt transitions we observe after depositions of substantial thicknesses; and, second, can we obtain any qualitative link between surface roughening rate and the measured epitaxial thickness.

2. Roughening during slow-diffusion deposition

A number of detailed models are available for surface roughening during growth under conditions where surface diffusion remains slow. Since these models are extensively discussed elsewhere in these proceedings, we will summarise only a handful of points here. Almost all models for roughening follow a slow thickness dependence in the asymptotic limit. Typically, $\Delta \propto \Delta_0 \cdot h^\beta$ with $0 < \beta < 0.5$. It should also be noted that high exponents (near 0.5) are associated with nearly diffusionless growth (i.e. when the exponent approaches 0.5 the roughness approaches $h^{0.5}$): this will turn out to contrast sharply with the observations. These slow-roughening models appear to be insufficient to explain the abrupt transition from epitaxy to amorphous growth (from defect-free epitaxy to fully amorphous over a distance from h_{epi} to $1.3 \cdot h_{epi}$). To achieve more rapid breakdown we can invoke nucleation of the amorphous phase at some large step "cluster": the probability of an n-step cluster then increases as the ity to the n'th power, or as the thickness to the (n/2)th power. However, the observed breakdown is so abrupt that the required thickness dependence remains inconsistent with any slow roughening. More rapid roughening is obtained from models that include macroscopic geometric shadowing. In particular, Bales and Zangwill [4] have modelled roughening and void formation due to geometric shadowing (of regions at the bottom of troughs) during sputter deposition of amorphous films. Since the roughening rate is now proportional to the degree of shadowing (i.e. to the instantaneous roughness) these models roughen very rapidly (near-exponential). However, the approach appears to be restricted to truly diffusionless growth, and seems unlikely to apply for epitaxy.

Tests for the models of roughening are few and far between. While the theory of kinetic roughening is well-understood in a variety of limits, experimental data on roughness increases during low-T growth are scarce. Experiments such as RHEED oscillation damping and LEED spot-profile measurements have been successfully interpreted to yield quantitative information on the initial stages of step accumulations during growth. The models, however, relate mainly to the asymptotic behaviour of the roughness in the limit of large thicknesses: this is also the range of most interest if roughness is linked to the breakdown of epitaxy.

3. Measurements of surface roughness in LT Si MBE

Experimentally, few techniques are available for the quantitative measurement of surface roughness. The approach we take here is to deposit a monolayer of Ge on the Si surface as a marker, and then continue to grow Si. The Ge marker layer can then be characterised *ex-situ* after growth of the layer is completed, using transmission electron microscopy of cross-section samples. Growth is carried out over a range of temperatures in which we see the transformation from epitaxial to amorphous after 200-10 00Å (i.e. from $\approx 250°C$ to $\approx 350°C$: below 200°C the h_{epi} transition is still well-characterised but the crystalline thicknesses are small, making our quantitative measurements more difficult). Typical MBE growth rates are used (1Å/s Si deposition, ≈ 0.1Å/s for Ge). The Ge monolayer is assumed to cover the surface conformally, with no tendency for the Ge to preferentially occupy either the hills or troughs on the Si surface. Our measurement averages over the thickness of the TEM sample ($\approx 200 - 500$Å) so that most surface roughness, being on lengthscales $\ll 500$Å, will give rise to apparent *broadening* of the Ge profile.

Fig. 1 shows a typical image of a typical sample with a sequence of Ge markers dispersed through the film. The micrograph is a bright-field image recorded at very large defocus (≈ 500nm) in the symmetric 3-beam (400) condition with a small objective aper ture. Under these conditions all interfaces should be vertical but at the same time diffraction from crystal planes is fairly weak. The large defocus greatly enhances the contrast visible from these small amounts of Ge. The measured width of all of the markers in the c-Si is 17Å, independent of position in the layer. This does not, however, imply a thickness-independent roughness of the c-Si surface, but is merely associated with the Fresnel fringe width of the Ge layer under these microscope imagin conditions. The resolution of the image being ≈ 17Å, a measured Ge marker with this width merely implies a surface roughness of ± 17Å/ or smoother.

The amorphous region also contains Ge markers, whose thickness is clearly substantially greater than that of the markers in c-Si. In a-Si, under the same imaging conditions, Ge marker widths are 45 ± 5Å/ throughout the a-Si film (i.e. independent of deposited thickness). We also observe void chains formed in the a-Si (arrowed), with the chains directed at the Si source. These observations, confirming a mechanism similar to that discussed by Bales and Zangwill, are not surprising: however the measurement of the Ge marker width at all points in the film represents the first quantitative determination of the limiting roughness in this growth mode.

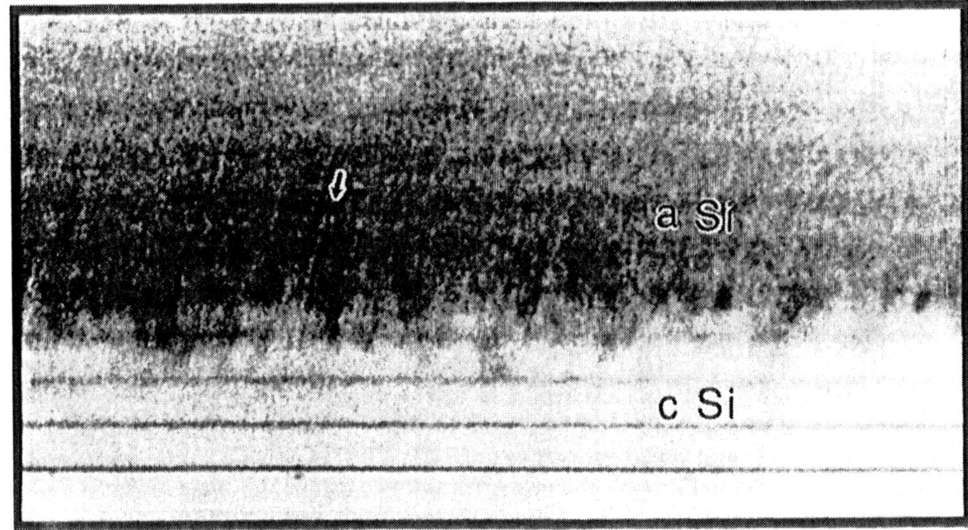

Figure 1. Typical micrograph of Ge markers in LT Si MBE before and after the epitaxial thickness transition. Note the much broader Ge lines in the amorphous portion of the film, and the decrease in intensity of successive markers in the crystalline region.

The striking feature of the crystalline regions in Fig. 1 is not the (thickness-independent) Ge marker width (which is determined by fitting Gaussians to the intensity profile of the image), but the intensity decrease of the dark Ge bands during growth of the Si layer. This intensity decrease can be seen more explicitly in Fig. 2 (taken from fits to a different image from Fig. 1), and turns out to give us an accurate measure of the Ge layer thickness. Under Fresnel imaging conditions the objective lens behaves as a narrow-pass filter for a particular spatial frequency (17Å in the image in Fig. 1): for a Ge profile whose width is considerably less than 17Å an increase in the width leads to a decrease in this low-frequency scattering potential, and thus to a decrease in image intensity. In detail, for a top-hat profile in the large-defocus limit under phase-object imaging conditions, we expect the intensity of the dark line due to the Ge to decrease *linearly* with increasing profile width. We now make a couple of approximations: that the initial Ge marker represents an abrupt profile (1Å or 1 monolayer wide), and that we can extend this large-defocus imaging regime out to large marker widths. This allows us to turn the Ge-marker-layer-intensity vs. layer position (from Fig. 2) into Si-surface-roughness vs. deposited-film-thickness (Fig. 3).

Fig. 3 represents a dramatic departure from our usual picture of asymptotic roughening, suggesting that at least in LT Si MBE the roughness may increase with some

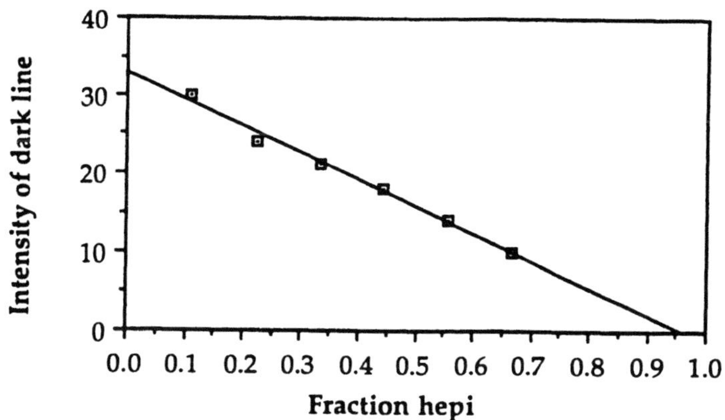

Figure 2. Fitted intensity distributions for a sequence of Ge markers in a film similar to that in Fig. 1 (but grown at higher T so that the epitaxy-amorphous transition occurs later).

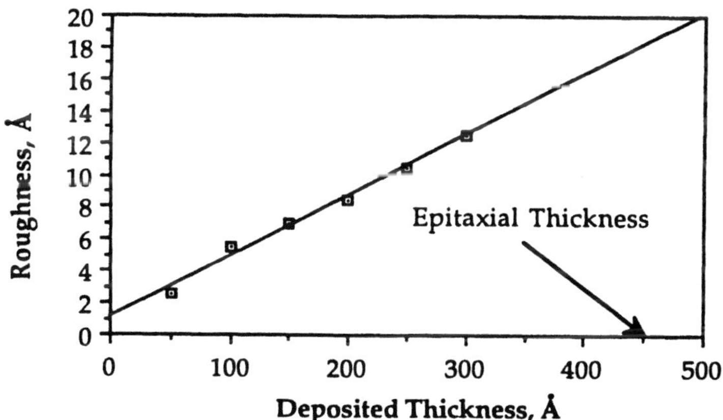

Figure 3. Measured increase in roughness during LT Si MBE, showing a linear thickness dependence.

large exponent in h ($\Delta \propto h$ in the above regression). This is particularly remarkable since the quantitative roughness is fairly small (in line with the expected large diffusion coefficients for this moderately high temperature): almost all models predict large roughening exponents only for large roughening (i.e. small diffusion rates). Thus the measured roughness differs wildly from that predicted by simple models.

4. Monte-Carlo simulations of roughening

In an attempt to address the inconsistencies between our observations and theories of
kinetic roughening, we have carried out a wide range of simulations. A wide variety of
previous studies have shown that exponents of 0.2-0.5 are obtained if simple assumptions are made. However, several authors have pointed out that more complicated
behaviour can be obtained if asymmetric diffusion or sticking at a step is incorporated.
In particular, continuum equations show [5] that some instability may develop when
diffusion from the upper terrace to the step is inhibited.

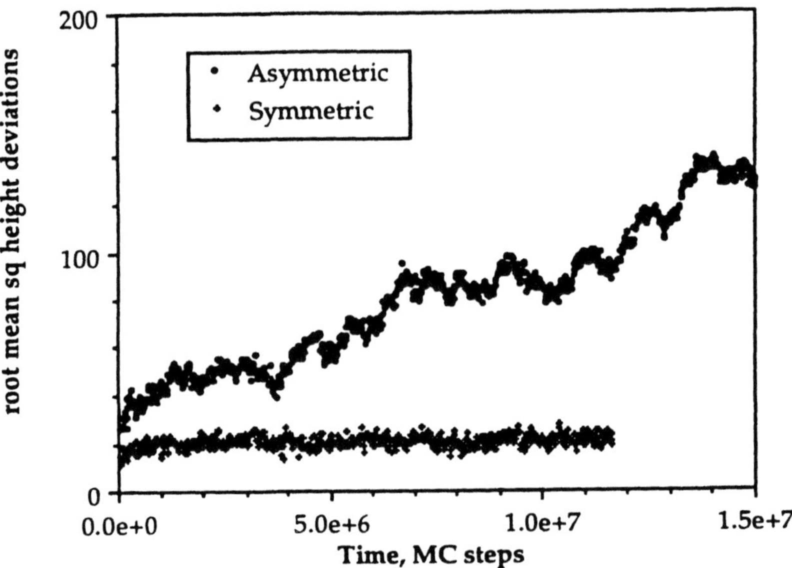

Fig. 4 Simulated roughness at long times in a Monte-Carlo calculation assuming
either symmetric accomodation at steps or asymmetric accomodation. Note the long-
time increase in rms height deviations caused associated with asymmetric processes at
steps.

In our simulations we treat a SOS model in which only adatoms diffuse (dimers
and step-edge atoms are assumed immobile). Asymmetry is introduced by reducing
the accomodation of atoms diffusing to a down-step with respect to accomodation at
an up-step: in the limit, up-steps are sinks, down-steps are mirrors. The consequences
of this are seen in Fig. 4: roughness accumulates over long time-scales even when there
is substantial diffusion (in Fig. 4 the ratio of the probabilities of a diffusion step to a
deposition step is 10^5:1, and on short time-scales this model carries out hundreds of

slightly-damped "RHEED" oscillations). This arises from the build-up of remarkable structures on the surface, where deep grooves appear at the intersections of large terraces. Growth of these structures can be seen as a surface-diffusion analogue of the Bales-Zangwill process. Most flux arriving at the surface arrives on a terrace, and the groove between the terraces is "shadowed" from this flux by its own down-steps.

REFERENCES

[1] J. A. Venables in *Epitaxial Growth, Part B* J.W. Matthews, ed., Academic Press, New York 1975 p. 389.

[2] H. Jorke, H.-J. Herzog and H. Kibbel, *Phys. Rev. B* **40** 2005 1989.

[3] D.J. Eaglesham, H.-J. Gossmann and M. Cerullo, *Phys. Rev. Lett.* **65** 1227 1990.

[4] G.S. Bales and A. Zangwill, *J. Vac. Sci. Tech.* **A 9** 145 1991.

[5] J. Villain, *J. de Physique I* **1** 19 1991.

SELF-AFFINE SCALING IN VAPOR-DEPOSITED METAL FILMS

J. Krim

Department of Physics, Northeastern University, Boston, MA 02115 USA

Abstract. Scanning tunneling microscopy, x-ray reflectivity and adsorption isotherm measurements have been carried out on vapor-deposited metal film surfaces in order to investigate whether self-affine fractal scaling is present. Self-affine surfaces with roughness exponents ranging from 0.5 to 0.96 have been observed. The physical conditions which produce these surfaces are outlined, and the relative strengths and weaknesses of the experimental probes are discussed.

1. Introduction.

This paper describes experimental measurements employing scanning tunneling microscopy (STM), x-ray reflectivity and adsorption to probe whether self-affine fractal scaling is present in certain vapor-deposited gold, silver and iron films. It summarizes the results of one recent publication [1] and four upcoming works.[2-4]

Self-affine surfaces are distinguished from self-similar ("genuine") fractals by an asymmetry in the scaling behavior perpendicular to the surface, generally manifested by an absence of surface overhangs.[5] A self-affine fractal can be described as a single-valued Gaussian rough surface $h(x, y)$ with average height h, and root-mean-square (rms) width $\sigma = < [h(x, y) - h]^2 >^{1/2}$. Such a surface is self-affine if $\sigma(L)$ increases with the horizontal length L sampled according to $\sigma(L) \propto L^H$, where $0 < H < 1$ is a scaling exponent which reflects the surface texture, or degree of height-height correlation.[6] (The letter w is also frequently used in the literature to denote the surface width.) Small H values are associated with jagged surfaces (anticorrelation), while large values are associated with well-correlated, smooth textured surfaces. (The parameter H is frequently represented by the letter α in the literature, and has also been denoted by χ, ζ, and h. It is variously referred to as the 'Hurst', or 'Hölder' exponent, the 'roughness parameter', and the 'static scaling parameter'.) All self-affine surfaces are characterized by an upper horizontal cutoff to scaling, or correlation length

ξ, beyond which the surface width no longer increases. For the case of film growth (or erosion), the time evolution of this maximum width is described by the scaling exponent β, according to $\sigma \propto t^{\beta}$ (where it is generally assumed that the film thickness h is directly proportional to the deposition time t). Film growth can alternatively be expressed in terms of a 'dynamic scaling exponent' $z = H/\beta$, which describes the time evolution of the correlation length ξ according to $\xi \propto t^{(1/z)}$.

Self-affine scaling in vapor-deposited films may be experimentally investigated at sub-micron levels by means of STM, x-ray reflectivity or adsorption. STM involves the monitoring of a tunneling current as a probe tip is scanned over the surface of the sample under study.[7] Direct topological information is in principle obtained, but STM measurements provide no information on surface overhangs and are also sensitive to tip curvature effects and chemical inhomogeneities in the substrate.

Adsorption measurements involve coating the sample with a liquid films which wets its surface. Surface roughness is probed by monitoring the quantity of adsorbed film particles required to form successively thicker coatings as the substrate roughness is progressively smoothed out.[8] Adsorption measurements provide a direct probe of surface area, including that due to pores and overhangs. The role of the liquid adsorbate surface tension complicates however the interpretation of the data.[9, 10]

Specular x-ray reflectivity, where the intensity of specularly reflected x-rays (angle of incidence equal to angle of reflection) is recorded as a function of incidence angle, is a widespread probe of σ.[11, 12] Diffuse reflectivity, where the x-ray detector is held fixed while the sample is rocked about the specular condition, is sensitive to surface height-height correlations. This technique has recently been developed as a probe of the parameters ξ and H.[11] X-ray measurements have the advantage of probing the entire film density profile. They are complicated however by scattering contributions from the interface between the vapor-deposited film and the substrate it is deposited on. They are also limited in application since diffuse scattering from surfaces with widths greater than about 50Å is negligible.

One or more of these three techniques has been employed to investigate the presence of self-affine scaling in certain vapor deposited silver, gold and iron films. Self-affine fractal scaling has been observed in one gold film test sample and three categories of films grown *in situ*. The measured values of H for these samples varies from ≈ 0.5 to 0.96.

2. Experimental Observations

2.1. Gold Film Test Sample: $H = 0.96 \pm 0.02$

The value $H = 0.96 \pm 0.02$ has been obtained for a gold film test sample which was studied in order to directly compare STM and adsorption measurements.[2] The sample is a gold elctrode of a commercial quartz crystal intended for use as a deposition rate monitor. It was selected on account of the visual self-similarity of STM images recorded at differing length scales. It was also selected on account of the fact that it was visually rough, indicating that the upper cutoff to scaling might fall well into the micron range. Figure 1 presents STM data recorded on this sample. Numerous scans, each of size L, were recorded at random places on the sample. The average surface width for each scan size (*determined after a plane fitting procedure has been carried out so as to make each image appear to be level*) was then plotted on a log-log scale versus L so that the slope of the line is the value H. The upper data set was recorded in air with two different scan heads (circles and squares) and several different scan tips. The lower data set (triangles) was recorded in ultra-high vacuum on a different instrument. The difference in absolute magnitude of the data sets is attributed to a calibration difference between the two microscopes. Self-affine scaling behavior is observed for horizontal lengths ranging from below 40Å up to about 40,000Å.

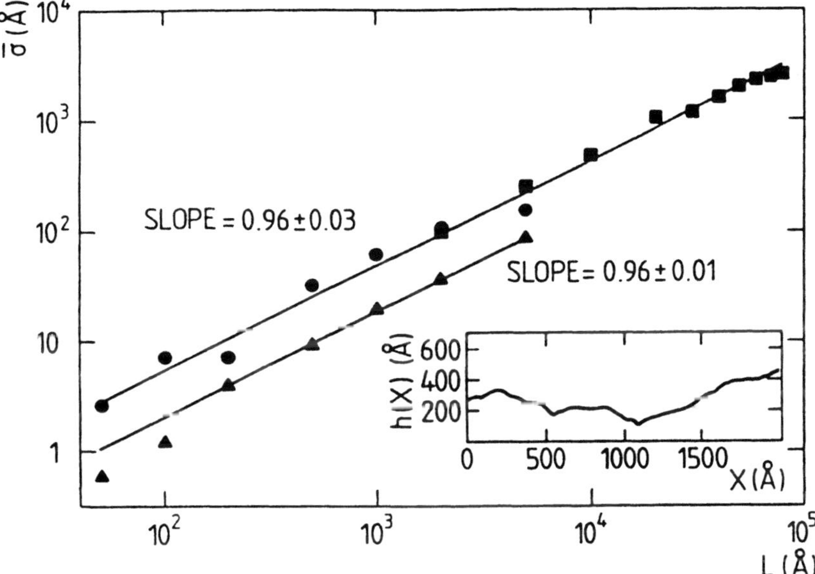

Figure 1. STM data recorded on the gold film test sample. (From Ref.[2])

Adsorption isotherms recorded on this sample yielded H values of either 0.98 ± 0.1 or 0.49 ± 0.1, depending on whether liquid surface tension effects were or were not neglected in the data analysis. [5] Only the analysis which nelgected surface tension is consistent with the STM result. This conclusion is strongly supported by the fact that both liquid nitrogen and liquid oxygen (which has twice the surface tension of liquid nitrogen at 77K) exhibited virtually identical adsorption behavior.

2.2. Silver Evaporated onto Polished Silicon at 300K: $H = 0.67 \pm 0.05; \beta = 0.26 \pm 0.05$

X-ray reflectivity measurements were performed *in situ* on a silver film deposited by thermal evporation at normal incidence onto a polished silicon substrate held at room temperature.[3] The diffuse data were fit by $H = 0.67 \pm 0.05$. Measurements were carried out for five film thicknesses ranging from 100Å to 1400Å. For these thicknesses, the surface width, as detemined by means of specular reflectivity, scaled as $\sigma \propto t^{0.26}$.(Figure 2).

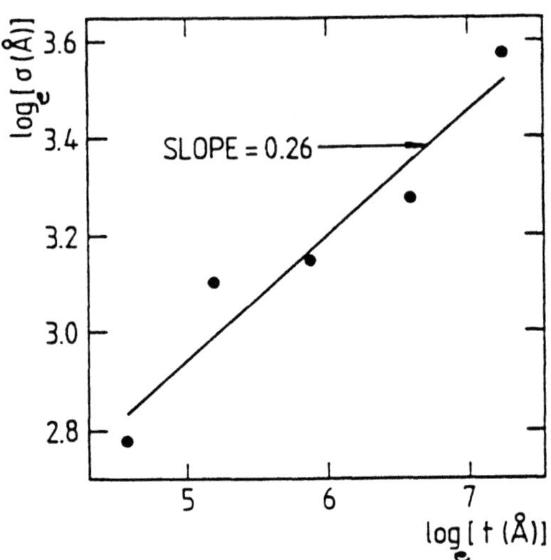

Figure 2. Width versus film thickness for silver deposited at 300K. (From Ref.[3])

Adsorption measurements were also carried out on silver films deposited onto optically polished quartz substrates under identical deposition conditions.[4] The adsorption data could not distinguish between this sample and a geometrically flat plane: the actual increase in surface area for films deposited in this manner is negligible.

2.3. Ion-Bombarded Iron Films Grown Epitaxially on MgO: $H = 0.53 \pm 0.02$

Figure 3 presents STM data recorded in ultra-high vacuum for an iron film sample.[2] The film, originally 2000Å thick, was grown epitaxially on a magnesium oxide substrate and then heated to $\approx 600°C$ in order to anneal and flatten the surface. The lower set of data points in Figure 3 (squares) was recorded after the annealing process. There is no linear region in the plot, so the surface cannot be considered as fractal in nature.

The annealed surface was then bombarded with 5-keV Ar$^+$ ions to probe whether self-affine scaling could result from an erosion process (Figure 3, circles). For scan sizes below 2000Å, the data fall on a straight line with slope $H = 0.53 \pm 0.05$. This line intersects with the curve for the annealed surface near $L = 2000$Å. Above this point the power law is evidently obscured, and the curve closely tracks that of the annealed surface. The observation of fractal scaling for this sample suggests that the applicability of scaling theory to describe erosion at sub-micron lengthscales may extend far beyond a previous case study involving erosion of a graphite surface.[13]

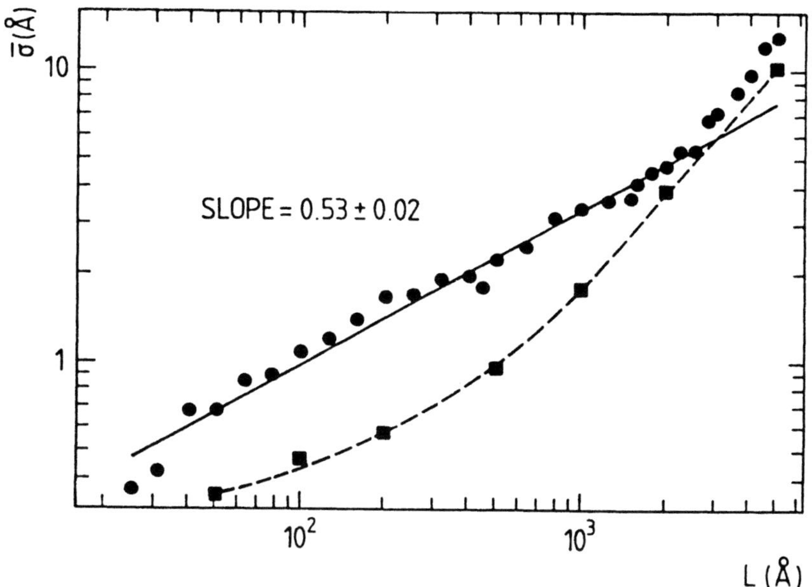

Figure 3. STM data recorded on the iron film sample. (From Ref. 2)

2.4. Silver Evaporated onto Polished Quartz at 80K: $H = 0.5 \pm 0.1$

Silver films evaporated at 5° off-normal incidence onto polished quartz substrates have been investigated by means of adsorption and x-ray reflectivity.[1] Diffuse x-ray data are fit by $H = 0.46 \pm 0.1$ for a 1100Å thick sample. Adsorption data recorded on films in this category indicate surface areas which are two to three times that of a geometrically flat plane. Liquid nitrogen and liquid oxygen exhibit strikingly different adsorption behavior on these samples (Figure 4),[4] presumably on account of the greater surface tension of liquid oxygen. Nitrogen adsorption meausurments yielded H values of either 0.64 ± 0.1 or 0.35 ± 0.1, depending on whether or not surface tension effects were nelgected in the analysis. Oxygen adsorption measurements meanwhile yielded the respective H values of 0.98 ± 0.1 and 0.49 ± 0.1. A value of $H = 0.5 \pm 0.1$ is consistent with the x-ray data and adsorption data, assuming that surface tension effects must be included in the oxygen adsorption data analysis.

Nitrogen adsorption measurments were carried out on 450-2000Å-thick silver films in order to investigate the thickness dependence of surface area (and therefore the thickness dependence of the surface width). While there is a clear trend to greater surface area for the thicker silver films, there is considerably more noise in the data than those recorded by x-ray reflectivity.

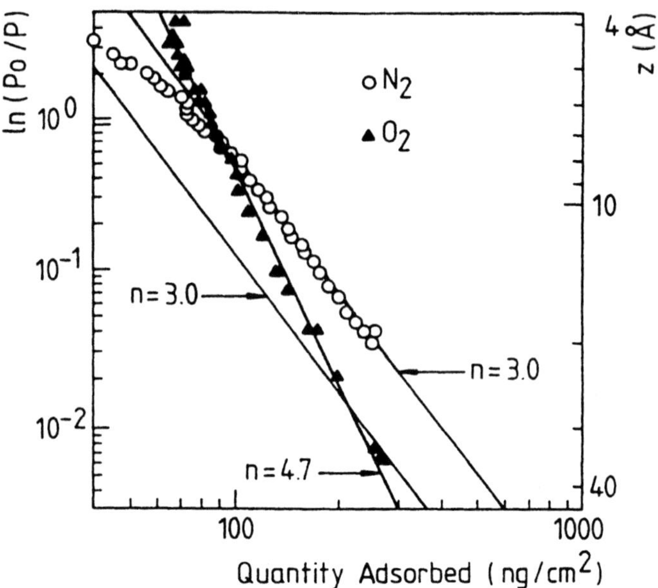

Figure 4. Comparison of liquid nitrogen and liquid oxygen adsorption data on cold-deposited silver. (From Ref.[4])

3. Discussion

Although self-affine scaling was observed in the samples described above, it was not observed in all of the metal film surfaces which were studied. Fractal scaling was most notably absent in film surfaces formed at the higher temperatures. Examples of this include gold films deposited onto polished quartz substrates held at 500K,[1] and the annealed iron film samples described above. Ag films deposited at room temperature exhibited fractal scaling, but very little increase in surface area. Ag films deposited at 80K (and then examined at room temperature) exhibited significantly increased surface area along with a lower fractal scaling parameter ($H = 0.5 \pm 0.1$) than that observed for films deposited at room temperature ($H = 0.67 \pm 0.05$). It is not unlikely therefore that the H value measured for Ag films deposited *and examined* at 80K will be even lower.

Two of the samples described above were probed by more than one experimental technique. In the case of the gold film test sample, the STM and adsorption measurements were in excellent agreement, with the STM data providing a more precise value for H. The excellent agreement can be attributed to the relatively high value of H for the sample under study: this smooth-textured sample provided a relatively uncomplicated geometry for the STM tip to track, while at the same time being less affected by the surface tension effects which complicate interpretation of adsorption data. Agreement between the x-ray and adsorption data was less perfect for the cold-deposited Ag sample, characterized by a significantly lower H value.

Among the various techniques, the surface width σ (and therefore the scaling parameter β) is most accurately measured by means of x-ray reflectivity, so long as the correlation length of the sample remains below that of the x-ray beam (≈ 1 micron). STM, if accurately calibrated, can also provide a precise value of σ, and in addition remains viable to much longer length scales (scan heads ranging up to 125 microns are now commercially available). The value of the correlation length ξ is best measured by STM, since no fitting is required in the data interpretation. The scaling parameter H appears to be most accurately measured by means of x-ray reflectivity and STM, but a detailed comparison of the two techniques has yet to be carried out. Adsorption meanwhile remains the only technique capable of revealing porosity in the outer surface topology, and the only technique which appears openly problematic when experimental complications are in fact present.

Acknowledgements

The following persons are collaborators in the work presented here: Y. Bruynseraede, R. Chiarello, Y.P. Feng, I. Heyvaert, G. Palazantzas, V. Panella, S.K. Sinha, C. Thompson and C. Van Haesendonck. JK is supported by grants from the NSF: DMR-8910315; DMR-8657211 and PRF: 22008-AC5, and a Fellowship from the Research Council of the Katholicke Universitiet Leuven.

REFERENCES

[1] R. Chiarello, V. Panella, J. Krim and C. Thompson, *Phys. Rev. Lett.* **67** 3408 1991.
[2] J. Krim, I. Heyvaert, C. Van Haesendonck and Y. Bruynseraede, preprint .
[3] J. Krim, C. Thompson, G. Palasantzas, Y.P. Feng and S.K. Sinha, manuscript in preparation .
[4] V. Panella and J. Krim, two manuscripts in preparation .
[5] F. Family and T. Vicsek, *J. Phys. A* **18** L75 1985.
[6] T. Vicsek *Fractal Growth Phenomena* World Scientific, Singapore 1985.
[7] M.W. Mitchell and D.A. Bonnell, *J. Mater. Res.* **5** 2244 1990.
[8] P. Pfeifer, Y.J. Wu, M.W. Cole and J. Krim, *Phys. Rev. Lett.* **62** 1997 1989.
[9] M. Kardar and J.O. Indekeu, *Europhys. Lett.* **12** 161 1990.
[10] M. Kardar and J.O. Indekeu, P. Pfeifer, M.W. Cole and J. Krim, *Phys. Rev. Lett.* **65** 663 1990.
[11] S.K. Sinha, E.B. Sirota, S. Garoff and H.B. Stanley, *Phys. Rev. B* **38** 2297 1988.
[12] A. Braslau, P.S. Pershan, G. Swislow and B.M. Ocko, *Phys. Rev. A* **38** 2457 1988.
[13] E.A. Eklund, R. Bruinsma, J. Rudnick and R.S. Williams, *Phys. Rev. Lett.* **67** 1759 1991.

TOWARD QUANTIFICATION OF THIN FILM MORPHOLOGY: REVISTED

Russel Messier
Pennsylvania State University, Materials Research Laboratory, University Park, PA 16802

Abstract. The quest for quantitative preparation-property relations must be preced in many instances by a quantitative thin film morphology model. Although thin film morphology has been recognized an important linkage, little practical progress has been realized over the last decade. An assessment of the past and present state of understanding of thin film morphology which results from low adatom mobility conditions will be discussed.

1. Introduction

If a thin film were nothing more than a thin slice of bulk material chemically bonded to a substrate across an atomic scale interface, their properties would be quite predictable. However, real thin films may have graded interfaces, may contain structural and chemical defects, grain boundaries, voids and void networks, etc. and often display a wide range of properties dependent upon their film characteristics. The film characteristics are highly dependent, in turn, upon the deposition parameters, and more importantly, the fundamental processes of temperature, bombardment, and chemistry as they affect adatom mobility. The good news is that such latitude gives the thin film engineer the ability to control and tailor the properties to the desired application. The bad news is that he/she also has enough rope to hang him/her self. The first question, therefore, is why?

In most thin film studies the atoms arriving at the growing film surface (adatoms) follow ballistic trajectories since the mean-free-path distance are much greater than the film morphology sizes. However, depending upon the nature of the deposition source (point, planar, etc.) and the degree of scattering in atom transfer between the source and the substrate, the average angle of arrival and the distribution of angles of arriving species can have significant effects when atomic level self-shadowing is considered [1,2]. In particular, depending upon the degree and nature of mobility of adatoms upon arrival at the growing film surface, a wide range of resulting film morphologies can result. These morphologies are typically anisotropic with the direction of anisotropy related in some way to average angle of incidence of the arriving atoms. Although the angle of the morphology was early on related to the incidence angle of the vapor through the tangent rule [3] this turns out to have neither theoretical nor experimental basis.

2. Review of thin film morphology

In order to gain a perspective on where this field is going, it will be important to first look at the progress of the last six years.

2.1. Experimental

In 1986 we presented a review of the status of both our and others' efforts toward quantification of thin film morphology [4]. The conclusions of this paper, references therein, and other papers from our group around this time period [5-17], are summarized below.
  1. Thin film characteristics and resulting properties are not the same as bulk materials. Thin film properties can vary by orders of magnitude.
  2. Quantitative preparation-property relations are often sought but seldom achieved.
  3. Thin film morphology often has a controlling influence on thin film properties.
  4. The structure zone models (SZM) of Movchan and Demchishin [18] and Thornton [19] are useful qualitative descriptors of morphology when, and only when, films of equal thickness are compared. Neither of these SZM's considered the implications of growth evolution. Although the SZM's indicate just several distinct zones, there is a continuum of morphologies and morphological changes. At low adatom mobility conditions for deposition a film is either rough, cauliflower-like Zone 1 morphology os smooth, very fine fine grain-size Zone T morphology. Such models have no hope of achieving quantitative thin film properties-morphology relations.
  5. At low mobility conditions (the primary emphasis of both the 1986 and this paper) clustering at the 5-30Å level is found for most materials. Both phase contrast TEM and field ion microscope techniques reveal this clustering, essentially independent of the atomic level structure. Micrographs of metals, semiconductors, and insulators as well as amorphous and nano-scale crystalline films display a striking similarity.
  6. Thin film morphology evolves. The evolution involves the competition for growth of power-law shaped cones. The power-law exponent varies with film preparation conditions. In general, all cones go through a growth and death stage since, in the limit, only one cone can survive under such conditions.
  7. Under certain conditions the power-law shaped conical units evolve toward a cylindrical shape. Such morphology has been referred to as needle-like or match-stick. It appears that the competition for growth decreases and a fibrous morphology ensues. The cylider size is typically 2-3 orders of magnitude larger in size than the initial cluster unit which indicates that competitive cone growth occured in the earlier stages of growth evolution.

8. Under conditions of higher substrate temperature or increased levels of ion bombardment during growth (but still in the low mobility regime $< 0.3 - 0.5T/Tm$) the morphology becomes fine-grained in cross-section and with a smooth top surface morphology, as viewed by the typical 20KX to 50KX magnification in SEM's. Close examination indicates that the clustering still occurs and that it is likely that the intense competition for cone growth leads to the surface smoothening.

9. The anisotropic columnar (not necessary cylindrical) morphology is defined by a void network structure. This void network appears to form a hierarchial structure in which the void regions form the boundaries of the columnal regions. This top surface cauliflower-like structure has been termed fractal-like. The three most extensively studied materials are SiC (amorphous and microcrystalline), Ge (amorphous) and pyrolytic graphite (microcrystalline).

10. Voids are missing atoms or small groups of missing atoms. A void region is typically what is viewed by high resolution microscopy and consists of voids in various distributions and on average have a density lower than that of the bulk. A void is the volume between clusters and groups of clustered units. Thus void regions are expected to have a wide range of sizes, densities, activation energies for diffusion, etc. A direct consequence is that a singular number defining the total void content of a material is of some interest but of little help in any detailed understanding of thin film morphology. Therefore, any general description of a thin film's void network structure is expected to contain the distribution functions of all aspects of this information.

11. In general the Zone T, very fine grained morphologies contain voids and void regions but no easily distinguishable void network. Often such films are termed "dense" with the implicit or explicit assumption that void regions are eliminated. This is not so and in one instance we have found a Zone T a-Ge film with a lower average density than a Zone 1, cauliflower-like, morphology film.

12. Initial substrate roughness, and in particular the distribution of the roughness, has a large effect on the growth morphology. In particular the tail of the distribution of roughness sizes controls the dominant growth morphology.

2.2. Modeling

In this same time period several approaches had been reported in which simple theory were related to the experimental studies. These models can be generally classified as molecular dynamics simulations (lots of physics), ballistic aggregation simulations (some physics), and geometrical, continuum simulations (very little physics). A good summary of the status is found in a series of articles published as part of a 1987 SPIE Topical Conference on "The Modeling of Optical Thin Films" [20]. In our paper on the computer simulation of the cross-sectional morphology of thin films [5] we summaried the status, which is given below in less detail.

1. Simple 2-D ballistic aggregation simulations by Dirks and Leamy [2] and 3-D simulations by Henderson [21] were the first to show the resulting clustering, columnar growth, the effect of the avera ge angle of incidence of the arriving vapor, and the consequence of atomic self-shadowing in these regards. Very large numbers

of particles can be considered which allows for the study of growth evolution 3-D over reasonable thicknesses. How to deal with local mobility, bond relaxation, cluster energy minimization, etc. can be difficult to do properly.

2. Molecular dynamics simulations are limited to relatively small numbers of particles. Especially the early studies of Muller [22] produce simulations with unrealistically high void concentrations and large voids.

3. The geometrical models of Yang et al. [5] allow experimental data on nucleation density, dominant columnar shapes, etc. to be input in a simple computer simulations model in which the competition for cone growth occurs by an intuitive set of rules. The main advantage is that very thick films can be studied and the ease of computation allows for many simulations.

It was expected that each of these approaches would have their greatest utility when used and analyzed in a coordinated manner along with experimental data.

3. Thin film morphology today

A very clear statement can be made regarding progress in the area of experimental quantification of thin film morphology: Minimal progress. Either the topic is uninteresting or the task is too formidable. The general conceptual model of a hierarchial, anisotropic structure arising out of the atomic self-shadowing induced clustering and resulting competition for cone growth seems simple. Furthermore, the commonality of the various low mobility morphologies at both the nm- and μm-level, which is the basis of the various SZM's, is a strong argument for a simple explanation. However, within this simplicity is a complexity in the fact that the resulting morphology contains a variable distribution of sizes, shapes, activation energies, etc. and that any quantifiable model must take into account this detailed information.

There have been no new studies reported on the quantitative image analysis of the cauliflower-like morphologies seen in SEM photographs. Likewise, no further detailed analysis has been attempted on higher resolution TEM and FIM micrographs beyond our earlier work [6]. In this same study we began to explore the use of scattering (x-ray and optical) studies for understanding the distribution of orphological units within a thin film, but not further studies have been reported. It might be expected that with the advent of the scanning tunneling microscopy (STM) and atomic force microscopy (AFM) that the potential exists for making progress in this regard, and there are indications that such work is in progress.

The area of spectroscopic ellipsometry has progressed to the point to where through effective medium approximation modeling we are able to develop detailed structural models which contain quantitative information such as nucleation density, interface structure and chemistry, surface roughness and roughness evolution, total void concentration and distribution with film thickness (evolution), and bandgap [23,24]. With the recent developments in the ability to perform real time spectroscopic ellipsometry (RTSE) we now have the ability to construct detailed morphological analysis of both a growing thin film as well as a thin film undergoing post-deposition processing, such as a dynamical property change (eg. electrochromic coloration by the reversible electroch emical diffusion of H^+ or Li^+ into and out of a WO_3 thin film). In both cases we can obtain quantitative information indirectly on both thematerial and its void network morphology. It must be recognized that such RTSE studies must be coordinated with other more direct morphological studies in order to be most effective.

The progress made in the area of modeling and theory of the low mobility morphology will be left for other participants in this workshop to evaluate in detail. From the point of view of an experimentalist, it does not appear that eny significant progress has been made in regard to a functional understanding which produces a step in the direction of a more quantitative model linking preparation-property relations under low mobility conditions.

4. Future directions

We conclude with a plea for a complex and yet simple model for thin film morphology. The complexity arises due to the need of introducing reality, and the reality is that average values are probably insufficient for describing thin film morphology. It is the continuum of morphological unit sizes and shapes that leads to the commonality of the resulting morphologies and the complexity in description. In many ways this is analogous to the description of crystalline and amorphous thin films in which single crystals are highly defined and have a precise free energy value while amorphous structures have a continuum and distribution of bonding configurations with a broad and continuous distribution of free energy states. The first question, then, is what experimental approaches have at least the potential of developing quantitative and realistic experimental data?

Because thin films are not simple materials it is more important to study systematic series of well conceived experiments, study the trends, and relate experiments to simulation models and theories. Since ion bombardment is known to have a large and often beneficial effect in preparation-morphology-property relations, the effect of varying levels of bombardment induced mobility should be telling. For instance, it appears that the bombardment increases the nucleation density of clusters, reduces the cluster size, the distribution of cluster sizes, and increases the competition for growth. Each of these suppositions needs to be tested and understood better. An example is use of MD to study the competitive growth from a small group of clusters under the influence of various growth trajectories and concommitant ion bombardment. Since competition is not an absolute but statistical in nature, many simulations may be required. Certainly any theoretical insights will be important.

In addition to better preparation experiments and relevant modeling studies, there will be a need for more quantitative morphology characterization methods. As indicated above AFM/STM studies of atomic level phenomena within and between clusters will be important since it could directly link real materials with computer simulations. The other technique which could become an important quantitative characterization tool is spectroscopic ellipsometry since it can be integrated into real deposition systems and can follow in real-time both growth processes as well as post-deposition modification. Thus RTSE can explore growth evolution and dynamical effects such as solid state diffusion in sufficient detail that it can be quantitatively related to both theory and models. In both bases there will be technique limitations, which cannot be addressed in this brief review, and thus the materials for study will have to be chosen wisely.

From our past and current studies we can suggest three model systems which clearly require further study: a-Ge, a-WO$_3$, and microcystalline pyrolytic graphite. Certainly other materials can be considered. In any case more time and effort needs to be taken in coordinating our experimental, modeling, and theoretical studies if we are to make significant progress over the next few years. This workshop is a good beginning.

REFERENCES

[1] A.I. Shaldervan and N.G. Nakhodkin, *Sov. Phys-Sol. State* **12** 1748 1971.

[2] A.G. Dirks and H.J. Leamy, *Thin Solid Films* **47** 219 1977.

[3] H.J. Leamy, G.H. Gilmer and A.G. Dirks, in *Current Topics in Materials Science*, 6 E. Kaldis, ed., North-Holland, Amsterdam 1980.
J.M. Nieuwenhuizen and H.B. Haanstra, *Philips Tech. Rev.* **27** 87 1966

[4] R. Messier, *J. Vac. Sci. Technol.* **A4(3)** 490 1986.

[5] B. Yang, B.L. Walden, R. Messier and W.B. White, *Proc. SPIE* **821** 68 1987.

[6] J.E. Yehoda and R. Messier, *Proc. SPIE* **678** 32 1986.

[7] J.E. Yehoda and R. Messier, *Appl. Surf. Sci.* **22/23** 590 1985.

[8] R. Messier, A.P. Giri and R.A. Roy, *J. Vac. Sci. Technol.* **A2(2)** 500 1984.

[9] R.A. Roy and R. Messier, *MRS Symp. Proc.* **38** 363 1985.

[10] R.A. Roy and R. Messier, *J. Vac. Sci. Technol.* **A2(3)** 312 1984.

[11] A.P. Giri and R. Messier, *MRS Symp. Proc.* **24** 221 1984.

[12] P.J. McMarr, J.R. Blanco, K. Vedam, R. Messier and L. Pilione, *Appl. Phys. Lett.* **49(6)** 328 1986.

[13] J.R. Blanco, R. Messier, K. Vedam and P.J. McMarr, *MRS Symp. Proc.* **38** 301 1985.

[14] B. Yang, L.J. Pilione, J.E. Yehoda, K. Vedam and R. Messier, *Phys. Rev.* **B36(11)** 6206 1987.

[15] L.J. Pilione, K. Vedam, J.E. Yehoda, R. Messier and P.J. McMarr, *Phys. Rev.* **B35(17)** 9368 1987.

[16] J.E. Yehoda, B. Yang, K. Vedam and R. Messier, *J. Vac. Sci. Technol.* **A6(3)** 1631 1988.

[17] R. Messier and J.E. Yehoda, *J. Appl. Phys.* **58(10)** 3739 1985.

[18] B.A. Movchan and A.V. Demchishin, *Phys. Met. Metallogr. USSR* **28** 83 1969.

[19] J.A. Thornton, *Annu. Rev. Mater. Sci.* **7** 239 1977.

[20] M. Jacobson, ed. *Modeling of Optical Thin Films* Proc. SPIE, **821** 1988.

[21] D. Henderson, M.H. Brodsky and P. Chaudhar, *Appl. Phys. Lett.* **25** 641 1974.

[22] K.H. Muller, *J. Appl. Phys.* **59** 2803 1986.

[23] Y.Cong, I. An, R.W. Collins, K. Vedam, H.S. Withma, L.J. Pilione and R. Messier, *Thin Solid Films* **193/194** 361 1990.

[24] R.W. Collins, Y. Cong, H.V. Nguyen, I. An, K. Vedam, T. Badzian and R. Messier, *J. Appl. Phys.* in press, 1991.

THE KPZ MODEL AND SPUTTER EROSION

R. Bruinsma

Department of Physics, University of California, Los Angeles, CA 90024

Abstract. A derivation of the KPZ model for sputter erosion is given and compared with recent experiments.

1. Introduction

Extensive theoretical work on kinetic roughening has stimulated parallel experimental efforts to verify the proposed scaling description, in particular on homo- and hetero-epitaxial growth and on imbibition fronts. With the exception of the recent study of Chevrier et al. [1] on hetero-epitaxy of Fe, the measured exponents so far appear not to agree with the KPZ theory however. The situation is different for sputter induced *erosion*[2]. A series of studies by Williams and collaborators [3] on the erosion of graphite by a beam of Ar ions found that at *lower* dosages the height height correlation function. measured by the scanning tunneling microscope(STM), is consistent with the scaling prediction

$$< |h(\mathbf{q})|^2 > \propto F(tq^z)/q^\nu \tag{1}$$

with $\nu = 2.9 \pm 0.2$ and $h(\mathbf{q})$ the height profile. This indeed would be inside the range of values for the height exponent α in $< h^2(L) > \propto L^{2\alpha}$ with $\alpha = \nu - 2$ found in numerical studies of theKPZ model in $2 + 1$ dimensions ($\alpha \approx .4$). Moreover, even though the incident beam made a 60° angle with the surface normal, the measured height-height correlation function was essentially isotropic – a feature which is characteristic of the KPZ model under modest anisotropy (see below).

Recently. these authors [4] extended their work to study very large radiation dosages (~ 4000 monolayers eroded) using the atomic force microscope (AFM). In the latter case. they found that the surface had lost its self-affine appearance and developed an anisotropic "ripple" structure with grooves along the beam direction and with a characteristic ripple spacing of order 0.1μ. The modulation showed up as a maximum in the height correlation function $< |h(\mathbf{q})|^2 >$ around $q \sim (0.1\mu)^{-1}$. The width of the surface was found to increase *linearly* in time (i.e. $\beta \approx 1$) whereas for KPZ, $\beta \approx 0.2$.

2. Sputter Yield and Growth Law

In these notes, I will give a (naive) derivation of the KPZ equation for the case of erosion to see what assumptions are made and how they could have failed. The starting point of the derivation is the *sputter-yield*, the numbers of atoms removed from the substrate per incident beam atom. It can be measured as a function of the angle θ between layer normal and beam direction. Typically, $Y(\theta)$ increases with θ until it reaches a maximum around 45°-60°and then drops rapidly. We assume here an amorphous surface so there is no dependence on the azimuthal angle. The yield also will depend on surface curvature κ: mountain peaks ($\kappa < 0$) should erode a little faster than valley bottoms ($\kappa > 0$) (due to shadowing of nearby parts of the surface). We will expand:

$$Y \approx Y_0 + Y_1\theta^2 - Y_2\kappa\delta \tag{2}$$

with $Y_0 \sim Y_1 \sim Y_2$ and positive while δ is an unknown length-scale which controls the erosion differential of mountains and valleys. We will later use the Williams et al. experiment [3] to estimate δ. Note that by symmetry only even powers of θ can appear in Eq.2.

If we are allowed to neglect redeposition and surface diffusion then the erosion velocity v_n is JYa with $a \sim 3\text{Å}$ or so the characteristic spacing of graphite layers and J the incoming flux. If $h(\mathbf{r},t)$ is the height profile then

$$\frac{\partial h}{\partial t} = V_n \sqrt{1 + (\nabla h)^2} \tag{3}$$

Eq.3 is only valid if the density of the film is not strictly fixed as it is in a crystal. If the Ar^+ beam has amorphized the graphite, as we will be assuming, then this condition is satisfied. The angle θ is recovered using $\cos(\theta) = \hat{\mathbf{n}} \cdot \hat{\mathbf{n}}^J$ with $\hat{\mathbf{n}}$ the layer normal and $\hat{\mathbf{n}}^J$ the beam direction:

$$\cos(\theta) = (n_z^J - \mathbf{n}_\perp^J \cdot \nabla h)/\sqrt{1 + (\nabla h)^2} \tag{4}$$

with n_z^J and $\mathbf{n}_\perp^J$ respectively the normal and inplane component of $\hat{\mathbf{n}}^J$. Combining Eqs. 2-4 and going to the limit of small θ and $(\nabla h)^2$ gives

$$\frac{\partial h}{\partial t} = -V_0 - V_1(\mathbf{n}_\perp \cdot \nabla h) - V_2(\nabla h)^2 + V_3(\mathbf{n}_\perp^J \cdot \nabla h) + \gamma\nabla^2 h + \eta(\mathbf{r},t) \tag{5}$$

with $V_0 \propto (JY_0 a)V_1 \propto (JY_1 a)V_3 \propto (JY_1 a)V_2$ a linear combination of V_0 and V_1, while $\gamma \propto (JY_2 a\delta)$. Finally, η is the shot-noise of the beam. By tilting the sample, i.e. by adding a linear function to $h(\mathbf{r})$, it is easy to remove the term $\mathbf{n}_\perp \cdot \nabla h$ from Eq.5. After setting $h \to -h$ we then recover the anisotropic KPZ model [5]. This model has the same scaling behavior as the KPZmodel provided the prefactors of $(\partial h/\partial x)^2$ and $(\partial h/\partial y)^2$ are of the same sign. This will be true for small $|n_\perp|$ in our case. Under these circumstances, numerical studies [6] of Eq.5 find an essentially isotropic height correlation function, in agreement with the experiments of ref. 3.

Even though the scaling behavior is correct, we must also estimate whether *the order of magnitude* of the coefficients agrees with experiment. First, if shot-noise is the roughening mechanism then we should epect that after N monolayers are eroded, typical height fluctuations δh should roughly be of order $\sqrt{N}a$ Indeed, for $N \approx 400$, the measured $\delta h \sim 100$Å while our estimate would be $\sqrt{N}a \sim 60$Å. Next, to find δ, we note that (for early times) in linear response [7] the length $\zeta \sim \sqrt{\gamma t}$ should be of the order of magnitude of the *width* of a mountain. Since $\gamma \propto (JY_2 a \delta)$, $\zeta \sim \sqrt{JY_2 a \delta t}$ or, with $\delta h \sim a\sqrt{JY_0 t}$,

$$\delta h/\zeta \sim \sqrt{Y_0 a/Y_2 \delta} \tag{6}$$

Eq.6 gives an independent check of the model: it predicts that the typical mountain slopes atearly times ought not to depend on beam flux J. Indeed over a remarkably large range of flux (and even dosage), $\delta h/\zeta \sim 0.1$ which would give, for $Y_0 \sim Y_2$, $\delta \sim 10^2 a$. Finally, the non-linearterm in Eq.5 becomes important when $V_2(\nabla h)^2 \geq \gamma \nabla^2 h$ or when $\delta h \geq \delta$ if we estimate $\nabla h \sim \delta h/\zeta$ and $\nabla^2 h \sim \delta h/\zeta^2$. For $\delta \sim 300$Å this would mean that we can expect significant non-linear corrections if we erode more than 10 layers or so, so non-linear corrections are in general significant and we should expect KPZ exponents rather than linear response, as is indeed the case in experiment.

3. Limitations of the Model

All these estimates based on Eq.5 appear to give very reasonable results. However, we saw that experimentally the scaling description actually does break down at late times [4]. We made a number of important assumptions along the way in deriving Eq.5: (i) we neglected surface diffusion – which is very reasonable for room temperature graphite, (ii) we neglected carbon-redeposition, (iii) we treated incorrectly shadowing as a *local* effect [7], and, (iv) we silently assumed chemical uniformity of the film. Starting with the second assumption, after an Ar^+ ion strikes the substrate, some of the re-emitted carbon atoms will skim along the surface, roughly along the beam direction, and be re-absorbed. This would amount to a very low-angle beam of C atoms spraying the eroded film and producing *re-growth*. The re-growth would be highly anisotropic along the beam direction. Such a low-angle growth mode is susceptible to growth instabilities – because of shadowing [7] – which may indeed explain the groove pattern at late times. Next, as mentioned, shadowing produces instabilities for *growth* (as it enhances the growth of protrusions) but for erosion, it is hard to see how it can lead to instabilities for. on the contrary, it would remove protrusions and thus smooth out the surface. Finally, chemical non-uniformity of the film also may well be compromised by long-time exposure to the Ar^+ beam. If the film is non-uniform (e.g. due to random local strains caused by radiation damage) then we would have to add *quenched* impurities to the right hand side of Eq.5. Such quenched-in noise will increase the surface roughness, as is known from the imbibition experiments, but it is unlikely to produce a width increasing *linearly* in time. In fact, the linear increase with time of the surface width late times by itself proves already that in general shot-noise *cannot* be responsible as for shot-noise the width $w(t)$ scales as t^β with $\beta \leq 1/2$ under rather general conditions. It is easy to see that very modest chemical alterations in the substrate indeed could produce linear growth instabilities: Imagine a ripple on the surface. There is actually no chemical "symmetry" between the C atoms at the maxima and at the minima of the ripple. Those minima that have just been exposed by the erosion process and are still surrounded by a matrix of amorphous carbon while the atoms at the mountain peaks are much more exposed and have been submitted to more radiation. This chemical asymmetry could lead to a different density and a different (local) bonding arrangement. This in turn could make the sputter yield at the maxima and minima different (apart from the shadow effects incorporated in Eq.2). If the erosion rate of the minima would exceed that of the maxima then this could effectively change the sign of γ and a growth instability would ensue with an amplitude proportional to time. It is easy to see that a ripple whose crests run along the beam direction would have a growth velocity greater than that of a ripple with crests perpendicular to the beam direction, because of the above-mentioned shadow-erosion. Qualitatively these features are indeed observed at late times but whether the above "chemical asymmetry" between the extrema exists is not known at present.

Acknowledgement

I would like to thank S. Williams for communicating his data before publication, R.Snyder for providing me with the videotape for the presentation, M. Paczuski and R. Messier for suggesting a derivation of Eq.5 based on the sputter yield curve, J. Krim for a critical reading of the manuscript, and D. Eaglesham for helpful discussions in particular on the redeposition process.

REFERENCES

[1] J. Chevrier, V. Le Thanh, R. Buys, and J. Derrien, *Europhys Lett.* **16** 737 1991.

[2] *Erosion and Growth of Solids Stimulated by Atom and Ion Beams* (G. Kirialudis, G. Carter, and J.L. Whiston, eds.) Nyhoff, Hingham, MA 1986.

[3] E. Eklund, R. Bruinsma, J. Rudnick, and R.S. Williams, *Phys. Rev. Lett.* **67** 1759 1991.

E. Eklund, R.S. Williams, and E.J. Snyder, *Mat. Res. Soc. Symp. Proc.* **157**, 305 1990.

E. Eklund, E.J. Snyder and R.S. Williams, *Surf. Sci.* (to be published).

[4] E. Snyder, D. Baugh, R. Williams, E. Eklund, and M. Anderson, (preprint) .

[5] D. Wolf, *Phys. Rev. Lett.* **67** 1783 1991.

[6] D. Wolf, (private communication) .

[7] G.S. Bales, R. Bruinsma, E. Eklund, R. Karunasiri, and A. Zangwill, *Science* **249** 264 1990.

FACETS AND STEPS

THE INITIAL STAGES OF GROWTH: PLATINUM ON PT(111)

**Georg Rosefeld, Andreas F. Becker, Bene Poelsema,
Laurens K. Verheij and George Comsa**
*Institut für Grenzflächenforschung und Vakuumphysik, KFA-
Forschungszentrum Jülich, Postfach 1913, D-5170 Jülich, Germany*

Abstract. Thermal Energy Atom Scattering (TEAS) has been used to study the
initial stages of Pt growth on Pt(111) at substrate temperatures between 200 K and
800 K. The cluster density at a given coverage decreases with increasing temperature in
accordance with nucleation theory up to 400 K. However, in contradiction to nucleation
theory, the cluster density increases by more than one order of magnitude between 400
K and 500 K. Above 500 K a high density of small, stable clusters (heptamers) is
observed up to coverages of ten percent of a monolayer.

1. Introduction

The growth of a thin film on a crystal surface by vapour deposition at a sufficiently
high supersaturation proceeds via nucleation of adatoms and the subsequent growth
of the nuclei to larger islands. In the case of homoepitaxial systems, we are dealing
with two-dimensional nucleation. Within the picture of nucleation, generally accepted
sofar, nuclei are formed in a short induction period, in most cases too short to be
accessible by experiment [1]. The rim of the growing stable nuclei is assumed to be
an almost perfect sink for adatoms. The accommodation of adatoms at the growing
nuclei reduces the supersaturation and thus the further formation of new nuclei is
largely suppressed. As a consequence, the island density during later stages of growth
is determined by the density of nuclei formed at the very first stages of monolayer
deposition, as long as effects reducing the island density like cluster coalescence and
Ostwald ripening can be excluded. Therefore, much effort has been made to understand
nucleation kinetics in order to control the morphology of ultrathin films [e.g. 2-4]. The
question arises what happens if the assumption, that clusters above a certain stable
size are perfect sinks for adatoms, is no longer valid. i.e. if adatoms are not efficiently

bound at these clusters. Then an interesting situation may occur: if the formation rate of stable clusters of a certain ("magic") size is higher than the rate at which they accommodate adatoms and grow to larger islands, an anomalously high density of these magic clusters is expected to build up during depostion of atoms onto the surface. From the experimental observations reported in this paper we conclude that this is indeed the case for the system Pt/Pt(111) at coverages below ten percent of a monolayer and substrate temperatures above 500 K.

2. Experimental Results

Thermal Energy Atom Scattering (TEAS) offers a unique way to monitor *in-situ* the evolution of the cluster (island) density on a crystal surface during vapour deposition. This is due to the large cross-section for diffuse scattering from various defects on the flat surface like adatoms, clusters and steps bordering larger islands [5]. For *in-phase* conditions, as used in the measurements reported here, scattering contributions from adjacent flat terraces interfere constructively and the attenuation of the specularly reflected He intensity is due to diffuse scattering alone [5,6]. For small islands (clusters) the cross-section for diffuse scattering is in a first approximation given by the overlap of the single adatom cross-sections, whereas for larger islands (> 12 Å in diameter) only the step atoms rimming the islands contribute to diffuse scattering [5,7]. Accordingly, the effective cross section for diffuse He scattering per island Pt atom, Σ', decreases with increasing island (cluster) size. Thus, at a given coverage, the total amount of diffuse scattering is the higher the smaller the structures on the surface are. The in-phase He intensity can be used to determine the mean cluster size at a given coverage or, equivalently, the cluster density at this coverage. The specularly reflected He intensity was recorded while depositing Pt atoms onto the Pt(111) surface at a rate of about 0.002 ML/s at a constant substrate temperature. The experiments were made at different substrate temperatures in the range of 200-800 K. The initial decay of the normalized He specular peak height was approximated by a straight line for each of the deposition curves (cf. inset in figure 1): $I/I_o = 1 - \Sigma' n_s \theta$, where θ denotes the coverage and n_s the number of lattice sites per unit area. The absolute value of the slope of this line, $\Sigma' n_s$, is a measure of the total amount of diffuse scattering at a given coverage and therefore, of the cluster density at this coverage. Figure 1 shows the initial slope, $\Sigma' n_s$, as a function of the substrate temperature. Between 200 K and 400 K, the initial slope decreases with increasing temperature, showing that in this temperature range, the cluster density at a given coverage decreases with increasing temperature in agreement with nucleation theory [1]. However, this trend is not continued above 400 K: the cluster density shows a dramatic increase between 400 K and 500 K. The absolute value of the initial slope, $\Sigma' n_s$, indicates that, above 500 K, we are dealing with a high density of small clusters with a mean size of approximately seven atoms (heptamers), as demonstrated in figure 1 by the dashed line [8].

These small structures have further been characterized by investigating their stability at the deposition temperature. The clusters were prepared by deposing Pt at a temperature of 600 K and stopping the deposition at five percent of a monolayer. The

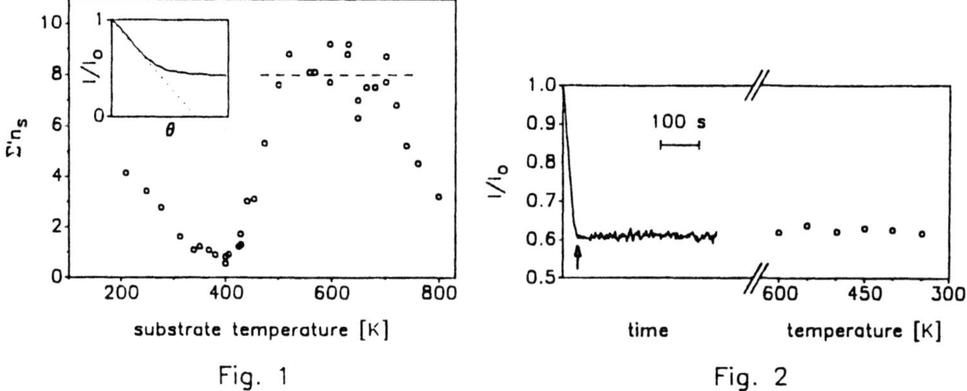

Figure 1. Initial slope of deposition curves as a function of the substrate temperature. Deposition rate: 0.0018-0.0023 ML/s. The dashed line corresponds to a situation where all clusters are heptamers. Inset: example for a deposition curve: the normalized He specular peak height versus coverage up to 0.15 ML.

Figure 2. Annealing behaviour after deposition of 0.05 ML (arrow) at a substrate temperature of 600 K. Left part: normalized He specular peak height versus time while the substrate is held at 600 K. Right part: normalized He specular peak height versus temperature while the substrate is cooled down.

He specular peak height was monitored further while keeping the sample at 600 K. As obvious from figure 2 (left part), no annealing is observed for more than 300 seconds, revealing that the small clusters are stable at a temperature of 600 K. Additionally, a possible contribution of free adatoms to the high amount of diffuse scattering above 500 K could be ruled out: the afore mentioned experiment was continued monitoring the He specular peak height while cooling the sample to 350 K (right part of figure 2). If there were a non-negligible adatom density at 600 K, a large fraction of the adatoms would undergo 2D- condensation as the sample is cooled and the He intensity should increase, correspondingly. The (Debye-Waller corrected) data in figure 2 (right part) show that this is not the case.

3. Discussion

The experimental observations described in the previous section indicate that we are dealing with a high density of small stable clusters formed during deposition at high temperatures. In the following, we propose a simple explanation of the observed phenomenon within an effective nearest neighbour model. The explanation is based on the assumption that the probability of simultaneously breaking two in-plane nearest neighbour bonds becomes a significant process at temperatures above about 400

K. This assumption is justified from recent findings during the investigation of later stages of growth ($\theta > 1$ ML) of Pt on Pt(111) [7]. Below this temperature two nearest neighbour bonds cannot be broken simultaneously and consequently, a triangular trimer is a stable cluster. All compact clusters consisting of more than three atoms are stable too, because each of the involved atoms has at least two nearest neighbours (see figure 3). Thus once a trimer is built, it will steadily grow to a large island by further accommodation of adatoms. This is the situation commonly described in nucleation theory and corresponds to our experimental findings up to temperatures of about 400 K.

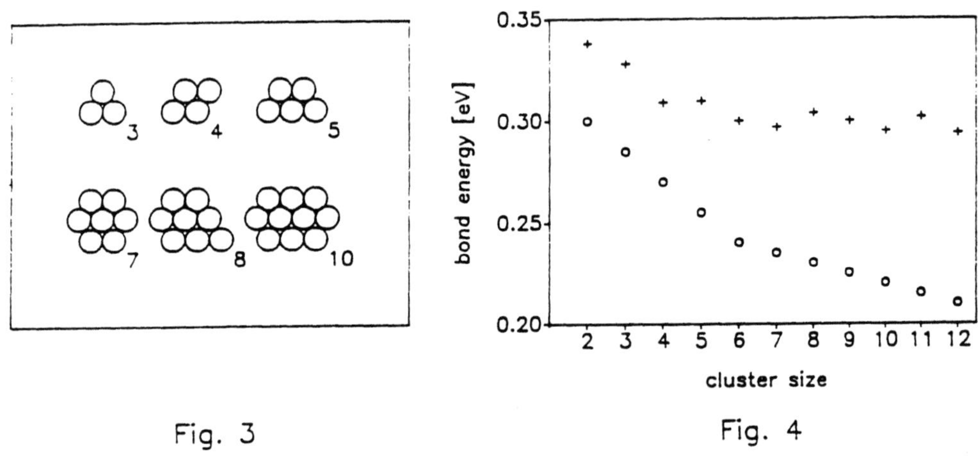

Figure 3. Examples for small 2D-clusters on Pt(111). These configurations were used in the calculations discussed in the text.

Figure 4. Effective dissociation energy per bond for Pt clusters on Pt(111). Crosses: results from calculations in the Effective Medium Theory [9]. Circles: values used in the rate equations (see text).

Above 400 K, where as assumed the simultaneous breaking of two nearest neighbour bonds becomes significant, the situation is different. Now a hexagonal heptamer (figure 3) is the smallest stable configuration, the next stable one being the compact decamer. The configurations in between contain one or two twofoldly bound atoms. Therefore, the steady growth of a heptamer to a large island is not at all obvious. If we assume for a moment (see below) that heptamers can be formed, our experimental results for temperatures above about 400 K lead to the following picture: heptamers formed by nucleation of adatoms are hindered from growing to larger clusters by the "stability gap" between heptamers and decamers. while the further deposition of atoms results in the production of new heptamers. Consequently, a high number of heptamers builds up, in line with the observed value of the initial slope, $\Sigma' n_s \approx 8$ (cf. figure 1). This explanation can only be true if the formation rate of heptamers is higher than the rate at which heptamers grow to decamers. In a simple nearest neighbour model with a nearest neighbour bond strength independent of cluster size, this condition is not fulfilled. Indeed, if two nearest neighbour bonds can easily be broken, the formation of a heptamer involves the successive formation of five unstable configurations (dimer to hexamer) whereas only two equally unstable configurations separates a decamer from a heptamer (see figure 3). This difficulty could be overcome if the two configurations between the heptamer and the decamer were less stable than the smaller configurations with likewise twofold bound atoms, i.e. if the effective nearest neighbour bond strength were decreasing significantly with increasing cluster size. Recent calculations by Norskov and coworkers indicate that this is indeed the case [9]. Figure 4 shows their results for the dissociation energy per nearest neighbour bond needed to remove an atom from a Pt cluster onto the terrace on the Pt(111) surface. We have checked whether the dependence calculated by Norskov et al (figure 4) allows for the formation of a high density of heptamers in terms of rate equations similar to those used by Venables et al [1]. Atom-atom interactions are modelled as effective nearest neighbour interactions and only adatoms are assumed to be mobile. The accommodation rate of adatoms of density n_1 at clusters consisting of k atoms (k-mers), with corresponding density n_k, is written as $\Gamma_{k,+} = \alpha_k n_1 n_k$, where α_k contains the diffusion constant and a capture number. The decay rate of k-mers is $\Gamma_{k,-} = \beta_k n_k$, where β_k is given by the binding energy of the most loosely bound atom in the cluster. The corresponding rate equations have been solved by a Monte Carlo approach. Using Norskov's values in the rate equations, the preferential formation of heptamers could not be modelled. However, for a stronger cluster size dependence of the strength per nearest neighbour bond, like that also shown in figure 4, the situation was more favourable: for a substrate temperature of 400 K, clusters quickly grew to larger islands, whereas for 500 K a high density of heptamers was generated up to coverages of a few percents of a monolayer [10]. We think that this result is quite promising for future theoretical work. A more sophisticated description of accommodation coefficients and decay constants is needed for the present case, where the structures involved consist of a few atoms only and small stable clusters are not perfect sinks for adatoms. From the experimental point of view, direct imaging techniques like Field Ion Microscopy and Scanning Tunnelling Microscopy are expected to contribute to a detailed knowledge of kinetic processes on an atomic scale.

Acknowledgement

We are deeply indebted to L.B. Hansen, P. Stoltze, K.W. Jacobsen and J.K. Norskov for the permission to use the results of their calculations in figure 4.

REFERENCES

[1] J.A. Venables, G.D.T. Spiller, M. Hanbücken, *Rep. Prog. Phys.* **47** 399 1984.

[2] G. Zinsmeister *Basic Problems in Thin Film Physics* R. Niedermayer, H. Mayer, eds. Göttingen 1966 pp.32-42.

[3] J.A. Venables, G.L. Price *Epitaxial Growth, Part B* J.W. Matthew, ed. New York 1975 pp.381-436.

[4] B. Lewis, J.C. Anderson *Nucleation and Growth of Thin Films* London 1978.

[5] B. Poelsema, G. Comsa *Scattering of Thermal Energy Atoms from Disordered Surfaces* Springer Tracts in Modern Physics 115, Berlin 1989.

[6] H. Xu, Y. Yang, Th. Engel, *Surf. Sci.* **255** 73 1991.

[7] B. Poelsema, A.F. Becker, G. Rosenfeld, R. Kunkel, N. Nagel, L.K. Verheij, G. Comsa, *Surf. Sci.* in press

[8] For this evaluation it has been assumed that the cross-section for diffuse scattering of Pt adatoms on Pt(111) is 120 Å^2, i.e. similar to a number of adsorbates like CO, NO and Xe for a room temperature He beam, incident at an angle $\vartheta_i = 40°$. Scaling this value to the present experimental conditions ($\vartheta_i = 68.5°$) and using the overlap principle [5], results in $\Sigma' n_s \approx 8$ for heptamers.

[9] L.B. Hansen, P. Stoltze, K.W. Jacobsen, J.K. Norskov, to be published

[10] Note that an adatom accommodation coefficient decreasing with increasing cluster size has a similar effect as the decrease of bond energy with increasing cluster size.

Surface Disordering:
Growth, Roughening and
Phase Transitions

STEP STRUCTURE ON SI(111) SURFACES

Ellen D. Williams and N.C. Bartelt

Department of Physics, University of Maryland, College Park, MD 20742

Abstract. We have studied the structure of steps on Si, with the goal of making detailed comparisons with statistical mechanical theories of stepped surfaces. By using STM to quantify kink structure, step correlations, and step-height mixtures at low temperatures, we find convincing evidence for energetic, in addition to entropic, interactions between steps. Using known values for the elastic constants of Si and the stress of the reconstructed surface, we can obtain a quantitatively consistent description of the observations assuming that the origin of the interaction between steps is elastic.

1. Introduction

The nature of steps on surfaces determines the morphology and many important properties of surfaces under both equilibrium and non-equilibrium conditions [1]. This importance has engendered a large theoretical literature, dating back to the early 1950's [2]. However, until very recently many of the crucial properties of steps, such as kink energies and step-step interactions, have largely been unquantified experimentally. The possibility of directly imaging steps at an atomic level with STM has changed this situation dramatically. In the following, we will describe observations (using STM and LEED) and analysis of the step behavior as a function of misorientation angle on these surfaces. we will concentrate on points in which simple behavior is demonstrated, and on points in which we are limited in our capability to extract fundamental information (usually about energetics) from the observations.

2. Background

Before attempting to compare measured step structure with theoretical models, it is crucial to establish whether the step structure is in thermal equilibrium. This, in general, requires a thorough understanding of the thermal behavior. Fortunately, Si is among the most studied materials in surface science, so we have a wealth of information in the literature on which to call for this purpose. Si(111) is known to form a complex (7×7) reconstruction, which disorders in a strongly first-order transition at about 850°C, leaving residual (2×2) order in the high temperature phase [3]. The vicinal surfaces of Si are all orientationally stable above this disordering temperature, and consist of periodic arrays of steps of single-layer height (3.1Å) [4,5]. Recent high-resolution X-Ray diffraction studies [6] have shown that these surfaces are rough, as expected for surfaces which are not facets on the equilibrium crystal shape. The formation of the (7×7) reconstruction is known to occur preferentially by nucleation at step edges, resulting in a specific registry of the reconstruction with respect to the upper step edge [7,8]. At low step densities corresponding to step separations of 1000Å or more, the single layer height steps become straighter, and wander less upon formation of the reconstruction [9]. However at larger step densities, corresponding to separations of 200Å or less, there are pronounced changes in the arrangements of the steps with a strong dependence on the orientation of the steps. The reversibility of the step rearrangements which occur on surfaces misoriented toward the $[\bar{2}11]$ and $[1\bar{1}0]$ directions indicate sufficient surface mobility for step equilibration down to approximately 750°C. Surfaces misoriented precisely along the $[\bar{1}\bar{1}2]$ direction are orientationally stable when the (7×7) reconstruction forms. On these surfaces, the formation of triple-height steps occurs reversibly with the reconstruction [10]. For 6° to 12° of misorientation, the diffraction signature of the surface indicates that nearly all the steps are triples at low temperature. However, at lower angles, the diffraction signature is intermediate between that expected for either singles or triples, indicating a mixture of step heights, with the fraction of triples decreasing with increasing step density.

The results listed above provide a cumulative body of evidence that indicate that step structure on Si surfaces can be equilibrated. The characteristics of the surfaces are complex, perhaps largely as a result of the complexity of the reconstruction. However, as is shown below, Si surfaces misoriented toward the $[\bar{1}\bar{1}2]$ direction have proven to be amenable to analysis based on simple descriptions of step behavior.

3. Characterization and Analysis

The experiments were performed using standard techniques. The samples used in this study were commercially obtained Si wafers nominally inclined by 1.3°, 2.5°, 3.8° and 6° towards the high symmetry $[\bar{1}\bar{1}2]$ direction. The samples were cleaned by flashing to 1250°C for 90s in ultrahigh vacuum [10,11]. During the flashing the background pressure did not increase above 4×10^{-10} Torr. After flashing, the samples were quickly cooled to 950°C and then slowly cooled to room temperature with a rate of less than 0.2°C/s. Diffraction measurements were performed using a high resolution LEED instrument (HRLEED) with an instrumental resolution of 500Å [12]. Scanning tunneling microscopy measurements [13] were performed in another UHV chamber with a base pressure below 1×10^{-10} Torr.

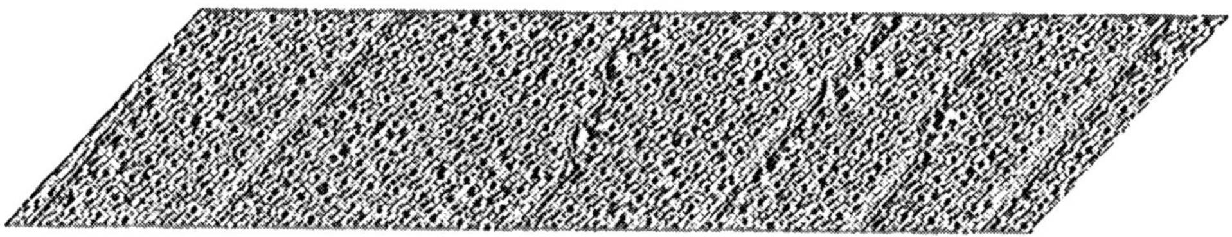

Figure 1. An STM image of a stepped Si surface

We discuss the observations in increasing order of complexity, beginning with the behavior of kinks, proceeding to the behavior of interacting single-height steps, and finally to the coalescence of steps to form triple-layer height steps.

3.1. Step Diffusivity and Kink Energy

The statistical behavior of steps is determined by the thermal formation of kinks, allowing step wandering. Step wandering can be characterized by the step "diffusivity," b, defined for an isolated step as:

$$b^2(T) = \langle (x(0) - x(y))^2 \rangle \frac{a_\parallel}{y}, \tag{1}$$

where $x(y)$ charts the course of the step has it travels along the y direction, defined as the net crystallographic direction of the step edge, and $a_\parallel$ is the smallest possible kink-kink separation. If the step is oriented in a high-symmetry direction, then its minimum energy configuration will have no kinks, and the diffusivity b^2 will be proportional to $e^{-\epsilon/kT}$, where ϵ is the kink energy [14]. Swartzentruber *et al.* [15] have effectively used this type of analysis to deduce kink energies of vicinal Si(001).

When the kink energies are very large compared to thermal energies, as they are on Si(111) at room temperature, one must use this result with some circumspection, however. This is because steps misoriented by some amount away from the high-symmetry direction will have an intrinsic kink density ρ. If the kinks are non-interacting, there is no energetically preferred configuration of kinks, and the resulting distribution of kink positions leads to a non-zero diffusivity $[(\rho - \rho^2)a_\perp^2$, where $a_\perp$ is the minimum kink depth] even at zero temperature. In the limit of low temperatures, the expression for the diffusivity in terms of the kink energy becomes

$$\frac{b^2}{a_\perp^2} = \frac{4z^2 - \sqrt{\rho^2 + 4z^2(1 - \rho^2)}}{4z^2 - 1} - \rho^2, \tag{2}$$

where z is $e^{-\epsilon/kT}$ [16]. Analysis of this function shows that the diffusivity is essentially constant at its zero-temperature value up to $kT \approx 0.2\epsilon$. Thus if a measurement is made at temperatures below this limit, there are almost no thermally excited kinks, and one obtains only a lower limit on the kink energy. These considerations can be used in evaluating the kink structure on Si(111). The structure of steps and kinks on Si(111)-(7×7) is illustrated in Fig. 1.

The registry of the (7×7) reconstruction with the step edge and with the kinks is immediately obvious. Analysis of many such images [17] has shown that the terraces are quantized in units of the reconstruction (the unit length perpendicular to the step edge is 23.3Å). Measurement of approximately 800 step segments similar to those in Fig. 1 has revealed only about 5% kinks, none of which have a depth greater than one unit. The steps are too close together to allow application of Eq. (1) to be valid. However, a naive estimate of the diffusivity and the kink energy based on this observation would result in a kink energy of about 0.3eV for the 7×3.84Å long kink edge, assuming equilibration at 800°C. This is an astonishingly low number for the kink energy, given other observations which indicate that kink formation is highly unfavorable on these steps [18]. However, if we note that for each series of steps, all the kinks were in the same direction, then it becomes apparent that Eq. (2) must be used with a value of ρ of about 0.05, leaving us able only to place a rough lower limit of 0.5eV on the kink energy.

3.2. Step-Step Distributions

Thermal wandering of steps on vicinal surfaces leads to a measurable distribution of step-step separations, which in turn (in principle) allows one to deduce the nature of step-step interactions. In the case where there are no energetic interactions between steps, there is still an effective repulsion due to the non-crossing condition. This leads to a broad distribution of step-step separations, which is peaked at the average separation l. If any additional energetic repulsive interactions are present, the distribution of separations becomes narrower. Specifically, for an interaction of the form

$$U = Ax^{-n}, \tag{3}$$

where x is the distance between steps measured perpendicular to the net step direction, the distribution of step separations is approximately a Gaussian of width [14]

$$w = \left[\frac{kTb^2(T)}{8n(n+1)Aa_{\parallel}} \right]^{\frac{1}{4}} l^{\frac{n+2}{4}}, \tag{4}$$

where l is the average step-step separation, b^2 is the diffusivity defined in Eq. (1), and $a_{\parallel}$ is the minimum kink-kink separation. It is thus apparent that measurements of the step distribution allow one to gauge the form of the interaction potential and, if the diffusivity is known independently, the magnitude A of the interaction. For the most commonly predicted form of step-step interactions, $1/x^2$, Eq. (4) reduces to a linear relationship between the average step separation and the width of the distribution. However, the relationship of the measurable value, which is the width of the distribution, to the unknowns through a power of four means that the relative experimental uncertainties will translate into much larger uncertainties in the values of the unknowns. We have used STM measurements such as those in Fig. 1 to assemble a statistical sample of the step-step separations on surfaces misoriented from 1.3° to 6°. The formation of triple-height steps complicates the analysis of the distributions, however on the two smallest misorientations measured the density of triples is small enough that large groups of pure single-height steps could be identified for measurement [13]. The results of the measured terrace distributions are shown in Table 1.

Table 1

Number of terrace widths, N, observed as a function of terrace separation, x, in units of half the (7×7) unit cell (23.3Å).

1.3°		2.5°	
x	$N(x)$	x	$N(x)$
1	0	1	7
2	0	2	73
3	3	3	213
4	12	4	127
5	31	5	46
6	37	6	7
7	33	7	0
8	20	8	0
9	6	9	6
10	3	10	1
11	1	11	3

The distribution of these data is considerably narrower than expected for entropic wandering. As shown in Fig. 2, the data fit a Gaussian form extremely well, indicating repulsive energetic interactions.

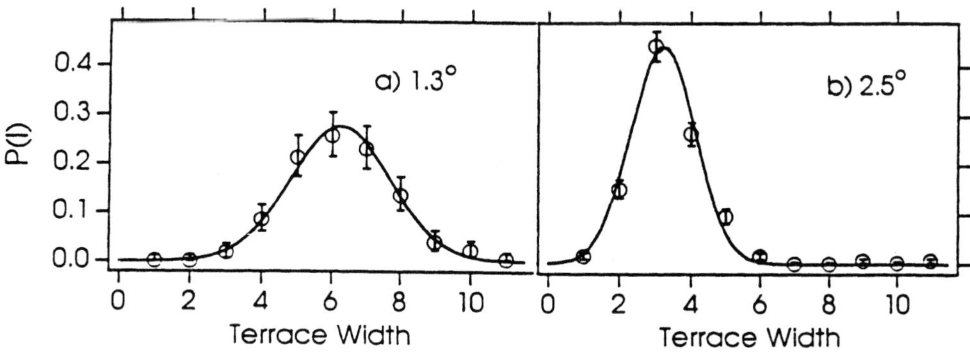

Figure 2. Terrace-width distribution for Si surfaces misoriented by a) 1.3°and b) 2.5°.

The average and widths of the two distributions are: for the 1.3° sample, $l = 6.2 \pm 0.1$ units and $w = 1.4 \pm 0.1$, and for the 2.5° sample, $l = 3.3 \pm 0.05$ units and $w = 0.9 \pm 0.04$ units. If we use assume a $1/x^2$ form for the step-step interaction, then a straight line fit of these data yields a slope of 0.26 ± 0.01. (If we allow the value of n to vary, we obtain a better fit with $n = 0.9 \pm 0.5$, however since there are only two data points, this cannot be taken too seriously.) Given the value of the slope of w vs. l, we can use our estimate of b^2 to deduce the value of the interaction energy between steps. The result is $A \approx 0.4 \pm 0.1$eV-Å, assuming a 20% uncertainty on our value of b^2, and an equilibration temperature of 800°C. This value can be compared with the value expected for elastic interactions between steps. Martinez et $al.$ [19] have measured the surface stress for Si(111)-(7×7) to be $s = 0.19$eV/Å^2. The elastic interactions between steps of height h are related to this stress via [20]:

$$A = \frac{2(1 - \sigma^2)}{\pi E} s^2 h^2 \tag{5}$$

where σ is Poisson's ratio ($\sigma = 0.31$) and E is Young's modulus ($E = 0.65$eV/Å^3). (As in Ref. [21], for simplicity, we only include the component of the elastic forces induced by the surface stress.) For the single step height of 3.14Å, this gives a predicted value of $A = 0.315$eV-Å, in good agreement with the estimate from the experimental values. A similar estimate of the value of A for the disordered surface using the data of Alfonso et $al.$ [21] yields $A = 0.22$eV-Å. This is quite comparable to the value of 0.14eV-Å predicted from Vanderbilt's calculations [22] of the surface stress for the (2×2) structure of Si(111).

3.3. Step-height mixtures

As mentioned in the background section, the steps on these surfaces tend to group to form triple height steps with increasing step density. This is illustrated in Fig. 3, which is an STM image showing two single and one triple-height step on a 1.3° misoriented surface.

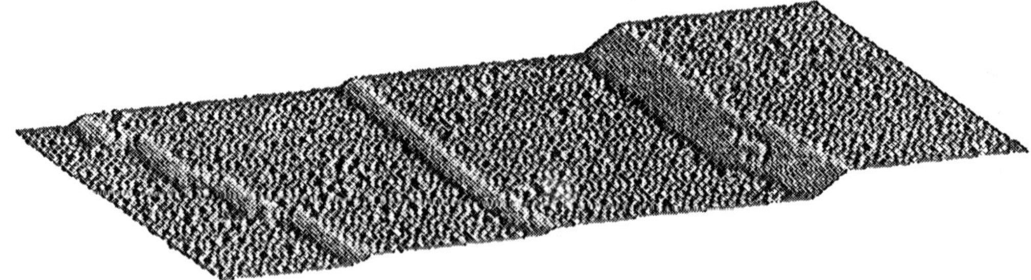

Figure 3. An STM image of a stepped Si surface with single- and triple-layer steps

The above discussion gives us confidence that it is reasonable to interpret our observations in terms of an equilibrium step structure in which there are repulsive energetic interactions between steps. Thus we can deduce from our observations that the formation energy for a triple height step from three singles ΔE must be quite comparable to the step-step interaction energy at small separations:

$$\Delta E = E_t - 3E_s \approx Al^{-2}. \tag{6}$$

In other words the energy difference appears to be less than a milli-electron volt! By compiling statistical information on the step separations and density of triples from STM measurements, we can determine whether the qualitative picture of step coalescence due to repulsive interactions is reasonable. The results of the measurements are shown in Table 2.

Table 2

STM data for the number of single-height steps, N, the number of triple height steps, M, and the average lengths of terraces between two single-layer steps l_{ss}, between two triple-layer steps l_{tt}, and between triple- and single-layer steps l_{st} as a function of misorientation angles. The lengths in the table are in units of half the unit cell width or 23.3 Å.

θ	N	M	l_{tt}	l_{st}	l_{ss}
$1.3 \pm 0.1°$	78	544	—	9.2 ± 1.6	6.9 ± 0.9
$2.7 \pm 0.3°$	222	1016	—	4.1 ± 0.4	3.3 ± 0.3
$5.2 \pm 0.4°$	257	340	3.4 ± 0.2	2.9 ± 0.1	2.1 ± 0.08
$6.6 \pm 0.4°$	282	113	2.9 ± 0.3	2.4 ± 0.2	—

The striking results of the measurements are the gradual increase in the fraction of triples with increasing angle, and the significantly greater average distance between triple-height steps and single-height steps.

These results can also be understood by assuming elastic interactions between steps. As shown in Eq. (5), the strength of the elastic interaction between two steps should be proportional to the product of the heights of the two steps. Thus the interaction between two triple height steps should be nine times stronger than that between two singles. Qualitatively, then one can expect differences in terrace widths as listed in Table 2. One can predict the relationship of the terrace widths quantitatively, at least for the ground state, by considering the Hamiltonian for an array of interacting steps, which can thus be written as

$$H = \frac{2(1-\sigma^2)}{\pi E} s^2 \sum_i h_i h_{i+1}/l_{i,i+1}^2 \tag{7}$$

where $l_{i,i+1}$ is the step separation between step i and step $i+1$, and we have assumed that surface stress is the only contribution to the torque at the step edges. In writing this equation we have neglected interactions beyond near-neighbor steps, as a first approximation. We use Eq. (7) to obtain the ground state structure for a given ratio of single and triple-layer steps by minimizing H with respect to step separation l (terrace lengths): with the constraint of fixed misorientation angle. This minimization yields a striking prediction for ratios of terrace lengths:

$$\frac{l_{tt}}{l_{st}} = \frac{l_{st}}{l_{ss}} = \left(\frac{h_t}{h_s}\right)^{\frac{1}{3}}, \tag{8}$$

where l_{st}, l_{tt} and l_{ss} are single-triple, triple-triple and single-single terrace lengths, h_s and h_t are the heights for single and triple-layer steps respectively. A comparison with the data presented in Table 2 shows that the measured terrace length ratios range from 1.2 to 1.3 with an uncertainty of about 20%, in good agreement with the predicted value of $3^{1/3} \approx 1.44$.

We can finally use Eqs. (6) and (7) to try to understand the variation of the density of triples with total step density. If we use the terrace width ratios of Eq. (8), then we can show that the ground state configuration based on Eq. (7) consists of a mixture of singles and triples with no triple-triple near neighbors when the fraction of triples is less than one half, alternating singles and triples when the fraction is equal to one half, and a mixture with no single-single pairs when the fraction is greater than one half. In each of these three regimes, the density of triples varies linearly with the density of steps with a different slope. The minimum angle θ_0 at which any triples should be observed is determined by the formation energy ΔE for a triple from three singles, according to

$$\tan \theta_0 = h\sqrt{\frac{\Delta E}{(12 - 6x)A}}, \tag{9}$$

where x is the cube root of the height ratio defined in Eq. (8), and A is the magnitude of the step-step interactions for single-layer height steps defined in Eq. (5). A comparison of the predicted variation of triple density with misorientation angle is shown in Fig. 4, for three different values of the triple-step formation energy ΔE.

When ΔE is zero, as would be the case for a solid-on-solid model with only near-neighbor interactions, the repulsive interactions drives the step configuration to pure triples at all step densities. A very small energy cost for forming a triple from three singles depresses the fraction of triples. There is qualitative agreement between the measured values and the calculated curves, indicating that the simple picture described above catches the main points of the behavior. We would expect entropic effects, which have been neglected in the description above, to round the calculated curve somewhat, which would improve the comparison with the data. Including the longer range interactions between steps which we have neglected, will tend to stabilize additional preferred step sequences [23], leading to more segments in the calculated curve, again with potential for improving the comparison with the data. Any alternative description in which kinetic trapping mechanisms drive the formation of triples leads to a much more rapid variation of triple density with angle beyond the minimum angle θ_0. We thus find an equilibrium description, in which a small energy cost for the formation of triples competes with repulsive step-step interactions, to be a satisfactory description of the observations.

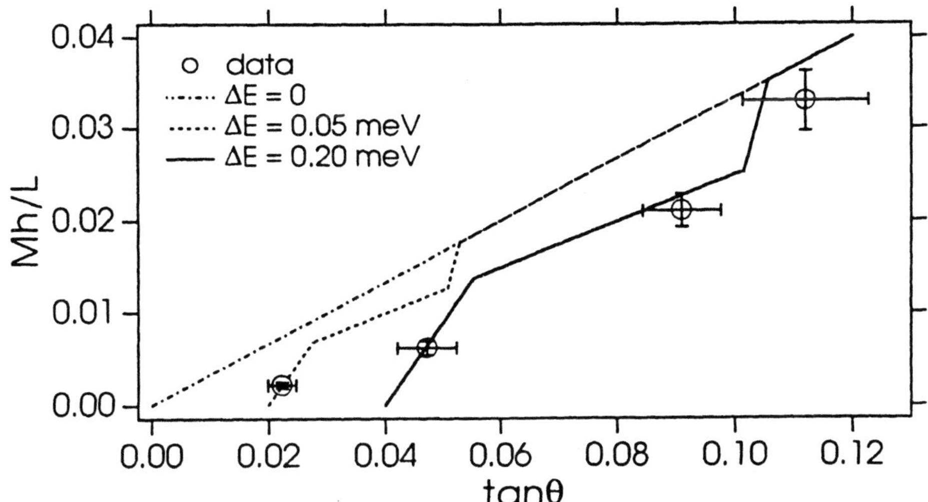

Figure 4. Density of Triple-layer height steps vs. misorientation angle. Circles are measured values, the three curves were calculated for $\Delta E = 0$, 0.05 meV/Å and 0.2 meV/Å.

4. Conclusions

We have developed what we believe to be a plausible and internally consistent picture of the energetic parameters governing the rather complex behavior of vicinal Si(111) surfaces. The simple theories of step interactions have proven to be a useful guide to understanding the observed behavior. In particular, the magnitude one expects for elastic interactions between steps satisfactorily accounts for the observed terrace-width distributions given the observed kink distribution. The converse problem, of trying to determine energetic parameters quantitatively from the experimental observations, is still limited by uncertainties in the measurements. However, as always, experimental observations reveal surprising behavior which stands as a challenge for future theoretical work. For example, it is puzzling why double-layer steps are unfavorable on Si(111), despite the fact that triple-layer height steps are evidently close in energy to three single-layer steps.

Acknowledgements

This work has been supported by the ONR under grant N00014-91-J-1401. We are very grateful for the experimental efforts of X.-S. Wang, J.L. Goldberg, and J. Wei whose respective theses provided the results discussed in this manuscript.

REFERENCES

[1] E.D. Williams and N.C. Bartelt, *Science*, **251** 393 1991.

[2] C. Herring, *Phys. Rev.*, **82** 87 1951.

[3] R.J. Phaneuf and E.D. Williams, *Phys. Rev. B*, **35** 4155 1987.

[4] N.C. Bartelt, E.D. Williams, R.J. Phaneuf, Y. Yang, and S. Das Sarma, *J. Vac. Sci. Technol A*, **7** 1898 1989.

[5] R.J. Phaneuf, E.D. Williams, and N.C. Bartelt, *Phys. Rev. B*, **38** 1984 1988.

[6] D.Y. Noh, K.I. Blum, M.J. Ramstad, and R.J. Birgeneau, *Phys. Rev. B* **44** 10969 1991.

[7] W. Telieps and E. Bauer, *Ber. Bunsenges. Phys. Chem.*, **90** 197 1986.

[8] R.S. Becker, J.A. Golovchenko, E.G. McRae and B.S. Swartzentruber, *Phys. Rev. Lett.*, **55** 2028 1985.

[9] N. Osakabe, Y. Tanishiro, K. Yagi, and G. Honjo, *Surf. Sci.*, **109** 353 1981.

[10] R.J. Phaneuf and E.D. Williams, *Phys. Rev. B*, **41** 2991 1990.

[11] B. S. Swartzentruber, Y.-W. Mo, M. B. Webb, and M. G. Lagally, *J. Vac. Sci. Technol.*, **7** 2901 1989.

[12] R.Q. Hwang, E.D. Williams and R.L. Park,, *Rev. Sci. Instrum.*, **60** 2945 1989.

[13] X.-S. Wang, J. L. Goldberg, N. C. Bartelt, T. L. Einstein, and E. D. Williams, *Phys. Rev. Lett.*, **65** 2430 1990.

[14] N.C. Bartelt, T.L. Einstein, and E.D. Williams, *Surf. Sci. Lett.*, **240** L591 1991.

[15] N.C. Bartelt, T.L. Einstein and E.D. Williams, *preprint*, 1992.

[16] B.S. Swartzentruber, Y.-W. Mo, R. Kariotis, M.G. Lagally, M.B. Webb, *Phys. Rev. Lett.*, **65** 1913 1991.

[17] J. L. Goldberg, X.-S. Wang, J. Wei, N. C. Bartelt, and E. D. Williams, *J. Vac. Sci.Technol. A*, **9** 1868 1991.

[18] J. Wei and E.D. Williams, *preprint*, 1992.

[19] W.E. Martinez, W.M. Augustyniak, and J. Golovchenko, *Phys. Rev. Lett.*, **64** 1035 1990.

[20] V. I. Marchenko and A. Ya. Parshin, *Sov. Phys. JETP*, **52** 129 1980.

[21] C. Alfonso, J.M. Bermond, J.C. Heyraud, and J.J. Métois, *Surf. Sci.*, **262** 371 1992.

[22] D. Vanderbilt, *Phys. Rev. Lett.*, **59** 1456 1987.

[23] E. Pehlke and J. Tersoff, *Phys. Rev. Lett.*, **67** 465 1991.

SCANNING TUNNELING MICROSCOPY AND X-RAY DIFFRACTION STUDIES OF GROWTH AND INTERFACIAL ROUGHNESS

D.E. Savage, E.J. Heller, Y.-H. Phang, M. Schacht, and M.G. Lagally

Department of Materials Science and Engineering, University of Wisconsin, Madison, WI 53706

Abstract. The analysis of film morphology is used to examine kinetic mechanisms in initial stages of film growth. Strategies are detailed by which information on specific atomic processes are obtained directly from STM measurements. Kinetic roughening at later stages of growth is quantified using height correlation functions. X-ray diffraction methods give evidence for interfacial roughness correlation.

1. Introduction

The morphology of growing films is governed by a competition between thermodynamics and kinetics. Consider, as an example, depositing a material onto a like substrate. Only under near-equilibrium conditions will the resulting morphology be determined by surface-free-energy considerations [1]. Generally, films are grown under conditions far from equilibrium; the resultant morphology will depend in a complicated way on the relative rates of atomic motion along a surface, incorporation into the growing surface, and the arrival rate of deposited atoms.

Turning this problem around, one can use measurements of surface morphology to make inferences about kinetic and thermodynamic processes involved during growth. In particular, by systematically evaluating the evolution of morphology on the atomic scale, one can obtain quantitative values for atomistic interactions and kinetic barriers. In the following sections we will show measurements, using scanning tunneling microscopy (STM), of growth and equilibrium island shapes, 2-d island densities, and denuded zones near steps to gain information on adatom diffusion, incorporation, and crossing of atomic steps. Examples will be drawn from the growth of Si on Si(001)[2] and GaAs on GaAs(001) [3].

113

As deposition rates are increased or substrate temperatures are reduced, film morphology is governed more and more by kinetics. Films may grow polycrystalline or amorphous and it becomes more difficult to separate out specific atomistic processes. Surface morphology measurements are still useful, but now one must measure changes in the average width of the growth front, the lateral scale of the roughness (the lateral correlation length), and the power spectrum of the surface (growth front) roughness. These measurements can then be compared with predictions of scaling obtained from various models of film growth [4]. We will show examples of growth front measurements for GaAs(001).

A more complicated situation arises when a material of a different chemical species is deposited. Here, film wetting behavior, chemical reaction, and interdiffusion must also be considered. If this process is repeated, alternately depositing different materials, heterogeneous multilayer films can be grown. For multilayer systems, film morphology may be coupled through the multiple interfaces. One way to evaluate multilayer morphology is again to examine the outer-surface roughness with STM. However, it is quite difficult to relate this measurement to the processes that have occurred during the growth of previously deposited layers. For multilayers, one can use x-ray diffraction to characterize buried interfaces [5], to obtain information on interface width, lateral scale, and correlation of roughness from interface to interface. Examples demonstrating interfacial morphology variations with growth parameters will be discussed for two very different types of multilayers, sputter deposited W/C [6] and MOCVD deposited $Al_xGa_{1-x}As$/GaAs [7]. We show that both types of films have significant amounts of vertically correlated roughness, but on very different lateral length scales.

We begin the next section with STM measurements and analysis of the initial stages of growth of Si on Si(001), followed by results of growth front morphology measurements on GaAs(001) films after many atomic layers have been deposited. Finally, we describe briefly x- ray characterization of multilayer interface morphology and relate these measurements to the nature of growth of heterogeneous layered systems.

2. STM Studies of Growth

Scanning tunneling microscopy is quite powerful for investigating morphology of the outer layers of a growing film, particularly during the initial stages of growth. In well defined situations, this morphology can be related to growth mechanisms and the kinetic barriers that are part of all growth processes.

2.1. Kinetic Mechanisms in the Initial Stages of Growth

In modelling growth, it is common to make various simplifying assumptions about the transport mechanisms involved. These include isotropic diffusion, atomic steps as either perfectly reflecting or perfectly transparent for "downward" diffusion, and atomic steps as perfect sinks for diffusing atoms on the down terrace. These assumptions can be evaluated experimentally using STM and the mechanisms quantified.

Figure 1 shows an STM micrograph of a vicinal Si(001) surface after 0.07 ML of Si has been deposited [2]. In Si(001), the surface layer forms dimers, giving a structure that has lower symmetry than the bulk and that rotates by 90° from layer to layer. This reduced symmetry has many consequences. Atomic steps are alternately rough and smooth, asymmetrically shaped Si islands form in the center of the larger terrace, and a large denuded zone (the area with a much lower island density) forms near the rough edge of the large terrace. Much can be inferred from this image: The asymmetry in the size of the denuded zone is evidence that atoms can stick more readily to the rough step (called S_B) than to the smooth step (called S_A). The shape of the islands is consistent with this interpretation; atoms attach more readily to the island ends, which are S_B type steps. The presence of the large denuded zone at the "up" side of an S_B step also implies that atoms must be able to diffuse readily downward over this step.

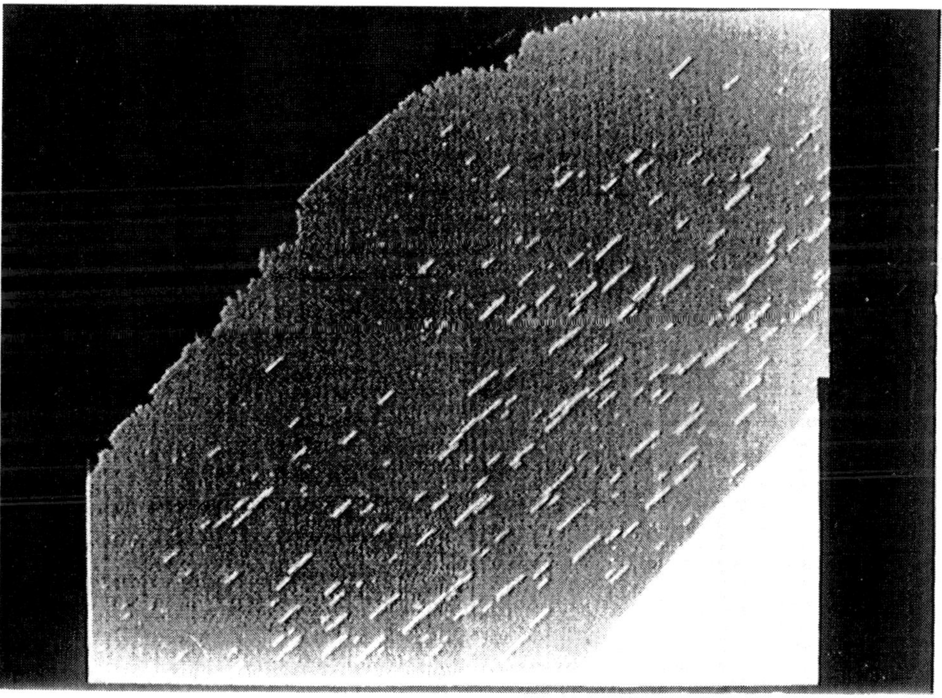

Figure 1. STM image of 0.07 monolayers (ML) of Si deposited on Si(001) with a deposition rate of 1/6000 ML/sec at a substrate temperature of 290°C. The scale of the image is $1\mu m \times 1\mu m$. Islands nucleate in the center of the large terrace and a denuded zone forms near the rough step. Five terraces are visible.

By measuring the island density in the central part of the terrace as a function of temperature, the diffusion coefficient of Si has been determined [2]. Because of the reduced surface symmetry, diffusional anisotropy can exist. The difference in the denuded-zone widths around a rough step was used to determine the amount of diffusional anisotropy, which for this system was $\sim 1000 : 1$ [2]. Finally, S_A steps were found to be nearly perfect barriers for atom crossing while S_B steps were found to be nearly transparent [2].

2.2. Growth Front Morphology

The deposition of GaAs(001) on GaAs(001) with MBE can be considered as the next level of complication in the study of growth. Here, the transport of two different types of atoms must be considered. In addition, perfect substrates can not be prepared by simply annealing a crystal, because Ga and As will evaporate at different rates, causing the surface to roughen. Typically, substrate roughness is reduced by growing a buffer layer. One simplification of the GaAs(001) surface compared with Si(001) is that, for As terminated surfaces, only double steps are present. As a consequence any anisotropies in atom incorporation or diffusion will be the same from level to level.

Initial results on morphological evolution of GaAs(001) films grown by MBE have been described recently [3,8]. Figure 2(a) shows a large-area STM scan of a $3000\text{Å} \times 2\text{mm}$ GaAs film deposited on a nominally flat substrate (no intentional substrate misorientation). There is an obviously large island shape anisotropy. This sample was grown at a very low deposition rate and high substrate temperature (610°C) followed by a 10 minute anneal, hence these are near equilibrium shapes. Using the same arguments as for Si [2], we suggest that the island shape anisotropy of $\sim 10 : 1$ reflects the ratio of the energies of steps terminating respectively the ends and sides of the terrace.

To quantify the growth front morphology shown in this image, we calculate the 1-d height-height correlation functions in the directions along and across the islands, as shown in Fig. 2(b). For line scans along an x direction the correlation function $C(X) = < z(x - X)z(x) >$, where z(x) is the displacement at position x from the mean (here 0) and X is the separation between two points along the line. The y-axis intercept in the figure gives the mean square roughness of the surface in the direction analyzed. For this example we see that roughness is anisotropic with a root mean square value of $2.6\text{Å} \times 2\text{mm}$ in the direction across the islands.

To investigate the initial stages of growth of GaAs(001) at conditions farther from equilibrium, one can continue to grow on the above surface, but at a lower temperature. Figure 3 is an image showing the film morphology after an additional 0.35 ML dose of GaAs was deposited at a substrate temperature of 500°C. Caution should be used in comparing island shapes to those in Fig. 2(a) because the GaAs(001) surface symmetry changes from a (4x4) reconstruction at 500°C to a (2x4) reconstruction at ~ 550°C. A striking feature in this image is the large number of small islands on islands, a clear indication of limited downward mobility of atoms over steps. In the limit of no downward crossing, the coverage at each atomic level can be described with Poisson statistics. For this example Poisson statistics predict the correct occupation in each level, implying a very small probability of any level crossing.

Figure 2. (a) STM image of 3000Å×2mm of GaAs deposited on GaAs(001) with a deposition rate of 0.05μm/hour at a substrate temperature of 610°C. The scale of the image is 6000Å× 6000Å. Large islands elongated along [1$\bar{1}$0] are observed and seven atomic levels are present.

3. X-ray Diffraction Studies of Interfacial Roughness

A traditional method for characterizing interfacial roughness in multilayer films is with x-ray diffraction. Typically, the specular intensity is measured with a $(\theta, 2\theta)$ scan and compared with a theoretical calculation [9]. While the specular intensity gives information on the average interfacial roughness, it gives no information on how the roughness is correlated within a given interface or from interface to interface. To gain this information the diffuse intensity (the intensity roughness removes from the specular direction and redistributes into other directions) must be probed. Methods for measuring and analyzing diffuse intensity profiles for multilayers have been outlined recently [5]. We illustrate the technique with examples from $Al_x Ga_{1-x}As/GaAs$ multiple quantum well structures and W/C multilayers (used for soft-x-ray mirrors).

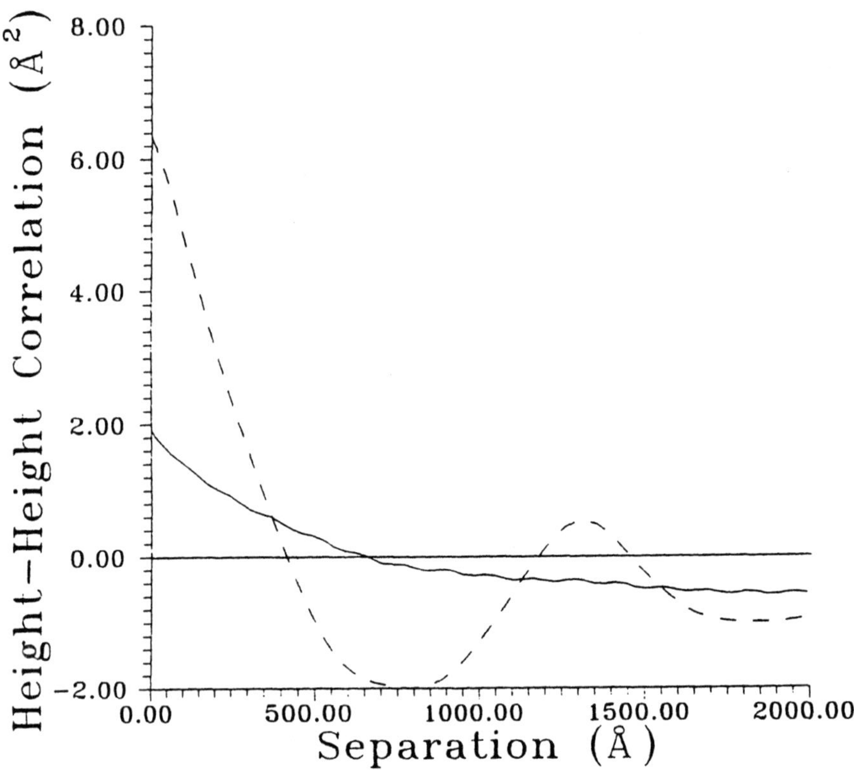

Figure 2.(b) Corresponding 1-d height-height correlation functions calculated along (solid) and across (dashed) the islands, with corresponding rms roughness and lateral correlation lengths of $\sigma = 1.4$Å, $\xi = 400$Å$\times2$mm and $\sigma = 2.6$Å, $\xi = 270$Å$\times2$mm respectively. The period of the damped oscillation observed in the dashed curve corresponds to the average island separation (~ 1300Å) in the direction scanned. Because STM has a limited field of view the roughness values obtained from the correlation functions are lower limits.

3.1. $Al_x Ga_{1-x} As/GaAs$ Multilayer Films

Epitaxial films develop the best structure and smoothest morphology if grown near equilibrium conditions. Films of GaAs/$Al_x Ga_{1-x}$As represent a heteroepitaxial system that should, at equilibrium, form perfectly smooth layers and interfaces. To determine the lateral scale of interfacial roughness correlated from layer to layer, we use rocking curves obtained with a narrow detector slit to probe the intensity distribution parallel to the surface. In Fig. 4 we show x-ray rocking curves through Bragg conditions for two nominally equivalent MOCVD deposited $Al_{0.7}Ga_{0.3}As/GaAs$ multilayers with only

118

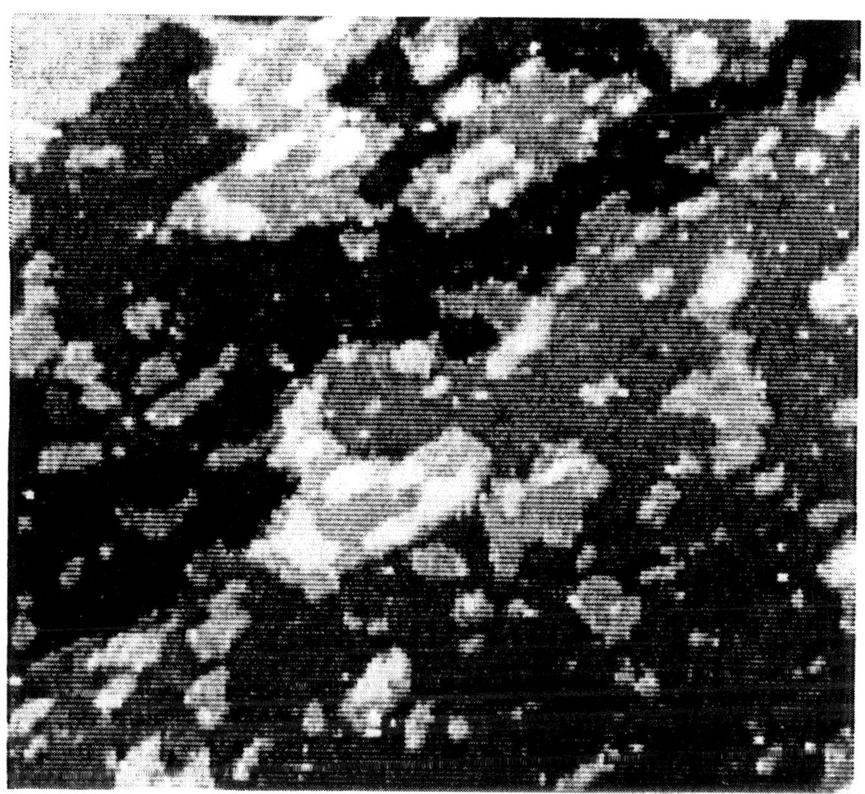

Figure 3. STM image of an additional 0.35 ML GaAs deposited onto a substrate prepared as in Fig. 2(a) and then cooled to temperature of 500°C before the added growth. The scale of the image is 1800Å× 1800Å. The many small islands on top of other islands imply limited level crossing of diffusing species.

the growth temperature changed. Each scan shows two components, a narrow central spike and a broader component (the diffuse background). For the sample grown at lower temperature, the diffuse component is relatively larger, a signature of greater interfacial roughness. Scans of the diffuse intensity distribution normal to the surface [7] indicate that roughness is replicated strongly from interface to interface.

A simple approach to fit the rocking scans treats interfacial roughness as if it were perfectly replicated through the film. One then has to specify the roughness of one interface to describe the multilayer. We use an isotropic correlation function of the form [10]

$$< [z(r - R)z(r)]^2 > = \sigma^2 \exp[-(R/\xi)^{2h}] \tag{1}$$

where σ is the rms roughness, ξ is the lateral correlation length, and h determines how jagged or smooth the surface is on length scales of the order of the lateral correlation

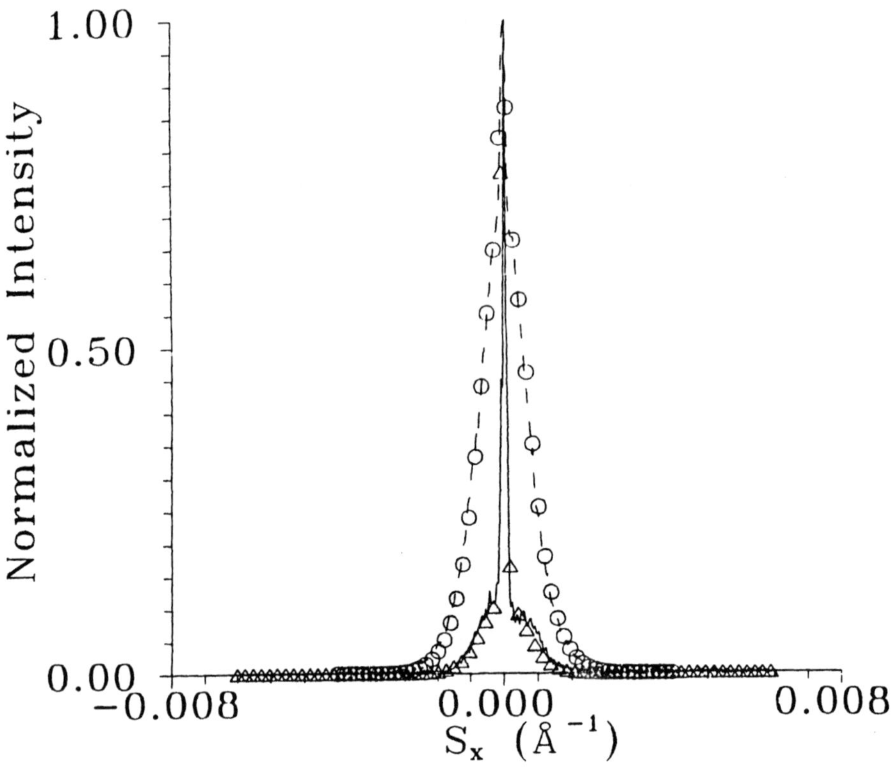

Figure 4. Rocking curves for two nominally equivalent MOCVD deposited $Al_{0.7}Ga_{0.3}As$ (80Å)/GaAs(50Å) multilayer films at different growth temperatures. For $T = 650°C$, the dashed curve is the measurement and the open circles are a fit using $\sigma = 8.7$Å, $\xi = 4300$Å, and $h = 1$. For $T = 750°C$, the solid curve is the measurement and the open triangles are from the model calculation using $\sigma = 3.9$Å, $\xi = 3000$Å, and $h = 1$. The 650°C rocking curve was taken through the 3rd Bragg condition while the 750°C rocking curve was taken through the 4th Bragg condition. A rocking curve through the 4th Bragg condition for the 650°C sample shows only the diffuse component.

length. The above correlation function can be thought of as a finite-film-thickness limit to a surface roughness correlation function that for infinite thickness would have a height difference correlation function of the form

$$< [z(r - R)z(r)]^2 >= AR^{2h} \tag{2}$$

The main difference between the two correlation functions is that for Eqn. 1 the roughness reaches a constant value for large separations. The exponent h can be equated with the exponent α used to describe the nature of a self-affine fractal [4].

The model correctly predicts a two-component diffraction profile. The fit to the data gives an rms roughness of $\sim 4°$ and lateral correlation length of $\sim 3000°$ for the sample grown at 750°C. The sample grown at 650°C exhibited a much greater roughness of ~ 9Å. For real multilayer systems, it is likely that larger-scale features are replicated better than smaller ones and thus one expects the lateral correlation length of interfacial roughness to be larger than the correlation length of the growth front. Growth front anisotropy can also be explored with x-ray diffraction by analyzing the diffuse intensity distribution in different directions. It is likely that different values for the lateral correlation length will be observed for such scans.

3.2. W/C Multilayer Films

W/C multilayers are different from $Al_x Ga_{1-x}As/GaAs$ films in that the former are grown far from equilibrium, while the latter are grown near equilibrium. Because W/C is a Volmer- Weber growth system, it would form 3D clusters if grown near equilibrium. Figure 5 shows a rocking scan of a 40Å-period, 40-bilayer sample. Again a diffuse component and central spike are observed, but this time on a very different lateral scale. The diffuse peak is more than an order of magnitude broader than was observed in the AlGaAs samples, giving a much shorter correlation length. The fit uses an exponent $h = 1/2$, which is predicted for growth of kinetically limited systems, an rms roughness of 1.4Å, and a lateral correlation length of 75Å. Note that the rms roughness of this kinetically limited film is lower than that of the equilibrium ones.

A more careful examination of the diffuse intensity distribution normal to the surface [11] shows that our approximation of roughness perfectly correlated from layer to layer is too simple and that interfacial roughness is frequently only partially correlated. We observe broadening along S_z (the normal component of the momentum transfer vector) compared with the specular intensity distribution and this broadening increases for scans taken at larger values of S_z. This shows that roughness is not perfectly replicated and that long-wavelength roughness is replicated better than short-wavelength roughness. The simple model treating roughness as perfectly correlated gives an upper limit to the lateral correlation length and a lower limit to the interfacial roughness. Partial correlation can be directly incorporated into a growth model in a phenomenological way [11].

Analysis of growth of W/C multilayers with various deposition parameter sets gives the following conclusions: [6] Average interfacial roughness increases only slowly as more layers are deposited. In contrast, increasing individual-layer thickness while keeping the total film thickness fixed increases the interfacial roughness. Finally, increasing thickness of a single layer of the film, i.e., a buffer layer, also increases the roughness. These observations suggest that the presence of interfaces retards the evolution of roughness for W/C by forcing the roughness to at least partially restart at each interface.

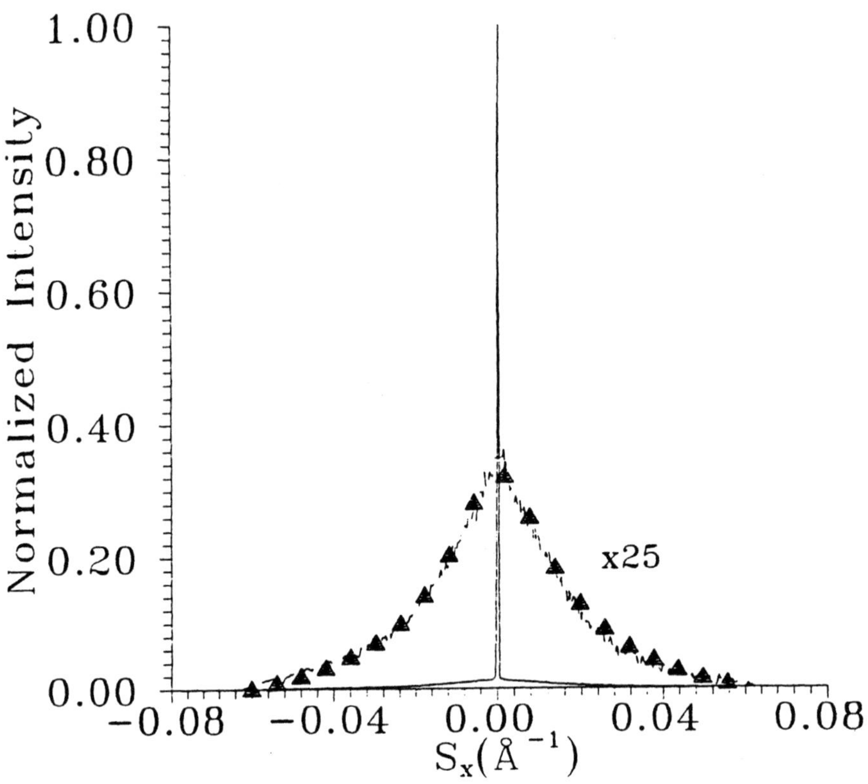

Figure 5. Rocking curve for a sputter deposited W(20Å)/C(20Å) 40-period multilayer film. The solid curve is the measurement, taken through the 5th-order Bragg condition. The diffuse component is magnified 25 times as is the model fit (solid triangles). The fit parameters are $\sigma = 1.4$Å, $\xi = 75$Å, and $h = 1/2$.

4. Summary

We have outlined how measurements of interfacial morphology can be used to obtain information about kinetic barriers limiting film growth. At the earliest stages of growth, STM is used to explore specific atomic mechanisms. At later stages, surface correlation functions are used to quantify STM images to obtain information on kinetic roughening, such as rms roughness and lateral correlation length. X-ray diffraction can be used to gain similar information about buried interfaces and specifically whether roughness is correlated from interface to interface. We have provided examples for both epitaxial (Si, GaAs, and AlGaAs/GaAs) and non-epitaxial (W/C) systems that are grown respectively near equilibrium and very far from equilibrium. In both cases, it is in principle possible to identify the mechanisms responsible for kinetic roughening and quantify individual aspects such as diffusion, transport over steps, and their anisotropies.

Acknowledgements

This research was supported in part by ONR and in part by NSF, Grant No. 89-18927. We acknowledge R. Kariotis and M.B. Webb for useful discussions.

REFERENCES

[1] See for example, E. Bauer, *Z. f. Kristall.* **110** 372 1958.
C. Herring, *Phys. Rev.* **82** 87 1951.
Structure and Properties of Solid Surfaces, R. Gomer and C.S. Smith eds., U. Chicago Press, Chicago 1953.
[2] Y.-W. Mo, J. Kleiner, M.B. Webb, and M.G. Lagally, *Phys. Rev. Lett.* **66** 1998 1991.
Y.-W. Mo, J. Kleiner, M.B. Webb and M.G. Lagally, *Surface Sci.*, in press.
[3] E.J. Heller and M.G. Lagally, *Apply. Phys. Lett.* in press.
[4] See, for example, F. Family, *Proc. of the Les Houches Workshop on Dynamical Phenomena at Surfaces, Interfaces, and Membranes* D. Beysens, G. Forgacs, and N. Bocorra, eds. 1991.
[5] D.E. Savage, J. Kleiner, N. Schimke, Y.-H. Phang, T. Jankowski, J. Jacobs, R. Kariotis, and M.G. Lagally, *J. Appl. Phys.* **69** 1411 1991.
[6] D.E. Savage, N. Schimke, Y.-H. Phang, and M.G. Lagally, *J. Appl. Phys.* **71** 1 1992.
[7] Y.-H. Phang, D.E. Savage, T.F. Keuch, M.G. Lagally, J.S. Park, and K.L. Wang, *Appl. Phys. Lett.* in press.
[8] G.W. Smith, A.J. Pidduck, C.R. Whitehouse, J.L. Glasper, A.M. Keir, and C. Pickering, *Appl. Phys. Lett.* **59** 3282 1991.
[9] E. Spiller and A.E. Rosenbluth, *SPIE* **563** 221 1985.
[10] This same form has been used to model surface roughness, see S.K. Sinha, E.B. Sirota, S. Garoff, and H.B. Stanley, *Phys. Rev.* **B 38** 2297 1988.
[11] Y.-H. Phang, R. Kariotis, D.E. Savage, and M.G. Lagally, *J. Appl. Phys.* submitted.

Surface Disordering:
Growth, Roughening and
Phase Transitions

LOW ENERGY ELECTRON MICROSCOPY STUDIES OF GROWTH PATTERNS ON SURFACES AND OF THIN FILMS

E. Bauer

Physikalisches Institut, Technische Universität Clausthal, D-3392 Clausthal-Zellerfeld, Germany

Abstract. The growth patterns appearing in first order phase transitions on Si(111) and Pb(110) surfaces and in Au films on Mo(110) and Si(111) surfaces, in the growth of Si(100) surfaces, in two-dimensional oxide growth on W(100) surfaces and in the early steps of deposition of Cu and Au on Mo(110) and Si(111) surfaces are studied by low energy electron microscopy.

1. Introduction

The dynamics of the morphological changes occuring in growth processes and phase transitions is one of the central topics of this workshop. While the theoretical treatment has reached already a high level of sophistication, there are only few experimental results on a microscopic scale for comparison with the theoretical predictions. Scanning tunneling microscopy (STM) would be an ideal tool to provide such information but at the present state image acquisition time and temperature variability are inadequate for dynamic studies. Low energy electron microscopy (LEEM) [1] however, though far inferior to STM in resolution, allows dynamical studies of surface processes over a wide temperature range. This is due to the high intensity available for imaging and the parallel image acquisition whichs allows image recording in real time. LEEM is particularly suitable for the study of morphological changes which are accompanied by changes in the crystal structure because the image contrast is predominantly caused by diffraction. In addition, geometric interference contrast, e.g. from monoatomic steps, plays an important role for the understanding of surface processes. In this contribution the information on growth dynamics which can be obtained with LEEM will be illustrated by a few examples. For information on the experimental method and its limitations the reader is referred to ref. 1 and the references therein.

2. The Si(111)-(1×1) ↔ (7×7) phase transition

When a Si(111) surface is cooled from high temperatures below about 1100 K it reconstructs into a (7×7) superstructure. Depending upon prehistory and cooling rate a wide range of morphological evolutions has been observed [2–5]. Some of them are the christmas tree, the compact and the chaotic morphology. The first two appear on surfaces with well-defined step structure, the last one on surfaces in which the step structure is obliterated by subsurface impurity segregation. When the supersaturation (undercooling ΔT) is small the (7×7) structure nucleates only in regions with high stress such at steps or the emergence points of screw dislocations. As soon as recognizable with the present lateral resolution (15 nm), they have triangular shape with the edges parallel to the $\langle 1\bar{1}0 \rangle$ directions, and the apices pointing in the $\langle 11\bar{2} \rangle$ directions.

Figure 1. Some typical growth morphologies of the (7×7) structure on Si(111). a) Christmas tree, b) compact and c) chaotic morphology. The sharp irregular line in the images is a crack in the channel plate image intensifier, the bar is 1 μm.

The apices are secondary nucleation sites which leads to the christmas tree morphology when growth is predominantly in a $\langle 11\bar{2} \rangle$ direction (Figure 1a). Supression of secondary nucleation, e.g. by slight contamination, leads to the compact (7×7) crystals (Figure 1b). The chaotic morphology caused by strong subsurface contamination is shown in Figure 1c. It should be noted that the diffraction pattern is identical in all three cases. Only the background is higher in the case of the morphology of Figure 1c and the transition occurs over a wide temperature range which gives it the appearence of a second order transition in diffraction studies. In contrast, in Figure 1a the transition is complete at an undercooling as low as $\Delta T = 3$ K within 1 min. For $\Delta T > 12$ K homogenous nucleation on the terraces occurs in addition to the hetergenous nucleation at steps.

3. The Pb(110)-(1×1) ↔ c(2×4) phase transition

A Pb(110) surface which in addition to its inherent surface stress experiences additional stress due to volume stress can develop a c(2×4) reconstruction [6] with a stress-dependent transition temperature ranging from 430 K to 465 K [7], that is in the temperature range in which already surface roughening occurs [8]. The growth and disordering pattern of the c(2×4) regions depends strongly upon the number of heating/cooling cycles and the maximum temperature [9].

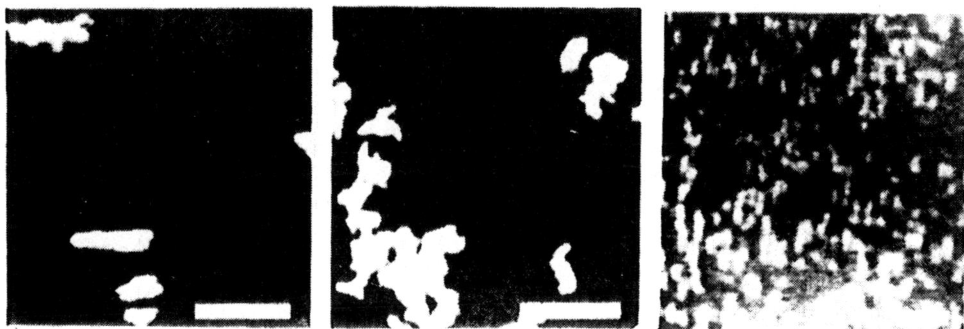

Figure 2. Growth patterns of the c(2×4) structure on Pb(110). a) Compact, b) dendritic and c) fine grained morphology.

On a freshly prepared surface, on which the reconstructed and partly stress-roughened surface layer has been removed by sputtering [7], the c(2×4) regions grow as compact islands with preferred [001] growth direction (Figure 2a). During the initial thermal cycles the nucleation rate is small and the morphology changes from compact to dendritic (Figure 2b). With increasing number of cycles — and increasing cycling temperature — the nucleation rate rises strongly and the islands grow preferentially in the [1$\bar{1}$0] direction (Figure 2c). The compact morphology can be regenerated by sputtering and annealing as mentioned above.

4. Phase transitions in submonolayer Au films on Mo(110) and Si(111) surfaces

Au grows in the submonolayer range on Mo(110) as a one-dimensionally misfitting, i.e. as an incommensurate layer, on Si(111) it forms several ordered phases ((5×1), ($\sqrt{3}\times\sqrt{3}$)R30° and (6×6)). Due to the incommensurability the two-dimensional (2d) Au "crystals" disorder easily into a 2d vapor/fluid so that the complete 2d phase diagram can be studied without desorption [10]. When the layer is cooled from the single phase region through the coexistence line into the two-phase regions a large number of 2d Au crystals form which are strongly elongated in the zero misfit directions (Figure 3a). There is no preferential nucleation at the substrate steps but each of the two equivalent orientations grows frequently preferentially in large substrate regions which suggests long-range interactions between the 2d crystals.

The ordered phases of Au on Si(111) and their transitions are completely different from those on Mo(110). The (5×1) structure which has a complicated atomic distribution [11, 12] disorders apparently continuously (in diffraction) into a (1×1) structure.

Figure 3. Growth patterns in phase transitions of two-dimensional layers. a) Au on Mo(110), b) (5×1) Au structure on Si(111) and c) (6×6) structure on Si(111).

LEEM quite clearly shows the first order nature of the transition with large surface regions transforming spontanously at each temperature step during cooling [13]. Figure 3b shows a partially converted surface which upon further temperature reduction completely transforms into the (5×1) structure. The (5×1) regions are laterally bounded by monoatomic steps. The high coverage (6×6) structure which always coexists with 3d Au crystals has a completely different growth morphology when it forms upon cooling from the high temperature $(\sqrt{3}\times\sqrt{3})$R30° phase (Figure 3c). Similar to the transitions on the Pb(110) surface, the nucleation rate also increases with increasing number of cycles through the transition.

5. Vapor growth of the Si(100) surface

The Si(100) surface has a (2×1) reconstruction consisting of dimer rows whose directions change by 90° from one monoatomic terrace to the next.

Figure 4. Growth morphology of Si on Si(100). a) Very low, b) low and c) large supersaturation.

This dimerization causes a strong anisotropy of the surface diffusion and of the step energies which has strong influence on the growth and equilibrium shapes of the 2d islands. Growth under quasi-equilibrium conditions produces almond-shaped islands (Figures 4a, b) from which a step energy anisotropy $\lambda_\parallel / \lambda_\perp \gtrsim 2.6$ can be derived, $\lambda_\parallel$ and $\lambda_\perp$ being the free energies of steps parallel and perpendicular to the dimer rows [14]. Growth at larger supersaturations causes rough 2d growth fronts (Figure 4c), mainly due to the strong anisotropy of the growth rate.

6. Two-dimensional oxide growth on the W(100) surface

The chemisorption of oxygen and the initial oxide growth on W(100) are complicated processes as numerous laterally averaging studies have shown. A brief laterally resolved study by LEEM combined with LEED [15] has shown that these processes are even more complicated than expected. This is illustrated in Figure 5 by the transition from the p(5×1) structure observed at a coverage of 1 monolayer (ML) at 1050 K to a "(2×2)" structure with increasing O_2 exposure. It is best developed at 1.25 ML and has tentatively been attributed to a 2d WO_3 layer [16]. The morphological evolution of this transition is shown in Figure 5. Initially small isolated bright regions form, apparently randomly without any correlation to the step structure of the substrate (Fig. 5a). These regions link up into chains which form a self-avoiding random walk pattern (Fig. 5b). Once the chains appear continuous they rapidly grow in width (Fig. 5c) until the complete surface appears bright at the electron energy selected for imaging and the "(2×2)" LEED pattern reaches saturation.

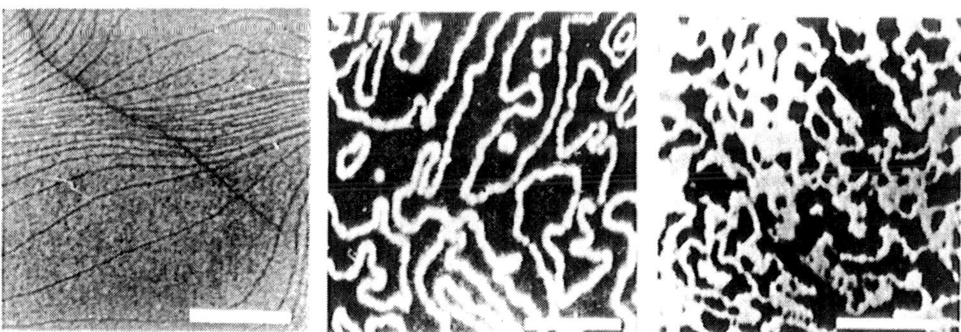

Figure 5. Two-dimensional oxide growth on W(100). a) Step structure of the clean surface, b) and c) at an early and a later stage of the oxide growth.

7. The initial stages of deposition of Cu and Au on Mo(110) and Si(111) surfaces

Cu and Au grow on Mo(110) surfaces without alloying while they react with the Si(111) surface and form 2d silicides. Consequently the growth patterns differ significantly. A

difference is, however, also found between Cu and Au on Mo(110) which is attributed
to the fact that the first monolayer of Cu grows commensurately while that of Au is 1d
incommensurate as already mentioned previously. The differences are seen in Figures
6a and 6b: the Cu layer grows by step flow growth with a smooth growth front [17], the
Au layer, however, with a rough growth front with same indication of the two preferred
directions seen in Figure 3a. Also the growth in the second layer differs significantly. An
interesting growth process has been observed in Cu films consisting already of several
monolayers at about 500 K: a considerable thickness difference between Cu layers on
different terraces builds up which is seen in LEEM by quantum size contrast (Fig. 6c)
[17]. Once nucleation has occured on a given terrace, the growth front spreads rapidly
across this terrace, so that it appears diffuse in the photograph, but does not transcend
the terrace boundaries. Obviously, diffusion is not limited by steps but growth is.

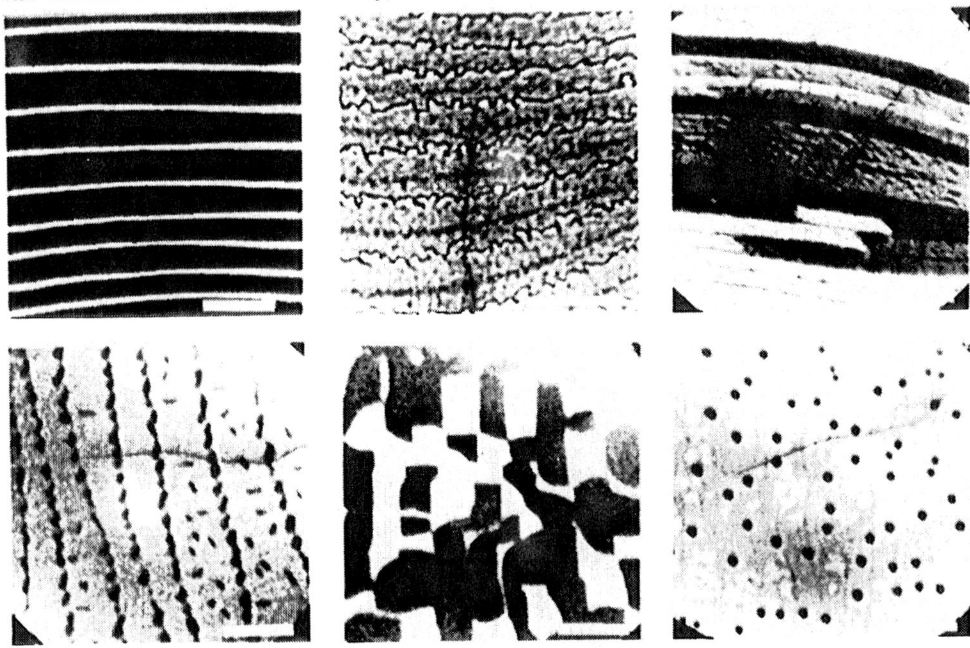

Figure 6. Growth stages of Cu and Au on Mo(110) and Si(111). a) Cu and b) Au
submonolayer growth on Mo(110), c) nucleation-limited roughening during further
growth of Cu on Mo(110), d) (5×1), e) ($\sqrt{3}\times\sqrt{3}$)R30° and f) (6×6) structure of Au
on Si(111).

On the Si(111) surface Cu forms, at least at elevated temperatures, initially a 2d
silicide before 3d silicide crystals begin to nucleate. The growth morphology of the
2d silicide is largely determined by the substrate step morphology: growth along the
steps in fast, normal to the steps slow [18]. This is understandable because the reaction
between Cu ans Si is easiest at steps. Similarly, the low coverage structure of Au on
Si(111), the (5×1) structure nucleates preferentially at steps but also at (7×7) domain

boundaries. Contrary to the incommensurate 2d Cu silicide layer the (5×1) crystals grow in well-defined geometric shape with three equivalent orientations (Fig. 6d) [13]. When they grow together they form much layer domains with the same orientation. After the surface has been covered completely with the (5×1) structure, nuclei of the $(\sqrt{3}\times\sqrt{3})$R30° form with a smooth irregular morphology (Fig. 6e). Their growth is limited by the step structure of the surface which changes considerable during the formation of the Au layer. When the surface is saturated with the $(\sqrt{3}\times\sqrt{3})$R30° structure 3d Au crystals form before the highest coverage 2d Au layer, the (6×6) structure, nucleates. The (6×6) regions grow in a rough irregular morphology (Fig. 6f). Thus, in one and the same system a wide range of morphologies is observed at constant temperature and incoming flux.

8. Summary

The examples reviewed here show the wide range of growth patterns which can occur on surfaces and in thin film growth. A full appreciation of the power of LEEM in this field can only be gained by viewing the video recordings of the growth process. The immense amount of data gained in such studies makes at the same time a quantitative analysis very time-consuming. This fact, together with insufficient control and measurement of temperature — and incoming flux in deposition experiments — are mainly responsible for the lack of a quantitative analysis of the data. Future experiments under well-defined experimental conditions will justify quantitative image analysis and allow to put the few speculations made here or in the original papers on firm grounds or to exclude them.

Acknowledgements

The results reviewed here have been obtained by my past coworkers W. Telieps, M. Mundschau, W. Swiech and M. Altman all of whom I wish to thank for their efforts. The work was supported in part by the Deutsche Forschungsgemeinschaft, in part by the Volkswagen Foundation.

REFERENCES

[1] E. Bauer in R. Vanselow and R. Howie, Edits., *Chemistry and Physics of Solid Surfaces VIII* Springer, Berlin 1990.
[2] W. Telieps and E. Bauer, *Surf. Sci.* **162** 163 1985; *Ber. Bunsenges. Phys. Chem.* **90** 197 1986.
[3] E. Bauer, M. Mundschau, W. Swiech and W. Telieps in D. Cherns, Edit., *Evaluation of Advanced Semiconductor Materials by Electron Microscopy* Plenum Press, New York 1989.

[4] E. Bauer, M. Mundschau, W. Swiech and W. Telieps in R.D. Bringans, R.M. Feenstra and J.M. Gibson, Edits., *Atomic Scale Structure of Interfaces* MRS Symp. Proc. 159 1990.

[5] E. Bauer, M. Mundschau, W. Swiech and W. Telieps, *J. Vac. Sci. Technol.* **A9** 659 1991.

[6] A. Pavlovska and E. Bauer, *Europhys. Lett.* **9** 797 1989.

[7] A. Pavlovska, H. Steffen and E. Bauer, *Surf. Sci.* **234** 143 1990.

[8] H.-N. Yang, T.-M. Lu and G.-C. Wang, *Phys. Rev. Lett.* **63** 1621 1989.

[9] M. Altman and E. Bauer, unpublished.

[10] J. Kołaczkiewicz and E. Bauer, *Phys. Rev. Lett.* **53** 485 1984; *Surf. Sci.* **151** 333 1985.

[11] E. Bauer, *Surf. Sci.* **250** L379 1991.

[12] Ch. Schamper, W. Moritz, H. Schulz, R. Feidenhans'l, M. Nielsen, F. Grey and R.L. Johnson, *Phys. Rev. B* **43** 12130 1991.

[13] W. Swiech, M. Mundschau and E. Bauer, *Surf. Sci.* **253** 283 1991.

[14] W. Swiech and E. Bauer, *Surf. Sci.* **255** 219 1991.

[15] M. Altman and E. Bauer, *J. Vac. Sci. Technol.* **A9** 659 1991.

[16] E. Bauer, H. Poppa and Y. Viswanath, *Surf. Sci.* **58** 517 1976.

[17] M. Mundschau, E. Bauer and W. Swiech, *J. Appl. Phys.* **65** 581 1989.

[18] M. Mundschau, E. Bauer, W. Telieps and W. Swiech, *J. Appl. Phys.* **65** 4747 1989.

LONG RANGE INTERACTIONS AND FACETTING IN BIDIMENSIONAL CRYSTALS

Cyrille Flament and Francois Gallet

Laboratoire de Physique Statistique de l'Ecole Normale Supérieure URA 1306 associée au CNRS et aux Universités Paris VI et Paris VII 24, rue Lhomond - 75005 Paris, France

Abstract. Assuming $1/R^n$ interactions between particles in a 2D homogeneous phase, we define and calculate the energy of the 1D phase boundary at equilibrium. We show that, for $n \leq 3$, a line tension cannot be defined. For a 2D crystal, a calculation to first order in the crystalline pinning potential U indicates that the interface is rough for $n > 2$, and facetted for $n < 2$, at any temperature. In the particular case $n = 2$ a roughening transition should occur at a finite temperature T_R that we explicitely calculate.

1. Introduction

It is usually admitted[1] that the 1D boundary of a 2D crystal is always rough, since the thermal fluctuations make its thickness $<y^2> \sim L \, k_B T / \lambda$ diverge with its length L at any temperature. This supposes the existence of a line tension λ (energy per unit interface length). However, there are experimental evidences for 2D facetted crystals, for example in the condensed phases of Langmuir films, made of amphiphilic molecules deposited at the free surface of water[2-6]. The behaviour of these films is usually dominated by long range electrostatic interactions between molecules. In this case, it is known that a line tension between two different phases may no longer exist[7-9]. The above argument must then be reconsidered, and the behaviour of the interface roughness has to be calculated through a direct integration of the interaction energy between molecules.

First we consider a 2D phase (crystalline or not) of average molecular density σ, and we calculate the total energy of a finite domain, at equilibrium, for an attractive pair potential $V_n(R) = - C_n/R^n$ $(n>0)$. This energy can be written as a sum of a surface term E_s and a boundary term E_c. For a straight interface of length $L \to \infty$, we show that E_c is proportional to L when $n > 3$ (in this case a line tension exists), to L^{4-n} when $n < 3$, and to $(L \ln L)$ when $n = 3$. A 2D crystal is modelized by a local potential U, of periodicity b in the direction normal to the boundary. The thickness $<y^2>$ of a free interface is found proportional to L for $n > 3$, to L^{n-2} for $2<n<3$, and remains finite for $n < 2$. In the particular case $n = 2$, $<y^2>$ diverges like $\ln(L)$. There, a first order calculation in U leads to a roughening transition at $2k_B T_R = b^2\sigma^2 C_2$. Finally we discuss possible experimental observations of facetted 2D crystals in Langmuir monolayers.

2. Line tension

We consider a collection of particles interacting through an attractive pair potential $V_n(R)$ $= - C_n/R^n$ only depending on their distance $R = |r_1-r_2|$, and supposed isotropic for simplicity. The energy E of a plane domain of area S, limited by a contour C, is :

$$E = -\frac{1}{2} \iint_S d^2r_1 \, d^2r_2 \; \sigma(r_1) \, \sigma(r_2) \, \frac{C_n}{|r_1 - r_2|^n} \tag{1}$$

Here $\sigma(r)$ is the particle density, assumed uniform on average. A cutoff a, of the order of the distance between neighbours, accounts for the short scale repulsions. We then write $\sigma(r_1)\sigma(r_2)= \sigma^2$ for $|r_1-r_2| > a$, and $\sigma(r_1)\sigma(r_2)= 0$ for $|r_1-r_2| < a$. It has been shown elsewhere[10] that the surface integral (1) can be transformed into a double integral along the contour C of the domain :

$$E = \oint \oint P_n(R) \, \vec{dl_1}.\vec{dl_2}$$

where $P_n(R)$ is such that $\nabla^2 P_n = -\frac{1}{2} \sigma(r_1)\sigma(r_2)V_n(R) = \frac{\sigma^2 C_n}{2 R^n}$ for $R > a$ and $\nabla^2 P_n = 0$

for $R < a$. A solution is :

$$P'_n(R) = A_n \left(\frac{a}{R}\right)^{n-2} \qquad \text{for } R > a \tag{2}$$

and $\quad P'_n(R) = A_n \left[1 - (n-2)\ln(\frac{R}{a})\right]$ for $R < a \qquad$ with $A_n = \dfrac{\sigma^2 C_n}{2 (n-2)^2 a^{n-2}}$

This function is continuous and derivable, and verifies the requested conditions everywhere except in $R = 0$ where $\nabla^2 P'_n = -2\pi(n-2)\, A_n \delta^{(2)}(R)$.

In the particular case $n = 2$, $P_2(R)$ is defined as :

$$P_2(R) = \frac{\sigma^2 C_2}{4}\left(\ln\frac{R}{a}\right)^2 \quad \text{for } R > a$$

and $\quad P_2(R) = 0 \qquad$ for $R < a \quad (A_2 = 0)$

The energy E can be written as the sum of a surface contribution E_s, proportional to the area S of the domain, and a boundary contribution E_c :

$$E = E_s + E_c = -2\pi (n-2)\, A_n S + \oint \oint P'_n(R)\, \overrightarrow{dl_1}.\overrightarrow{dl_2} \tag{3}$$

To evaluate the asymptotic behaviour of E_c for a large domain, we consider a straight boundary of length $L \to \infty$. Replacing P'_n by its analytical expression, the integration of (3) leads to :

$$E_c = A_n\left[\frac{2(n-2)^2}{(n-3)}La + \frac{2}{(n-3)(n-4)}\left(\frac{L}{a}\right)^{2-n}L^2 - \frac{(n-2)^2}{2(n-4)}a^2\right] \tag{4}$$

The asymptotic behaviour of E_c depends on the sign of $(n-3)$:

i) for $n > 3$, $E_c \sim L$. Then, one can define a line tension $\lambda = \sigma^2 C_n/(n-3)a^{n-3}$ for the boundary. The interactions between particles are short range enough, so that one expects the boundary to remain always rough.

ii) for $n < 3$, $E_c \sim L^{4-n} \gg L$ when $L \to \infty$. The boundary energy is no longer proportional to its length, and a line tension cannot be defined. A direct calculation of the fluctuations amplitude will be done in the next section to determine the boundary state.

iii) - $n = 3$: $\quad E_c \sim \sigma^2 C_3 L \ln\frac{L}{a}$. No line tension exists for n=3 included.

$\qquad$ - $n = 2$: $\quad E_c \sim \sigma^2 C_2 \left(\frac{L}{2}\ln\frac{L}{a}\right)^2$. This last case will be further examined in the next section.

3. Interface thickness and roughening transition

The smoothness or roughness of a crystal interface at rest results of the competition between its thermal fluctuations, which tend to delocalize it, and the restoring force due to the crystal periodicity, which tends to pin the interface along a given orientation.

Quantitatively, for an interface parallel to Ox in average, this restoring force can be modelized by a potential U(y), periodical in the Oy direction normal to the boundary[11-12]. The pinning energy E_p, in the presence of a given interface deformation y(x), can be written :

$$E_p = \int dx\, U(y)$$

U results of short range anisotropic interactions between particles (others than $1/R^n$ interactions). Several 3D calculations have shown that the interface state does not depend on the details of U, and we take for simplicity $U(y) = U_0 \cos(2\pi y/b)$ (Sine-Gordon model). Here b is the reticular spacing in the Oy direction, or the height of kinks on the interface. The interface state is determined by the value $<E_p>$ averaged over the amplitude y of line fluctuations. If $<E_p>$ diverges when $L \to \infty$, the pinning is relevant on a macroscopic scale, and the interface is facetted. On the contrary, if $<E_p>$ averages to zero, the long wavelength fluctuations are free, and the interface is rough. To first order in U, $<E_p>$ is equal to :

$$<E_p> = \int dxdy\, U(y)\, p(y) \tag{5}$$

where $p(y) \sim \exp(-\Delta E_c(y)/kT)$ is the probability for a free thermal fluctuation y(x). Let us first calculate the mean square thickness $\Delta = (<y^2>)^{1/2}$ of a free portion L of straight boundary. Supposing small fluctuations (dy/dx « 1), one has :

$$E_c(y) = \int_0^L \int_0^L dx_1\, dx_2\, P'_n(|r_1 - r_2|)\left(1 + \frac{dy_1}{dx_1}\frac{dy_2}{dx_2}\right)$$

Assuming a boundary condition y(L) = y(0), some transformations lead to $E_c = E_{c0}(y=0) + \Delta E_c(y)$, with[10] :

$$\Delta E_c(y) = \frac{1}{2}\int_0^L dx_1 \int_0^L dx_2\,(y_1 - y_2)^2\, \nabla^2 P'_n = \frac{\sigma^2 C_n}{2}\int_0^{L-a} dx_2 \int_{x_2+a}^L dx_1\, \frac{(y_1 - y_2)^2}{(x_1 - x_2)^n} \tag{6}$$

Picking up a given Fourier component $y(x) = y_k \exp(ikx)$, the corresponding value $\Delta E_c(y_k)$ is :

$$\Delta E_c(y_k) = \sigma^2 C_n y_k^2 k^{n-2}(kL\, J_n - J_{n-1})$$

with $\quad J_n(ka, kL) = \int_{ka}^{kL} \frac{du}{u^n}(1 - \cos u)$

At large kL and small ka, $\Delta E_c(y_k)$ behaves in the following way :

i) $\Delta E_c(y_k) \sim \sigma^2 C_n y_k^2 \dfrac{1}{(n-1)(n-2)} \dfrac{1}{L^{n-2}}$ for $n < 1$

ii) $\Delta E_c(y_k) \sim \sigma^2 C_n y_k^2 k^{n-1} L J_n$ for $1 < n < 3$ ($J_n = J_n(0, \infty)$ is then finite)   (7)

iii) $\Delta E_c(y_k) \sim \sigma^2 C_n y_k^2 \dfrac{1}{2(n-3)a^{n-3}} k^2 L$ for $n > 3$

It can be shown that the fluctuation modes of different wavevectors k are independant. Assigning an energy $\langle \Delta E_c(y_k) \rangle = k_B T / 2$ to each mode, one finds :

$$\langle y^2 \rangle = \sum_{k=-\pi/a}^{\pi/a} \langle y_k^2 \rangle \sim \frac{k_B T}{2 C_n \sigma^2} f_n(L) \qquad (8)$$

with $f_n(L) = (n-1)(n-2) \dfrac{L^{n-1}}{a}$ for $n < 1$

$f_n(L) = \dfrac{1}{(2-n)\pi J_n} \left(\dfrac{\pi}{a} \right)^{2-n}$ for $1 < n < 2$

$f_n(L) = \dfrac{1}{\pi J_n} \left(\dfrac{L}{2\pi} \right)^{n-2} \zeta(n-1)$ for $2 < n < 3$ (with $\zeta(n) = \sum_{p=1}^{\infty} \dfrac{1}{p^n}$)

$f_n(L) = (n-3) a^n \dfrac{3L}{6}$ for $n > 3$

Notice that $\langle y^2 \rangle$ diverges like a power of L for $n > 2$, but remains finite for $n < 2$. In the particular case $n = 2$, $f_2(L) = 2/\pi^2 \ln (L/2a)$.

Now, the integral (5) defining $\langle E_p \rangle$ can be explicited. Since $p(y)$ is a gaussian function of the deformation y, one has[12] :

$$\langle U \rangle = \int dy\, U_0 \cos (2\pi \tfrac{y}{b}) p(y) = U_0 \exp (-2\pi^2 \frac{\langle y^2 \rangle}{b^2})$$

and then :

$$\langle E_p \rangle = U_0 L \exp \left(- \frac{2\pi^2 k_B T f_n(L)}{b^2 C_n \sigma^2} \right)$$

Replacing $f_n(L)$ by its value, one observes that $\langle E_p \rangle$ is zero at large L for $n > 2$; the pinning potential U vanishes on large scale and the interface is always rough. On the contrary $\langle E_p \rangle$ diverges when $n < 2$: U is relevant and the interface is always facetted.

The case $n = 2$ is the marginal case :

there $\langle E_p \rangle \sim 2a\, U_0 (\dfrac{L}{2a})^{\alpha(T)}$ with $\alpha(T) = 1 - \dfrac{2 k_B T}{b^2 C_2 \sigma^2}$ (9)

The behaviour of $<E_p>$ then depends on the temperature T. A roughening temperature T_R appears, for which the interface equilibrium state changes from smooth to rough. In our first order development in U, T_R is given by :

$$2 \, k_B T_R = b^2 \, \sigma^2 \, C_2 \tag{10}$$

Like for a 2D interface[13], this value is universal in the sense that it does not depend on the details of the periodic crystalline potential U(y), but only on the amplitude C_2 of long range interactions.

It is interesting to notice that the above predictions are identical to Kjaer and Hilhorst[14] results concerning a unidimensional chain of N particles, interacting through a discrete gaussian hamiltonian $H = 1/2 \, \Sigma_{i \neq j} \, V(i\text{-}j) \, (y_i - y_j)^2$, where the potential V behaves like $1/(i\text{-}j)^n$ for (i-j) » 1. They found that, if a roughening transition exists, it should occur for n=2 at the temperature $k_B T_R = 1$. This is not surprising, since the unidimensional hamiltonien H is a discrete equivalent, within a factor of two, to our definition (6) of the fluctuation energy ΔE_c.

4. Experimental observations and conclusion

The possible existence of facetted bidimensional crystals has got renewed interest since the observation of ordered domains in Langmuir monolayers. These bidimensional systems are made of amphiphilic molecules deposited at the free surface of water. Since the beginning of the century, they are known to present several phase transitions[15], leading to plateaus or slope breaks on the 2D isotherms (surface pressure versus molecular area). Various techniques have now made possible the characterization of these monolayers : for instance fluorescence microscopy, spectroscopy, X-rays or electron diffraction, atomic force microscopy. Recently, the different phases have been classified, in a way analogous to the various smectic phases of liquid crystals[16] : the densest phases are expected to be ordered, and to behave like solids. Moreover, a few examples of crystalline order are known in these films[3], for which electron diffraction patterns present high order peaks.

A common feature to all the materials used to make Langmuir films is the existence of long range electrical interactions between molecules : the amphiphilic behaviour of these molecules is due to their polar head, which carries either a permanent dipole moment, or a finite electrical charge. In this sense, these monolayers are good candidates for the observation of facetted 2D crystals, and eventually of a 2D roughening transition. Indeed, dipole-dipole interactions behave like $1/R^3$, while charge-charge interactions behave like 1/R. However, this monopole-monopole interaction is generally screened by the ionic strength of water, on a distance of the order of the Debye-Hückel length. This length varies with the ionic concentration, but should never be larger than its value in pure water, of the order of $1\mu m$.

Recent observations under an optical microscope have clearly shown that the solid domains may be facetted in some Langmuir monolayers[2-6]. Technically, the visualization is made by introducing a small amount of fluorescent dye in the film. Actually, two different situations may occur : sometimes the facetted shape is observed during the domain growth[2], but this dynamical facetting is related to the anisotropy of the growth kinetics, rather than to the state of its boundary at equilibrium. On the other hand, some experiments are performed close enough to equilibrium to consider the shape as an equilibrium one[3]. In this reference[3], the film is made of two components of opposite charge : one is an insoluble amphiphilic molecule, spread on the water surface, and the other is a soluble surfactant acting as a counterion under the surface. This film displays a crystalline phase with facetted domains. A similar situation is described in reference[5]. Although these mixed films could be good candidates for a further study of facetting, a full description of their molecular interactions is still a non trivial problem.

As a final remark, it is instructive to estimate the roughening temperature T_R from formula (10). In the dense phases of Langmuir films, typical values are $\sigma \sim 1/30$ molecule/Å^2, $b^2 \sim 1/\sigma$. Assuming that one can achieve a $1/R^2$ potential by an interaction between a typical charge $q = \alpha\, e$ and a typical dipole $p = \beta\, e.\text{Å}$, we calculate $C_2 = q.d/4\pi\varepsilon_0$ (α and β are numerical constants of order unity). This leads to $T_R \sim 2800\, \alpha\beta$ (K). Reasonnable values for α and β are between 0.1 and 1. Then, if a roughening transition exists in Langmuir crystals, the corresponding temperature should be in the range of experimental investigations.

REFERENCES

[1] See for instance P. Nozières, Lectures at the Beg-Rohu summer school, C. Godreche ed., Cambridge University Press, 1992.

[2] J.M. Flesselles, M.O. Magnasco and A. Libchaber, *Phys. Rev. Lett.* **67** 2489 (1991)

[3] S. Kirstein, H. Möhwald and M. Shimomura, *Chem. Phys. Lett.* **154** 303 (1989) ; S. Kirstein and H. Möhwald, *Chem. Phys. Lett.*, **189** 408 (1992).

[4] L. F. Chi, R.R. Johnston and H. Ringsdorf, *Makromol. Chem. Macromol. Symp.* **46** 409 (1991).

[5] D. Ducharme et al., *ibid.* 97

[6] D. Möbius, private communication.

[7] D. Andelman, F. Brochard and J.F. Joanny, *J. Chem. Phys.* **86** 3673 (1987).

[8] H.M. McConnell and V.T. Moy, *J. Phys. Chem.* **92** 4520 (1988).

[9] P.Muller and F. Gallet, *J. Phys. Chem.* **95** 3257 (1991).

[10] C. Flament and F. Gallet, submitted to *Europhys. Lett.*

[11] S.T. Chui and J.D. Weeks, *Phys. Rev. Lett.* **40** 733 (1978).

[12] P.E. Wolf et al., *J. Physique* **46** 1987 (1985) ; P. Nozières and F. Gallet, *J. Physique* **48** 353 (1987).

[13] C. Jayaprakash, W.F. Saam and S. Teitel, *Phys. Rev. Lett.* **50** 2017 (1983).

[14] K.H. Kjaer and H.J. Hilhorst, *J. Stat. Phys.* **28** 621 (1982).

[15] For a recent review about monolayers properties, see for instance C. M. Knobler, *Adv. Chem. Phys.* **77** 397 (1990).

[16] A. M. Bibo, C. M. Knobler and I. R. Peterson, *J. Phys. Chem.* **95** 5591 (1991).

KINETICS OF GROWTH AND ORDERING OF CU ON CU(100) INVESTIGATED WITH HELIUM ATOM BEAM SCATTERING

H.-J. Ernst, F. Fabre and J. Lapujoulade
CEA Saclay, Srsim, 91191 Gif Sur Yvette, France

Abstract. We present experimental results on homoepitaxial growth and ordering processes on a single crystal surface. The activation energy for migration is determined to 0.28 eV. The early time exponent, which characterizes the evolution of the width of the interface before saturation, is found to be 0.4 at temperatures, for which thermally activated diffusion is highly operative. The ordering process from a kinetically rough state towards equilibrium of half a monolayer (representing a system undergoing "Spinodal Decomposition" in two dimensions) is shown to evolve in a selfsimilar fashion. Ordering exponents close to 1/4 are in accord with ideas of corrections to the Lifshitz Slyosov growth law due to excess mass transport along domain boundaries.

1. Introduction

Ordering and growth processes are very common and general phenomena in nature. The kinetics and dynamics involved address fundamental questions of nonequilibrium thermodynamics, but at the same time these processes are of prime interest in technology as well. For example Molecular Beam Epitaxy (MBE) has been proven to be a powerful tool in the growth of crystals and interfaces, for both scientific and technological use. While the basic processes involved are conceptually simple (impingement, adsorbtion, migration and incorporation of adatoms into the growing surface), a comprehensive understanding in the sense that the outcome of an MBE experiment can be reliably predicted, is not yet available. One reason for this can certainly be attributed to the difficulties in handling far from equilibrium phenomena with many different energetic and kinetic processes competing. This considerable task can be made somewhat easier, if one deals with homoepitaxial growth on a nonreconstructed surface. Then, the growth mode is exclusively determined by kinetic parameters.

The close connection between diffusion and growth is obvious. One expects that if the impinging flux density is very high or the substrate temperature too low to allow for sufficient mobility, even a homoepitaxial system will not grow in a layer by layer fashion: the system becomes "kinetically rough". While this state of the interface is in general not of great technological use for the fabrication of electronic devices (however it might be for example for catalysis), it has recently attracted much attention from a more fundamental point of view.

Theoretical considerations and computer simulations suggest that the morphology of a kinetically rough interface can be characterized by two characteristic length scales only: the growing surface develops heightfluctuations up to a characteristic length parallel to the surface. This length increases as time evolves, and simultaneously, the amplitude of these height fluctuations increases as well. The morphology of the surface is found to exhibit dynamical scaling; i.e. the heightfunction looks the same for all times, provided the lengths perpendicular and parallel to the surface are appropriately rescaled. This behaviour is reminiscent to ordering processes. Here as well dynamical scaling is expected to hold, with however the major difference being that ordering (of N dimensional structures in N dimensions) can be characterized by one single length scale only. In both cases, the lengthscales increase according to an algebraic law, and in analogy to equilibrium phase transitions, the values of the exponents are expected to characterize certain universality classes.

Our contribution addresses these aspects from an experimental point of view. In the following, we present the results of our He Scattering experiments on the growth and ordering of Cu on Cu(100). First, we discuss the nucleation and island formation of adatoms deposited at comparatively high temperatures and for a coverage below 1 ML. From these experiments the migration coefficient is infered. In the second paragraph, we explore the decay of intensity oscillations in multilayer deposition experiments and its relation to the height exponent (β), which characterizes the width of the growing interface before saturation. Finally, we report the results of the experiment that simulates "spinodal decomposition" in 2D. A system with equal volume fraction of two phases and conserved order parameter is realized by deposition of half a ML Cu on Cu(100).

2. Filling the first layer

The migration of adatoms on surfaces is a fundamental problem in surface science and has attracted attention for many years and still does. However, despite its importance, techniques for determining the pure migration coefficient are rare. Our approach is based on the close relation between migration, nucleation and growth, and relies on measuring in situ and in real time the separation between large islands formed on individual terraces during a submonolayer deposition. Nucleation theory relates this quantity to the migration coefficient.

Fig. 1 shows a series of angular distributions, taken in the out of phase condition, as a function of time during deposition of Cu on Cu(100) for a coverage up to 0.5 ML. The specular intensity decreases, while at the same time additional diffraction peaks appear. A simple simulation of the diffraction pattern within kinematical scattering theory reveals that the position of these diffraction peaks reflect the separation L of large islands formed during deposition. At a flux of 1270 sec / ML, these "sidebands" can be easily separated from the specular peak up to 250K; below about 160K, only a broad base below the specular is observed, which is presumably related to the filling of higher levels due to reduced mobility of adatoms. Fig. 2 shows the separation L as a function of inverse temperature. Between about 160K and 250K, L follows and Arrhenius type behaviour, the slope and the intercept being 70 me V and 5×10^3 Å, respectively.

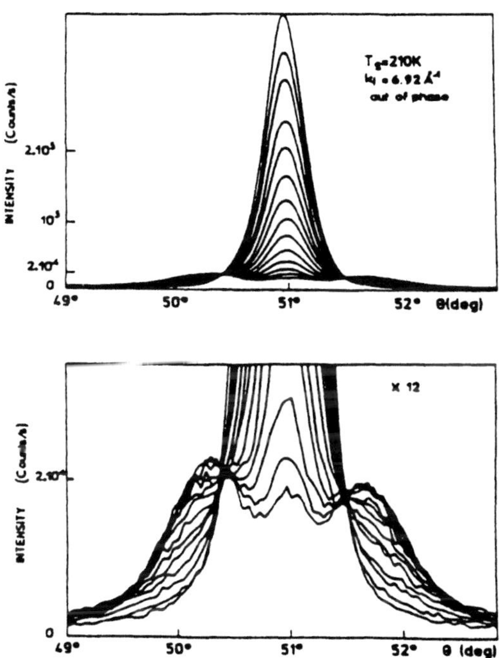

Fig. 1: Series of angular distributions, taken in the antiphase condition, during deposition of Cu at a rate of 2700 sec/ ML. The specular intensity ($\Theta = 51°$) decreases, while additional diffraction peaks appear. The diffraction peaks reflect the separation between large islands formed during deposition.

Rate equation theories [1], which are currently in use and believed to give a rather detailed picture of the nucleation process, relate these quantities to the preexponential factor and the activation energy for migration. For isotropic diffusion, and when the stable nucleus is formed by two adatoms, the parameter dependence of the total density of nuclei is expressed as $L \propto (D/R)^p$. Herre R is the arrival rate of adatoms, and $D = D_0 \exp(-E/kT)$ the migration coefficient. We have determined p by varying, at fixed substrate temperature, the arrival rate of adatoms and find, within experimental

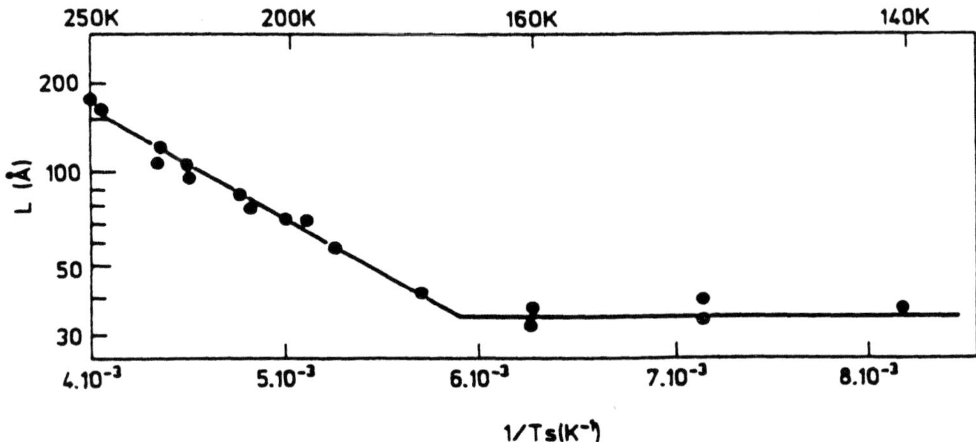

Fig. 2: Mean separation L between large islands as a function of inverse temperature. The data were taken with an incident flux of 1270 sec/ ML.

error, $L \propto R^{-1/4}$. This leads to $E = 0.28 \pm 0.06$ eV and $D \approx 10 - 5 \, \text{cm}^2/\text{sec}$ for the migration of Cu on Cu(100).

This result necessitates some comments. The finding $p = 1/4$ is at first sight sursprising. In fact, the rate equation approach predicts two nucleation regimes, with different values of p. At the very beginning of the nucleation process, in the transient regime, the number density of adatoms increases steeply and so does the the density of stable nuclei N. Thereafter, in the steady state regime, adatoms are also captured by already existing nuclei, and new ones are formed only at a lower rate. P takes the value $1/2$ in the transient regime, but $p = 1/3$ in the steady state regime, i.e. with $L = 1/N^{1/2}$, $p = 1/4$ and $p = 1/6$, respectively. The crossover is parameter dependent, and in our case it should take place at coverages far below 0.1 ML. At a coverage of 0.5 ML, we must have reached certainly the steady state regime, and $p = 1/6$ should apply. We note that this reasoning accounts for the total density of islands formed, regardless of their size. The position of the observed diffraction peaks from which we infer L, reflect however only the separation of the largest islands, and these islands are formed at the very beginning of the nucleation process [1,2] and there $p = 1/4$ applies. This explains as well, why we find the position of the diffraction peaks independent of coverage up to about 0.7 ML.

Our analysis is based on the assumption that a dimer forms already a stable nucleus. Molecular Statics calculation [3] reveal that the binding energy of a dimer on this surface is at least 0.3 eV. If we apply the rate equation formulas to our data under the assumption that a trimer forms a stable nucleus, the migration energy amounts to less than 0.15 eV. In view of existing data for other materials, this seems to be too small for self-diffusion on a fcc (100) surface.

Finally, we note that recent calculation by Hansen et al. [4] gives a value of 0.23 eV for the activation energy. Interestingly. they found that this value corresponds to an exchange diffusion mechanism; a bridge hopping process needs nearly double the energy. By comparison with this theory, our data provide an experimental indication for the exchange diffusion path.

3. Multilayer Deposition

The observation of anti-phase diffraction peak oscillations as a function of time in multilayer deposition experiments are taken as a signature of well ordered growth, often called loosely "layer by layer" growth. Intuitively one expects well ordered growth at high deposition temperatures and moderate fluxes. Based on the previously determined migration coefficient of Cu on Cu (100), we can determine the time needed to change one lattice site: this value amounts to about 10^{-5}sec at 240K, but 10^4sec at 100K. However, this simple reasoning based on horizontal mobility parameters only might be quite misleading, as recently demonstrated by Comsa et al. for Pt on Pt(111), when asymmetries in the perpendicular and horizontal mobilities come into play.

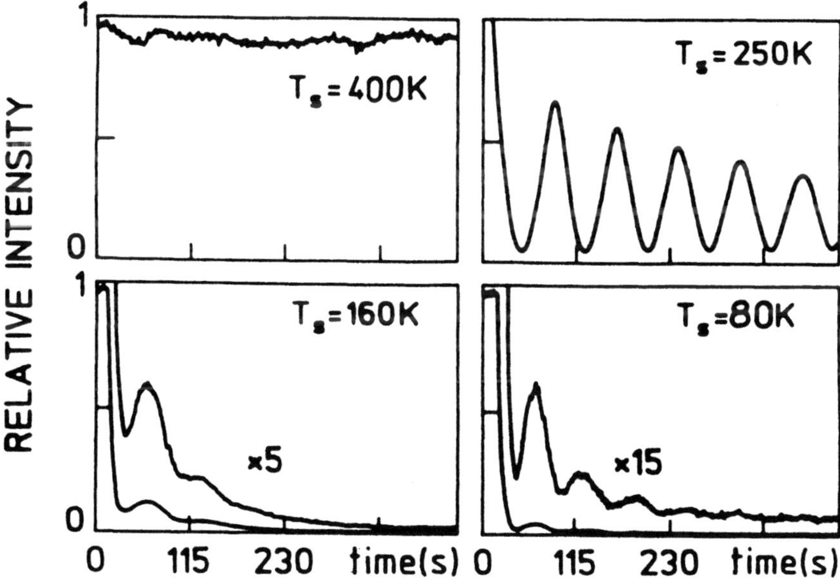

Fig. 3: Normalized specular intensity in the antiphase condition vs. deposition time for the temperatures indicated. The incident flux is 65 sec/ML.

Fig. 3 shows the specular intensity as a function of deposition time. Above 400K. no change in intensity is observed. At that temperature, growth proceeds without island formation, but via propagation of residual steps. Pronounced intensity oscillations are observed in the temperature range around 220K. Here, growth proceeds via nucleation and growth of islands. However, the maxima never reach the original value. This is a clear indication that growth is actually not ideally layer-by-layer; more and more levels are only partially filled as time evolves, and the width of the interface increases continuously. This damping is more pronounced at lower temperatures. However, even at 100K, in the absence of thermally actived mobility, a couple of oscillations are still observed, which means that the interface does not grow in a purely random mode. Is this expected?

Truly random deposition appears actually only when interlayer mass transport is completely inhibited. However, under the assumptions that the sticking coefficient of Cu on Cu (100) is one, and atoms adsorb in fourfold hollow sites (both have recently been confirmed in MD simulations[5]), the geometry of a fcc(100) structure imposes a natural correlation, and growth cannot be random in the sense defined above. This is in contrast to a sc lattice, for which atoms in the absence of mobility stick to on top sites and all columns can grow independently according to a Poisson statistics.

The decreasing amplitude of the oscillations is a signature of increasing interface width, and in principle, it should be possible to determine the relationship between width and time (or height). However, for quantitative analysis the measured antiphase intensity has to be corrected, for two reasons. First He has a large cross section for scattering from island edges or adatoms. Second, the scattered intensity in the in phase condition consists in general of a coherent "δ"-contribution, surrounded by a broad incoherent distribution. It is the centrl spike which contains information on the height difference correlation funcion at infinity, i.e. on the width of the interface. We have accounted for this by measuring the intensity slightly off the specular direction in the antiphase condition. Within Gaussian approximation, it follows that the normalized corrected intensity is then related as $I_c/I_0 = \exp(-\pi^2 w^2)$ to the width w of the interface.

Fig. 4 shows the result of this procedure for three different deposition temperatures and a coverage up to about 30 ML. The width of the interface is expected to behave as $w^2 \propto t^{2\beta}$ at the beginning of the deposition process. At 240K this exponent amounts to $\beta \approx 0.39$, while at 100K and 160K, the width of the interface increases with $\beta \approx 0,5$ at the beginning, but seems to saturate for longer times. At present, it is not clear, if the saturation appears only within the accuracy of the experiment, or if the data represent actually a crossover to a logarithmic behavior. In fact, there is a priori no reason, why the width of an interface should saturate in a (quasi-)infinite system, represented by a real crystal. Only the transfer width of the instrument imposes some threshold, up to which correlations can be readily detected, thus simulating a finite system. But the transfer width of our instrument is estimated to be about 1000Å, which implies that an apparently saturating width should have been detected for a height of the deposit much larger than about 10 ML.

On the other hand, a logarithmically growing interface would be compatible with the predictions of the EW model[6], with or without diffusion to kink sites, in 2+1 dimensions. However, it seems that the ν term in the EW equation, associated in the original model with gravity, is considered as irrelevant for MBE. We note, however, that a mechanism coined "funneling" [6], which boosts the filling of fourfold hollow sites on deep levels, could take the role of gravity.

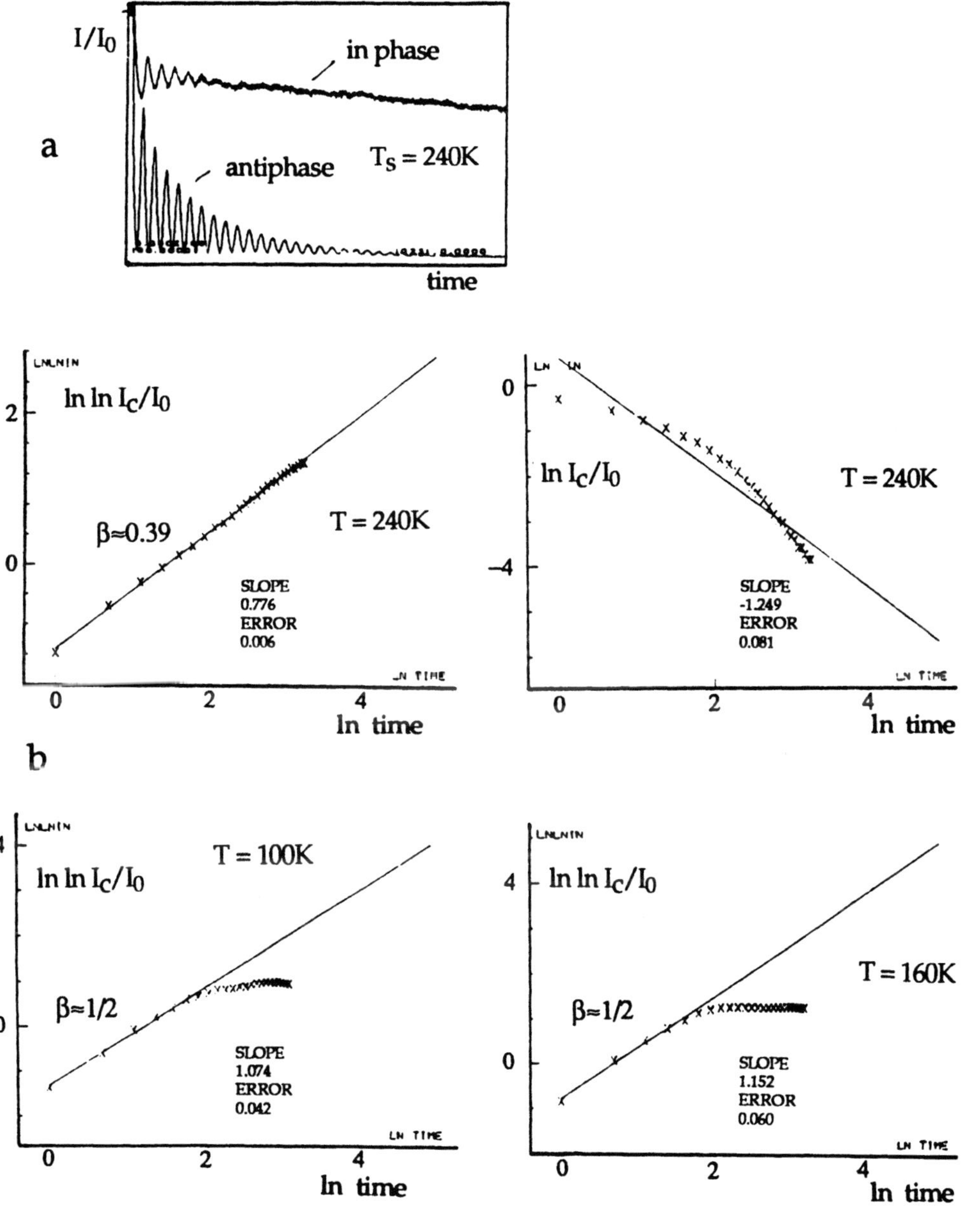

Fig. 4: a) Normalized specular intensity under in phase (upper curve) and antiphase condition for a deposition of about 30 ML at 240K.
b) Corrected normalized intensity (see text) for three different deposition temperatures. Maximum time corresponds to about 30 ML.

147

The finding $\beta \approx 0.39$ at 240K, for which no doubt thermally activated diffusion is highly operative, is not predicted so far by any of the existing theoretical models that might describe MBE in 2+1 dimensions.

4. The ordering process of half a monolayer Cu on Cu(100)

The ordering pattern of 2D structures on a 2D substrate can be characterized by a single length scale $L(t)$. The time dependence of this characteristic length often obeys an algebrais law, $L(t) = A(T) * t^x$, where A isthe growth rate constant, depending on temperature T, and x the growth exponent. There are suggestions from the theoretical side that only two different exponents exist, depending only on whether the order parameter is conserved or a nonconserved quantity [7]. However, for the case of conserved order parameter, which is experimentally represented for wxample by system undergoing phase separation, this is not yet completely settled, in particular in 2D.

Lifshitz and Slyosov examined the case of widely spaced domains growing in the matrix of the second phase. Under these conditions, ordering is mediated by evaporation of monomers from small domains, long range diffusion and condensation onto larger domains. In that case, the theory reveals an algebraic growth law with $x = 1/3$. It is a priori not clear, if this result still holds in a situation, of which the volume fractions are comparable, as it is realized in spinodal decomposition. A near critical quench usually results in a geometry, where one has not isolated domains, but instead an interpenetrating and interconnected domain structure. This opens the possibility for additional activated ordering mechanisms, for example for diffusion along boundaries.

In our experiment, a system in 2D with conserved order parameter an equal volume fractions of the two phases is realized by adsorbing 0.5 ML Cu on Cu(100). The initial state is prepared by deposition at 100K. At that temperature, thermally activated diffusion is virtually zero, and atoms are, on the average $L = 1/0.5^{1/2} = 1.4$ lattice constants apart, resulting in a structure, which consists of a network of interconnected domains, thus simulating "spinodal decomposition". The surface is subsequently raised to and held at higher temperature, and the kinetics of domain growth is monitored by measuring diffraction profiles as time evolves. Fig. 5 shows typical angular distributions for an upquench to 235K. The diffraction peaks move towards the specular, become narrower and increase in intensity as time evolves. We use the position Q_{max} of the diffraction peaks as a measure of $L(t)$. In order to check if dynamical scaling holds, we have plotted the shape $S(Q)$ of the diffraction peaks according to $S(Q,t) = (Q_{max}(t))^{-D} F(Q/Q_{max}(t))$ [8], where D is the dimension of the system (here D=2) and F a time independent master function. That is, although the characteristic length changes with time, the functional form of the domain pair correlation function does not, and all diffraction scans taken at different times must collapse to one single curve. Fig. 6 shows that dynamical scaling is followed to a very good approximation. Similar results have been obtained for other temperatues, except for the highest. However, it is possible that this does not represent a true deviation from scaling, but is due to the influence of the instrument, which becomes noticeable for large correlations.

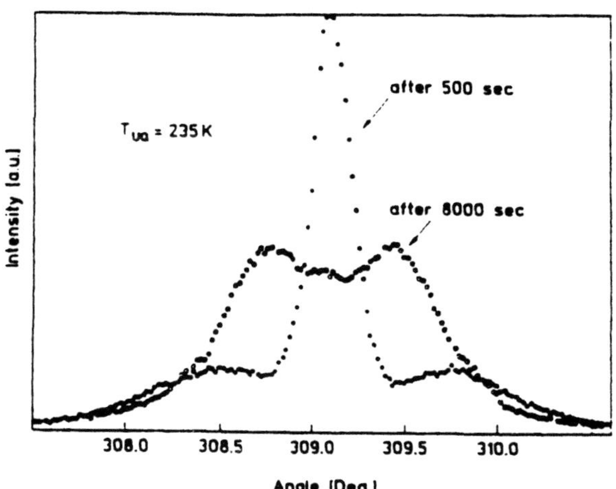

Fig. 5: Angular distributions for an upquench to 235K, after $t = 500$sec and $t = 8000$sec.

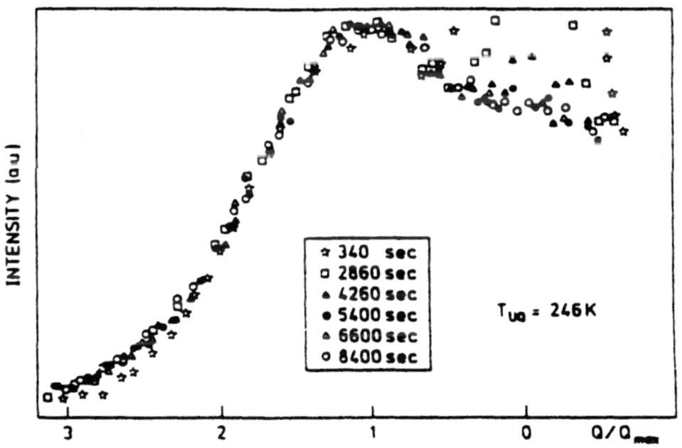

Fig. 6: Demonstration of scaling. Angular profiles have been transformed into momentum space and scaled according to the equation given in the text. The specular peak has not been removed from the data set. Therefore, intensity is not zero at $Q/Q_{max} = 0$ and apparent deviations from scaling are observed.

This dynamical scaling is observed, although the growth exponent is found to be far from the Lifshitz-Slyosov value. Fig. 7 shows the evolution of the characteristic length L as a function of time. When fitted to an algebraic growth law, we obtain exponents confined between 0.12 and 0.25 in the temperature and time range covered by our experiment.

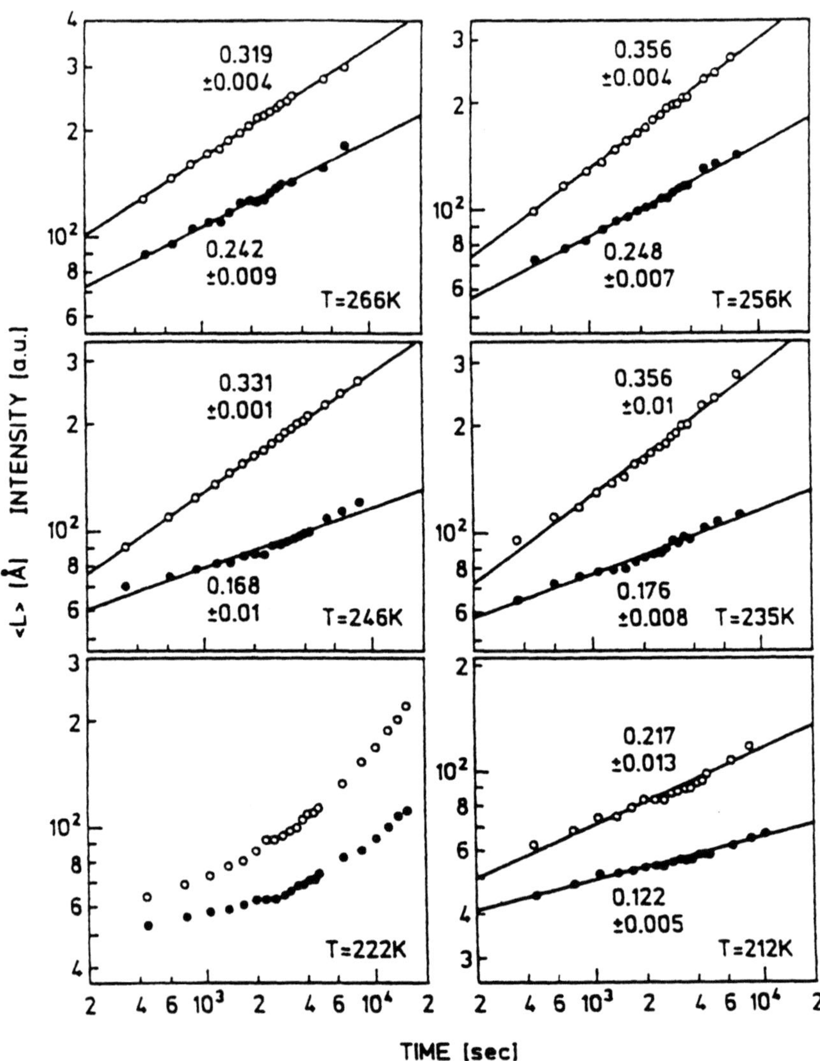

Fig.7: Variation of the characteristic length L ($\bullet$) and of the peak intensity ($\circ$) vs. time for different upquench temperatures. Lines are fits to an algebraic growth law. The numbers indicate power law exponents.

Comparable low ecponents have been found in MC simulations on spinodal decomposition in 2D as well. Huse [9] explained these findings by the fact that masstransport along domain boundaries contributes in parallel with the evaporation–condensation mechanism to ordering. The relative rate of these processes must depend on the total length of domain boundaries in the system, hence, he suggests a correction of the LS exponent of the form $x = 1/3(1 - L_0(T)/L(T,t))$, where L_0 depends on temperature roughly as $\exp(E/kT)$, E being the difference in activation energies for both mechanisms. A quantitative comparison of measured exponents and L reveals that L_0 is smaller than L, but of the same order of magnitude. This suggests that in our experiment ordering is actually dominated by mass transport along domain boundaries. It is likely, that Huse's correction is not valid in this case. In fact, based on dimensional arguments, Mullins [10] predicts that interface diffusion by itself leads to an algebraic growth law with exponent 1/4. This value is in good agreement with the observed exponent at the highest temperatures. The observed lower exponent at lower temperature should presumably be interpreted as a correction to this growth law due to an hitherto unidentified kinetic process.

That in our experiment ordering is dominated by interface diffusion is supported by experiments, for which we have adsorbed only 0.3 and 0.1 ML, followed by upquenches to 252K. If the evaporation mechanism were operative, we should approach, within the spirit of the LS theory, $x = 1/3$. Instead, we find $x = 0.08$ for 0.3 ML, while ordering effectively stops after a few minutes for 0.1 ML.

In general, we expect the growth law with exponent 1/4 to hold only, when the evaporation mechanism is in fact strictly zero. But because both mechanisms are thermally activated processes, the 1/3 growth law will eventually always catch up at only sufficiently long times. But from a realistic point of view, i.e. in view of the limited measuring time, the exponent 1/4 obtained in this particular situation of fully connected domains and deep quenches might be regarded as asymptotic.

REFERENCES

[1] S. Stoyanov and D. Kashiev, *Curr. Top. Mat. Sci.* **7** 69 1981.
    J.A. Venables, G.D.T. Spiller, M. Hambrücken *Rep. Prog. Phys.* **47** 399 1984.
[2] G. Zinsmeister, *Thin Solid Films* **7** 51 1971.
[3] N. Papanikolau, V. Pontikis,   to be published.
[4] L. Hansen, P. Stoltze, K.W. Jacobsen, J. Norskov, *Phys. Rev. B* **44** 6523 1991.
[5] J.W. Evans, D.E. Sanders, P.A. Thiel, A.E. dePristo, *Phys. Rev. B* **41** 5410 1990.
[6] D.Wolf and J.Villain, *Europhys. Lett.* **13** 389 1990.
    S. Das Sarma, P. Tamborenea, *Phys. Rev. Lett.* **66** 325 1991.
[7] O.G. Mouritsen in *Kinetics of Ordering and Growth at Surface* ed. by M.G. Lagally (NATO ASI Series B, Vol. 239) *Plenum‹NY.*
    1990
[8] J.D. Gunton, M. San Miguel and P.S. Sahni in *Phase Transition and Critical Phenomena*, Vol. 8. (ed. by C. Domb, J.L. Lebowitz Academic, NY 1983.
[9] D.A. Huse, *Phys. Rev. B* **34** 7845 1986.
[10] W.W. Mullins, *J. Appl. Phys.* **59(4)** 1341 1986.

FRONT PROPAGATION, PINNING AND UNCONVENTIONAL NOISE

KINETIC ROUGHENING WITH MULTIPLICATIVE NOISE

T. Vicsek

*Department of Atomic Physics, Eötvös University, Budapest, P.O. Box 328,
1445 Hungary and Institute for Technical Physics, Budapest, P.O. Box 76,
1325 Hungary*

Abstract. In this paper a new interpretation of the role of inhomogeneities in the experiments on kinetic roughening is introduced by considering a stochastic differential equation for the surface development with a *multiplicative noise*. The main motivation is to make assumptions which are as close to the experimental conditions as possible. By numerically integrating the equation we have obtained i) surfaces remarkably similar to those observed in the experiments and ii) a scaling of the surface width as a function of time with an exponent being in an excellent agreement with the measured value. Variations of the model, crossovers and questions concerning the applicability of additive noise to wetting experiments are also discussed.

1. Introduction

One of the exciting recent questions regarding the growth of rough interfaces [1] is the apparent discrepancy between the experimental results and the corresponding predictions based on the most general theoretical approaches and related simulations. In particular, for the 1+1 dimensional case the renormalization group treatment of the KPZ equation [2] and large scale numerical studies of the simplest aggregation models [3] give $\alpha = 1/2$ and $\beta = 1/3$, where these exponents describe the scaling of the width of the surface [4]

$$w(x,t) \sim t^\beta f(t/x^{\alpha/\beta}) \tag{1}$$

as a function of time t and the linear extension x of the surface over which the width is calculated.

On the other hand, the existing experimental estimates obtained for the interface of viscous flows and the surface of bacteria colonies range between 0.63 and 0.81 [5-7]

155

for α and give $\beta \simeq 0.65$ [6]. These values are in clear conflict with the predictions 1/2 and 1/3.

Recently a few specific models have been proposed to eliminate the above mentioned disagreement. Results in part consistent with the experiments have been obtained by assuming power law distributed noise [8-12], studying a simplified KPZ type equation with quenched noise [13], and by modifying the growth rule in various growth models [14-17] in order to better account for the physical conditions determining the process of kinetic roughening.

The purpose of the present paper is to introduce a new concept of studying the actual physical situation by considering a stochastic differential equation for the surface development with a multiplicative noise. Our goals are i) to make assumptions which are as close to the experimental conditions as possible, ii) to numerically investigate the resulting equation and iii) to compare the obtained behaviour with that observed in the experiments.

2. Multiplicative noise

We propose that the development of the interface $h(x,t)$, e.g., in the experiments on quasi 1+1 dimensional viscous flows should be described by the equation

$$\frac{\partial h}{\partial t} = \left(\nabla^2 h + v(1 + (\nabla h)^2)^{1/2}\right)(p + \eta), \tag{2}$$

where $p > 0$ is some constant, v is the normal velocity and the term $\eta > 0$ corresponds to quenched noise with no correlations, i.e.,

$$\langle \eta(x, h)\eta(x', h')\rangle = C\delta(x - x')\delta(h - h'). \tag{3}$$

We *do not assume* that the distribution of the noise amplitudes is Gaussian with a zero mean; it would be equivalent to supposing that flat parts of the interface would move backward at places with $\eta < -p$ (this question will be discussed in more details in the next section). Rather, we shall assume that η follows some other simple distribution, e.g., the Poisson distribution.

In order to give the interpretation of the form in which the noise term enters (2) let us consider the experiment on the two phase flow of viscous fluids in porous media. We are interested in the case when the more viscous, wetting fluid advances due to the presence of capillary forces and the interface exhibits kinetic roughening. Under such circumstances the system can be considered as a network of randomly interconnected channels of widely distributed sizes and geometry. The motion of the wetting fluid is determined by the simultaneous effects of surface tension, capillary forces and local flow properties (permeabilities of the channels). The advancement of the interface at a given point is proportional to the local driving force and the permeability; just as the electric current $\vec{j}$ is proportional to the conductivity σ and the electric field $\vec{E}$, $\vec{j} = \sigma\vec{E}$. In our case the the driving forces are i) the capillary force which would produce a velocity v for unit permeability and, ii) the forces due to the surface tension which are represented by the term $\nabla^2 h$. Thus, eq. (2) is equivalent to

$$v_s = \epsilon F, \tag{4}$$

where v_s is the velocity of the surface in the vertical direction, ϵ is the randomly changing local permeability and F denotes a general driving force.

In addition to the above discussion of the origin of the multiplicative nature of noise in (2) we would like to make two short comments on the other aspect of the proposed equation. i) The term $v(1 + (\nabla h)^2)^{1/2})$ is included in its full form (instead of its linearized version used in the KPZ equation), because in the actual experiments at the majority of the points along the interface the condition $\nabla h \ll 1$ is not satisfied. This statement becomes very relevant when pinning forces are present and the interface develops deep valleys with $\nabla h \gg 1$ playing a determining role in the process of roughening. ii) Naturally, eg. (2) can be extended to include additive noise by adding a new term ζ which can be independent of or proportional to η, and a term $\lambda(\nabla h)^2$ instead of $(\nabla h)^2$ only (λ is a parameter). In this case (2) reads as

$$\frac{\partial h}{\partial t} = \left(\nabla^2 h + v(1 + \lambda(\nabla h)^2)^{1/2}\right)(p + \eta) + \zeta. \tag{5}$$

Now we are in the position to describe the development of the interface in terms of kinetic roughening dominated by pinning forces. At places where $p + \eta \ll 1$, the motion of the interface slows down dramatically (see section 4). These points can be considered as temporarily pinned. However, like in all of the existing experiments (with no evaporation), after some time the surface passes by this place or region of low permeability and advances further without a complete stop.

An interesting special case of the noise is when η depends only on x (this possibility is also discussed in Ref. 14, for additive noise). An existing aspect of the physics is reflected in this choice: the motion of the interface is determined not only by the conditions at the surface, but also by the permeability of regions already left behind (which may partially block the supply of additional fluid).

As will be briefly demonstrated in section 4., the numerical studies of eq. (2) result in surfaces very similar to the experimental ones and exhibit a rich behaviour including a non-trivial crossover in time depending on the value p. Preliminary results concerning the model with x dependent η will also be presented in section 4.

3. Some remarks on the applicability of the KPZ approach to the experiments

According to the KPZ equation the development of the surface is descibed by

$$\frac{\partial h}{\partial t} = \nabla^2 h + \lambda/2(\nabla h)^2 + v + \eta(x, t), \tag{6}$$

where λ is a parameter which for wetting flows is larger than 0. As was pointed out by Kessler et al [13], for the interpretation of the experiments it is more appropriate to use a quenched noise in (6), $\eta(x, h)$, and this is the version we shall discuss below.

The applicability of (6) to the wetting experiments can be questioned on the following basis. The anomalous experimental values for the exponents are likely to be due to the fact that the motion of the interface is dominated by pinning forces. At the places where the surface is locally almost pinned (slowed down) $\partial h/\partial t \ll 1$. On the other hand, at the same locations $\nabla^2 h + \lambda/2(\nabla h)^2 \gg 1$. According to (6) this can hold only if $-(v + \eta) \gg 1$ in these points. We argue that large negative values of the noise η are not physical, because this would mean that a flat surface in the given point would move with a large velocity in the direction *opposite* to the growth. Since the fluid is wetting, its spontaneous motion cannot be the one suggested by this approach.

Nevertheless, the equation (6) with quenched noise can be solved numerically making various assumptions for η. We find that i) to obtain surfaces as rough as the experimental ones we have to use η values distributed in a rather wide range (e.g., between -20 and 20). This case corresponds to very small surface tension and the surface becomes extremely ragged, unlike those observed in the experiments. ii) For less widely distributed η the surface width is much smaller then in the experiments. Note that in figure 2 of Ref 13 the axes are not isotropic; if the scales along the axes were the same (as in the experimental pictures) the calculated surface would look almost like a straight line. It has to be pointed out that the choice for the units in the horizontal and the vertical directions cannot be arbitrary in the present case.

Finally, we would like to mention three recent models designed to simulate growth in the presence of pinning forces. In the very interesting approach by Parisi [14] the noise is additive, and the above questions have to be considered when discussing its applicability to wetting experiments. The growth models proposed in Ref. 15 and 16 are somewhat closer to the approach introduced in this paper. For example, the blocking sites of Ref. 15 correspond to places with zero $p + \eta$. However, the actual mechanisms of growth in the two pictures are rather different.

4. Results

Since the main goal of this work is to understand what are the most relevant factors determining the behaviour of experimental surfaces, we have numerically studied eq. (2) for times and system sizes compatible to those which have been realized in Refs. 5 and 6. It is an almost trivial task to integrate (2) numerically, the associated questions are briefly discussed in Ref. 13 (quenched case) and Refs. 18 and 19 (for time dependent noise). For simplicity we assumed that η was distributed uniformly on $(0,1]$. Fig. 1 shows the calculated surfaces for the following set of parameters: system size $L = 800$, $p = 0.0001$ and $v = 0.5$.

Next we investigated $w(L,t)$, the time dependence of the width of the entire system. The preliminary results shown in Fig. 2 demonstrate that at early times there exists a non-trivial scaling

$$w \sim t^\beta \tag{6}$$

with $\beta \simeq 0.65$ in a surprisingly good agreement with the only available experimental result. This value also agrees with the corresponding estimates obtained in Refs. 15 and 16, but is different from 3/4 published in Ref. 14. The crossover to a behaviour

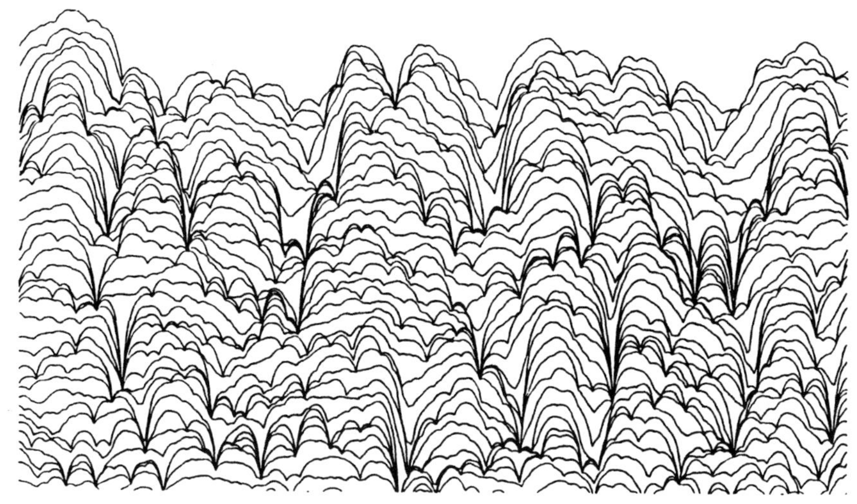

Figure 1. Subsequent "snapshots" of the evolving surface obtained by numerically integrating eq. (2) for $p = 0.0001$ and $v = 0.5$ for a system of linear size $L = 800$.

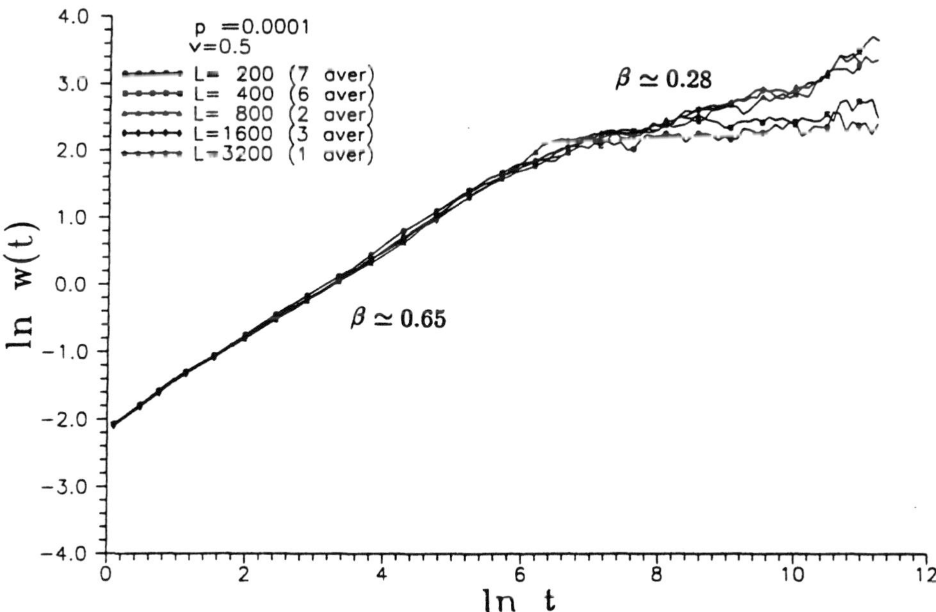

Figure 2. The time dependence of the surface width w for various system sizes. There is a well defined crossover at a time independing on L from a scaling according to an exponent $\beta \simeq 0.65$ to a scaling with β about 0.28.

described by a smaller exponent $\beta = 0.28 \pm 0.05$ is well pronounced and according to our simulations the crossover time depends only on p.

The situation is less clear concerning the spatial scaling of w. Our preliminary calculations indicate that, at least in the first approximation, there exist no well defined straight parts in the $\log w$ versus $\log x$ plots. Further studies are needed to clarify the scaling behaviour of the surface with as a function of x.

Finally, we briefly discuss the case when η depends on x only. Fig. 3 shows a typical series of surfaces for $L = 800$, $p = 0.0001$ and $v = 0.5$. In this model there is well pronounced scaling both in time and space, with numerically determined exponents close to 1 ($\alpha \simeq \beta \simeq 0.95 \pm 0.06$).

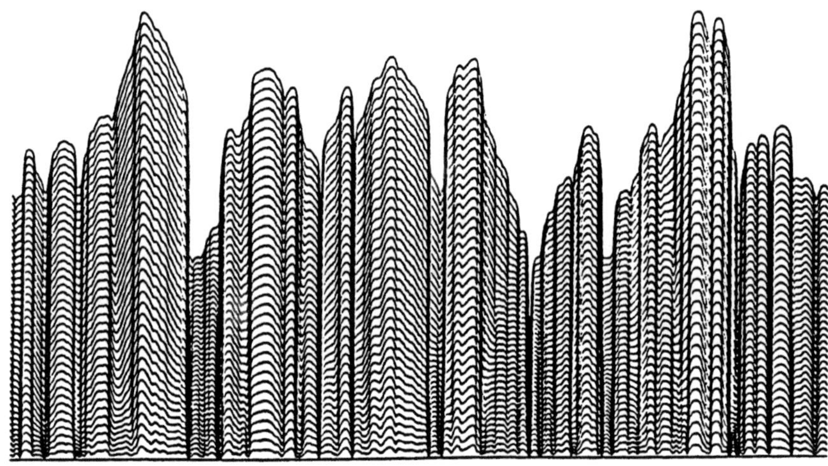

Figure 3. Series of surfaces obtained as a function of time for the model with η depending on x only (see the text).

A comparison of Figs. 1 and 3 with the experimental interfaces published in Ref. 6 suggests that, as expected, the experimental conditions could be best described by a combination of the two models.

5. Conclusions

In summary, we have proposed an approach which is intended to take into account the experimental conditions during two-phase fluid flows in inhomogeneous media as well as possible. By introducing a multiplicative noise into the stochastic partial differential equation describing the development of the interface we have been able to obtain i) surfaces remarkably similar to those observed in the experiments ii) a scaling behaviour of the surface width with an exponent being in an excellent agreement with the measured value.

Approaches based on equations or models with additive noise raise questions when their application to wetting experiments is considered. Finally we have also studied a model in which the noise depended only on x.

Acknowledgements: The author thanks E. Somfai and M. Vicsek for assistance in carrying out the simulations reported here. Useful discussions with J. Kertész are acknowledged.

REFERENCES

[1] *Dynamics of Fractal Surfaces* F. Family and T. Vicsek eds., World Scientific, Singapore, 1991.

[2] M. Kardar, G. Parisi, and Y.-C. Zhang, *Phys. Rev. Lett.* **56** 889 1986.

[3] T. Vicsek, *Fractal Growth Phenomena* World Scientific, Singapore, 1992.

[4] F. Family and T. Vicsek, *J. Phys.* **A 18** L75 1985

[5] M. A. Rubio, C. A. Edwards, A. Dougherty, and J. P. Gollub, *Phys. Rev. Lett.* **63** 1685 1989.

[6] V. K. Horváth, F. Family, and T. Vicsek, *J. Phys.* **A 24** L25 1991.

[7] T. Vicsek, M. Cserzö, and V. K. Horváth, *Physica A* **167** 315 1990.

[8] Y.-C. Zhang, J. de Physique **51** 2113 1990.

[9] J. Amar and F. Family, *J. Phys.* **A 24** L79 1991.

[10] S. V. Buldyrev, S. Havlin, J. Kertész, H. E. Stanley, and T. Vicsek, *Phys. Rev. A* **43** 7113 1991.

[11] S. Havlin, S. V. Buldyrev, H. E. Stanley and G. H. Weiss, *J. Phys.* **A 24** L925 1991.

[12] V. K. Horváth, F. Family and T. Vicsek, *Phys. Rev. Lett.* **67** 3207 1991

[13] D. A. Kessler, H. Levine and Y. Tu, *Phys. Rev.* **43** 4551 1991.

[14] G. Parisi, *Europhys. Lett.* **17** 673 1992.

[15] A-L. Barabási, S. Buldyrev, E. Caserta, S. Havlin, H. E. Stanley, and T. Vicsek, preprint

[16] L-H. Tang and H Leschhorn, preprint

[17] M. Cieplak and M. Robbins *Phys. Rev.* **B41** 11508 1990.

[18] K. Moser, J. Kertész and D. Wolf, *Physica* **A178** 215 1991.

[19] J. Amar and F. Family, *Phys. Rev.* **A41** 3399 1990.

Surface Disordering:
Growth, Roughening and
Phase Transitions

INTERFACE FLUCTUATIONS AT DEPINNING

Giorgio Parisi

*Dipartimento di Fisica, II Università di Roma, Tor Vergata and INFN, Sezione
di Roma, Tor Vergata Via della Ricerca Scientifica, Roma 00133, Italy*

Abstract.   In this talk I will present some evaluations of the critical exponent β
characterizing the broadening of the interface width at the depinning point. Both
numerical simulations and simple analytic computations are presented.

1. Introduction

The subject of this talk will be the study of the fluctuations of an interface near
the depinning transition. Before discussing the pinning of a real interface it may be
convenient to recall some recent work I did with Luciano Pietronero on the depinning
of charge density waves [1,2], because this case is better understood and some of the
techiques used are in common. Indeed it turns out that the behaviour of a fluctuating
interface is very similar to that found for charge density waves.

2. Charge density waves

In the model for charge density waves I consider here the dynamical quantities are the
fields $\phi_i(t)$, which satisfy the following differential equation [3]:

$$\dot{\phi_i} = E + \Delta\phi_i - sin(\phi_i - h_i), \tag{1}$$

where $\Delta\phi_i$ is the lattice Laplacian and i denotes a point of a d dimensional lattice. We
are interested to know the time dependance of the current $J(t) = \frac{1}{V}\sum_i \dot{\phi_i}(t)$, when
we start with a simple boundary condition at time equal to zero (e.g. $\phi_i(0) = 0$).

For small values of E, J goes to zero at large times, while for large values of E J goes to a non zero value at large times. We expect therefore the existence of a critical value of E (i.e. E_c) which separates the two regimes. It is reasonable to believe that, just at the critical point (i.e. $E = E_c$), the current decays as a power law:

$$J(t) \sim t^{-\omega}. \tag{2}$$

One of the goals of our study was to compute the value of ω, although we were mostly interested in understanding the stretched exponential decay observed in numerical simulations for $J(t)$ in the low E region [4].

An other quantity we can compute are the fluctuations of ϕ, defined by

$$W(t)^2 = \frac{1}{V} \sum_i \dot{\phi}_i^2 - (\frac{1}{V} \sum_i \dot{\phi}_i)^2. \tag{3}$$

At the critical point one expects that

$$W(t) \sim t^\beta, \tag{4}$$

where $\beta = 1 - \omega$.

3. The coarse-grained model

We can argue that the main effect of the fields h_i is to pin the values of ϕ more or less strongly (strongly if the values of the h's in a given region are similar, weakly if they differ of about π so that their effects cancel ones with the others). We finally arrive to an other model [5] which is supposed to describe the large scale behaviour of charge density wave:

$$\dot{\phi}_i(t) = F\theta(F) \quad F = \Delta\phi_i(t) + p + \eta_i, \tag{5}$$

where η is a random uncorrelated Gaussian noise and p is a controll parameter which plays the same role of E. One can easyly convince ourself that the critical value of p is just $p = 0$.

Our task is now to compute the large time behaviour of $J(t)$ at $p = 0$. Let us assume that the previous equations have a solution in the continuum limit, where they become

$$\dot{\phi}(x,t) = F\theta(F) \quad F = \Delta\phi(x,t) + p + \eta(x), \tag{6}$$

where the correlations among two η's are given by

$$\overline{\eta(x)\eta(y)} = \delta(x - y). \tag{7}$$

In this case we can use dimensional analisys. We can measure everyting in dimensions of length ($[x] = 1$). If we do so we find:

$$[\eta] = -d/2 \quad [t] = 2 \quad [\phi] = (4 - d)/2 \quad [J] = -d/2. \tag{8}$$

Dimensional analysis implies that

$$\omega = d/4 \quad \beta = (4 - d)/4. \tag{9}$$

The crucial assumption is the existence of the the solution of eq.(6) in the continuum. In many case stochastic differential equations with a dimensionless coupling are not well definined directly in the continuum limit. If a lattice spacing is introduced at a preliminary stage, one may finds an addictional dependence of the observables on the lattice spacing. Naive dimensional counting cannot be used and the renormalized operators (which have a finite expectation value in the continuum limit) have an effective dimension which differs from the naive dimension. The previous analysis assume the absence of anomalous dimensions and it corresponds to a mean field approach.

The absence (or the presence) of divergences in the continuum limit should be studied using renormalization group techniques. This has not yet been done; however accurate numerical simulations in 1 and 2 dimensions show that the prediction $\omega = d/4$ is satisfied with an accuracy less than 1%, so that it seem rather reasonable (although we lack a formal proof) that eq. (9) is an exact statement.

4. A model for interface pinning

A fast moving interface in a random medium is usually described by the following equation [6]:

$$\dot{h}(x,t) = \Delta h(x,t) + p\left(1 - 1/2(\nabla h(x,t))^2\right) + \eta(x,t), \tag{10}$$

where the noise η is Gaussianly distributed; its correlation functions are

$$\overline{\eta(x,t)\eta(y,t')} = \delta(x - y)\delta(t - t'). \tag{11}$$

However, if the interface is slow moving, equations (10) and (11) are no more correct and they must be replaced by [7-9]:

$$\dot{h}(x,t) = \Delta h(x,t) + p\left(1 - 1/2(\nabla h(x,t))^2\right) + \eta(x,h(x,t)), \tag{12}$$

$$\overline{\eta(x,h)\eta(y,h')} = \delta(x - y)\delta(h - h').$$

These equations describe the movement of an interface in a medium with random friction. They should be modified in the case of an interface in a random potential [10] (e.g. a magnetic interface) by using the appropriate form of the correlation of the η's.

Near the depinning transition the velocity is very slow and the non linear term is likely irrelevant, so that we can restrict ourselves to the simpler equation:

$$\dot{h}(x,t) = \Delta h(x,t) + p + \eta(x,h(x,t)). \tag{13}$$

A slightly different equation, which should be equivalent near the pinning and describes slightly better the random friction case, is

$$\dot{h}(x,t) = F\ \theta(F) \quad F = \Delta h(x,t) + p + \eta(x, h(x,t)). \tag{14}$$

This last equation differs from the previous one only by forbidding the backward movement of the interface. It also differs from the one studied in the previous section only in the form of the noise.

For this equation (similar results shouls also hold for the other equations (13) and (14)) we expect that for $p < p_c$, when the time goes to infinity, $h(x,t)$ has a finite limit (which may depend on the initial conditions).

A possible approach to the study of the large time behaviour of $h(x,t)$ consists in computing the number $\mathcal{N}(p,\eta)$ of solutions of the stationary equation

$$0 = \Delta h(x,t) + p + \eta(x, h(x,t)) \tag{15}$$

inside a region of size $V = TL^d$.

It is reasonable to assume that the function

$$f(p) = \frac{1}{V}\overline{ln(\mathcal{N})} \tag{16}$$

is positive only for $p < p_c$ and vanishes exactly at $p = p_c$. Therefore the properties of the pinned interface and the related static exponents could be computed by considering a generic solution of eq.(13). The replica formalism can be used to compute the average number of solutions but such a computation has not yet done.

5. A simple theoretical analysis

After the succes of the naive dimensional analysis in charge density waves, we could try to use the same method here; one easyly finds that:

$$[\eta] = -d/2 - [h]/2 \quad [t] = 2 \quad [h] = (4 - d)/3, \tag{17}$$

where d is the number of transverse dimensions. One consequently would conclude that the interface width $W(t)$ scales as t^β, with

$$\beta = (4 - d)/6. \tag{17}$$

In particular one would conclude that $\beta = 1/2$ for in two dimensions ($d = 1$). This result is in variance with numerical simulations which show without doubts that in two dimensions β is very near to $3/4$, with an error less or equal to 3% [9,11].

It is quite amazing that this result is exactly the same as the one obtained in the previous section with a time independent noise. In order to understand the origine of this discepancy we can concentrate our attention on the fluctuations of the friction field η. We define the quantity

$$E(L,T) = \int d^d x dt \; \eta(x, h(x,t)), \tag{18}$$

where the integral is done over a region of size L in space and size T in time.

We are interested in understanding the behaviour of $< E^2 >_c = < E^2 > - < E >^2$. Naively we would expect here that

$$< E^2 >_c = \int d^d x dt \; \left(\frac{\partial h}{\partial t}\right)^{-1} \sim \frac{L^d T^2}{< h(T) - h(0) >}. \tag{19}$$

This kind of extimate is at the basis of the previous dimensional analysis. This result obviously differs from the one obtained for an h- independent noise, where

$$< E^2 >_c \sim L^d T^2. \tag{20}$$

In order to derive eq.(19), one must assume that $\left(\frac{\partial h}{\partial t}\right)^{-1}$ does not strongly fluctuate from point to point. i.e. the movement is smooth. Now near the depinning this assumption is not correct and the surface growth looks more similar to a sequence of avalanches, well localized in space and time (surfacequakes). Between surfacequakes $\frac{\partial h}{\partial t}$ is pratically zero, so that there are regions in which the noise becomes time independent, just because h is time independent. In this situation equation (20) is again correct and the exponent should be the same as for the time independent noise. We find again the prediction eq. (9) for β.

In other words near the depinning most of the interface is not moving, and the regions which are already pinned give a contribution to $< E^2 >_c$ which scales as T^2. Naive dimensional analysis does not work here and if we use the equation (20) for extimate the effective dimension of the noise η, we find again $[\eta] = -d/2$. One is consequently lead again to the prediction eq. (9) for the value of β. In this way we are able to understand the origine of the puzzling result found in $1 + 1$ dimensions.

Obviously extra work is necessary to understand better the origine of this prediction, in particular from the numerical side it would be very interesting to compute the probability distribution of $h(t + \tau) - h(t)$ near the depinning transition for different choices of τ and to characterize better the probability distribution of avalanches. A detailed study of the tail of the probability distribution of $h(t + \tau) - h(t)$ is needed in order to understand the presence (ar absence) for not too small τ of a power law behaviour.

The reader should notice that this analysis can be applied only to models with random friction or random forces. The situation in models with moltiplicative noise [12] may be quite different; these models may belong to a different universality class, where different critical exponents are present.

REFERENCES

[1] G. Parisi and L. Pietronero, *Europhys. Lett.* **16** 321 1991.
[2] G. Parisi and L. Pietronero, *Physica A* **198** 666 1991.
[3] H. Fukuyama and P. A. Lee, *Phys. Rev. B* **17** 535 1978.

[4] A. Erzan, A. Veermans, R, Heijungs and and L. Pietronero, *Physica A* **166** 447 1990.

[5] L. Mihaly, M. Crommie and G. Gruner, *Europhys. Lett.* **4** 103 1987.

[6] M. Kardar, G. Parisi and Y.-C. Zhang, *Phys.Rev. Lett.* **56** 889 1986.

[7] R. Bruinsma and G. Aeppli, *Phys. Rev. Lett.* **52** 1547 1984.

[8] J. Koplik and H. Levine, *Phys. Rev. B* **32** 280 1985.

[9] D. A. Kessler, H. Levine and Tu Y., *Phys. Rev. B* **43** 4551 1991.

[10] M. Dong, M. C. Marchetti, A. Alan Middleton and V. Vinokur, *Syracuse University preprint* 1991.

[11] G. Parisi, *Europhys. Lett.* **17** 673 1992.

[12] T. Vicsek *These proceedings* 1992.

PINNING RELATED TO DIRECTED PERCOLATION

Lei-Han Tang

*Institut für Theoretische Physik, Universität zu Köln, Zulpicher Str. 77,
D-5000 Köln 41, Germany*

Heiko Leschhorn

*Fakultät für Physik und Astronomie, Ruhr-Universität Bochum, Postfach
102148, D-4630 Bochum, Germany*

Abstract. The roughening behavior of a driven interface between two fluids in a two-dimensional porous medium is examined. It is argued here that the unusually large roughness exponent observed in recent experiments is a characteristic of a critical depinning transition related to directed percolation.

1. Introduction

One of the recent experiments introduced to test the modern theory of kinetic roughening[1,2] is the displacement of air by water in a two-dimensional cell of randomly distributed glass beads fixed between two glass plates[3,4]. In this experiment, water is filled in slowly from one end and drawn into the cell by capillary forces at the interface. When the flow rate is low, it is reasonable to assume that the pressure field behind the water-air front is uniformly distributed (i.e., the viscous effects are weak). In this case the motion of the interface is, to a good approximation, isolated from that of the bulk. Surprisingly, measurements of the transverse interface fluctuation w as a function of window size L yield a power-law $w \sim L^\zeta$ with an exponent ζ in the range 0.73–0.81 [3,4], much larger than the value $\zeta = 0.5$ given by either linear[5] or nonlinear[1] theories of kinetic roughening.

For a static interface at equilibrium, the equation which governs the physics on the scale of pore size is the Laplace formula $\Delta P = P_{air} - P_{water} = 2\gamma/R$, where ΔP is the pressure difference across the interface, γ is the air-water surface tension, and R is the mean radius of curvature of the interface[6]. In addition, the interface must intersect the glass beads at a contact angle θ as determined from the Young's equation. It is our daily experience that the capillary rise of water at ambient pressure is higher in a narrow glass tube than in a wide one. Thus, for a given ΔP, it is easier for the water to flow through narrow channels than the wide ones in the glass-bead experiment.

To study the large scale roughness of the interface, it is useful to construct a continuum description of the interface dynamics on a length scale larger than that of the pore size. Rubio et al.[3], and Zhang[7] have suggested that the Kardar-Parisi-Zhang equation[1]

$$\frac{\partial h}{\partial t} = \nu \nabla^2 h + \frac{\lambda}{2}(\nabla h)^2 + v_0 + \eta(\mathbf{x}, t) \tag{1}$$

may be applicable, with a noise term η which is either correlated or power-law distributed so as to account for the anomalously large exponent ζ mentioned above. However, for interface motion in an isotropic medium, the parameter λ in (1) should be proportional to the mean interface velocity v_0[1], thus the nonlinear term is absent in an experiment conducted at a low flow rate. In addition, the disorder that acts on the interface is quenched rather than fluctuating in time. The main physical mechanism for interface roughening in the quasistatic limit should be pinning by quenched random forces rather than nonlinear growth with a fluctuating noise.

Koplik and Levine[8], and later Kessler, Levine and Tu[9], studied the following equation of motion for the interface profile $h(\mathbf{x}, t)$,

$$\frac{\partial h}{\partial t} = \nu \nabla^2 h + F + \eta(\mathbf{x}, h) \tag{2}$$

where F represents an average driving force and $\eta(\mathbf{x}, h)$ is a quenched random force from the porous medium. This equation also arises in the study of domain wall motion in a random-field Ising model[8,10-12]. As shown by several authors[8,11,12], for interface dimension $d < 4$, a moving solution to (2) exists only for F larger than a threshold value $F_c > 0$. The precise nature of the transition at F_c has not been discussed in detail analytically.

The purpose of this note is to present a simple mechanism for interface pinning in (1+1)-dimensions. At the present level of our understanding we can only provide a few plausible arguments as to why such a mechanism could be relevant for the anomalous roughening behavior observed in the glass-bead experiment. Detailed predictions of this pinning mechanism remain to be confirmed by future experiments.

2. Interface Pinning: A Simplistic View

When the pressure difference ΔP between air and water is small, only sufficiently wide channels can sustain a stable meniscus. A necessary condition for the water-air front to be in equilibrium is that the interface runs through only these wide channels. The existence of a percolating cluster of wide channels, however, is in general not a sufficient condition for pinning since there are usually more than one channel that lead to the same pore.

As emphasized by Martys et al. [10], an important effect which must be included in discussing pinning in wetting-invasion is the interference between neighboring menisci which leads to an "overlap instability". This effect eliminates sharp corners on a pinned interface. More importantly, it creates a dynamic correlation between adjacent growth events. Due to this collective effect, an initially planar interface is able to maintain its orientational order during growth. As a consequence, an interface element is more likely to invade a pore from a particular direction. Pinning takes place when the wide channels oriented around this direction form a connected path across the cell.

The mathematical problem of finding such a path is that of directed percolation[13,14]. Consider for instance a square lattice with a given density q of pinning sites, as indicated by the crosses in Fig. 1. A pinning site A is effective only if it is joined by another one B on the side or C or D along the diagonals. It is known that, for $q > q_c \simeq 0.539[15]$, the effective pinning sites percolate through the system.

Although pinning of the whole interface takes place only for $q > q_c$, there exist clusters of effective pinning sites even below q_c. The typical size of these clusters is given by $\xi_\parallel$ along the interface and $\xi_\perp$ perpendicular to the interface, as shown in Fig. 2. The two lengths diverge as the percolation threshold q_c is approached,

$$\xi_\parallel \sim |q - q_c|^{-\nu_\parallel}, \qquad \xi_\perp \sim |q - q_c|^{-\nu_\perp}. \tag{3}$$

Latest series calculation gives $\nu_\parallel = 1.733 \pm 0.001$ and $\nu_\perp = 1.097 \pm 0.001[16]$. As we shall see below, these clusters present a strong resistance to fluid flow close to the depinning transition.

Equation (3) has a physical meaning for $q > q_c$, too. In this case the percolating directed paths organize themselves into a lattice-like structure, with a typical distance $\xi_\parallel$ between the nodes parallel to the interface and $\xi_\perp$ perpendicular to the interface[17]. At each node two or more percolating paths join each other, as shown in Fig. 2. The total number of percolating paths grows exponentially with $V/(\xi_\parallel \xi_\perp)$ where V is the volume of the system.

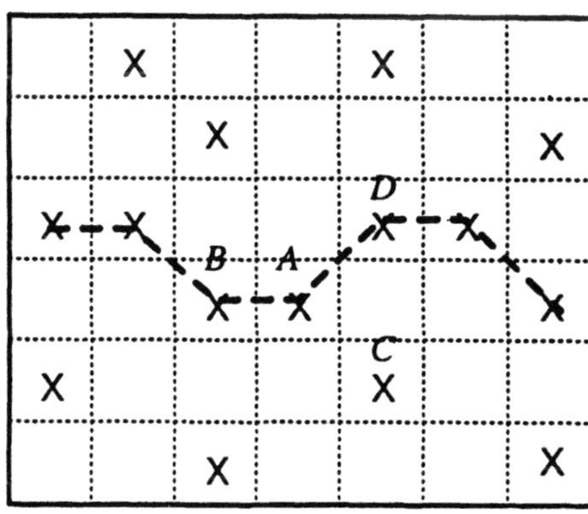

Figure 1. Directed percolation. The dashed line indicates
a percolating path across the system.

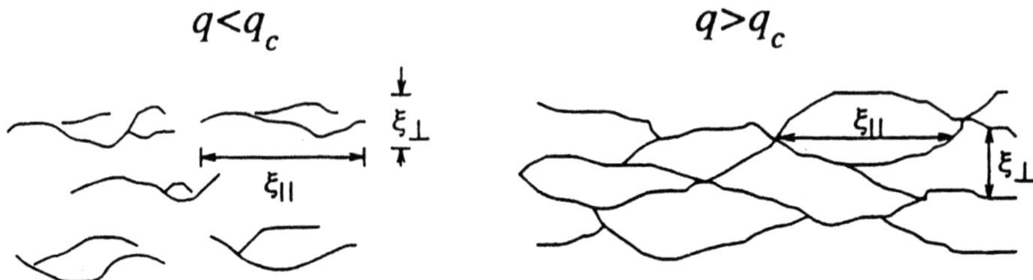

Figure 2. Percolating clusters above and below the threshold.

As one approaches the depinning threshold q_c from above, the number of stationary configurations decreases until only one is left. Although we are not aware of any explicit calculation of the roughness of the last percolating path, it appears plausible that this path is self-affine with a roughness exponent $\zeta = \nu_\perp/\nu_\parallel \simeq 0.63$. One argument in favor of this conjecture is that pieces of this last (or first) percolating cluster should behave the same way as those paths which define the lattice-like structure above q_c, and as those finite clusters which exist below q_c. In both cases the transverse fluctuation of a path is of the order $\xi_\perp$ across a distance of order $\xi_\parallel$.

3. A Lattice Model

The lattice model we consider here is motivated by Eq. (2). The smoothening effect of the Laplace term in the continuum equation is represented by a down-hill current which transports height from higher to lower parts of the interface[5]. The disorder of the medium is modelled by a random pinning force $f(\mathbf{R})$ for each site $\mathbf{R}$, uniformly distributed in the interval $[0,1]$. During growth, a column i is randomly selected. If the interface element in this column at height h_i is higher than either of its two neighbors by two or more units, i.e., $h_i \geq \min\{h_{i-1}, h_{i+1}\} + 2$, the lower of the two neighbors is made to advance by one unit, *irrespective* of $f(\mathbf{R})$. In the opposite case, $h_i < \min\{h_{i-1}, h_{i+1}\} + 2$, the interface element in column i is made to advance by one unit if the pinning force $f(i, h_i)$ is less than a given applied pressure p, but otherwise it remains at its original position. The critical height difference for height transport is set at two rather than one to allow for a certain amount of "self-sustained" roughness on short length scales. This is usually necessary for pinning to take place.

It turns out that pinning of the interface in this model occurs if and only if sites with $f(\mathbf{R}) > p$, connected by nearest neighbor horizontal bonds or next-nearest neighbor $\pm 45°$ bonds, percolate through the system[18]. As we mentioned earlier, directed percolation in this geometry takes place when the density q of sites with $f(\mathbf{R}) > p$ exceeds the critical value $q_c \simeq 0.539$. Figure 3(a) shows the mean interface displacement $H(t) = \langle \overline{h_i(t)} \rangle$ as a function of time t in a Monte Carlo simulation of

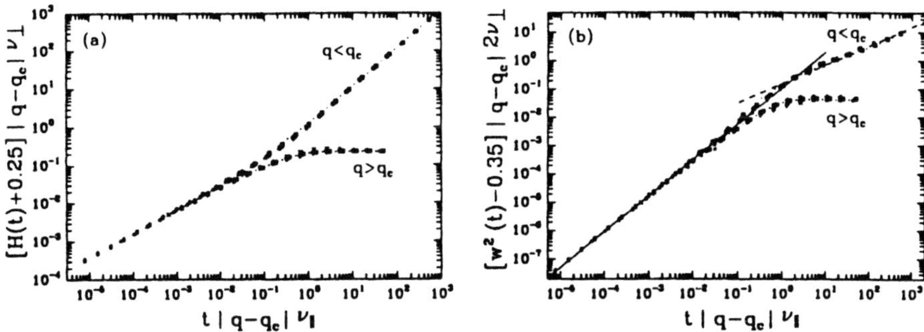

Figure 3. Scaling plot of (a) average interface displacement
vs. time. (b) Mean-square interface width vs. time.

the model for many different values of q around q_c. The initial interface configuration
at $t = 0$ is flat. Periodic boundary condition is used. After scaling $H(t)$ by $\xi_\perp$ and
t by $\xi_\parallel$, data at a wide range of q values fall on two universal curves. (A correction-
to-scaling term 0.25 is included in making the scaling plot.) The lower branch is for
$q > q_c$ (pinned phase), which reaches a plateau for $t \gg \xi_\parallel$. The upper branch is for
$q < q_c$, which crosses over to a line of slope one on the log-log plot, yielding a finite
steady-state velocity

$$v(q) \sim \xi_\perp/\xi_\parallel \sim (q_c - q)^{\nu_\parallel - \nu_\perp}, \tag{4}$$

where $\nu_\parallel - \nu_\perp \simeq 0.636$. In the critical regime $t \ll \xi_\parallel$, the average displacement of the
interface grows as a power-law of t, $H(t) \sim t^{\nu_\perp/\nu_\parallel}$.

The mean-square width of the interface $w^2(t) = \langle \overline{[h_i(t) - H(t)]^2} \rangle$ exhibits a similar
scaling behavior as $H(t)$. Figure 3(b) shows the scaled data from the simulation. For
$t \ll \xi_\parallel$, a power-law behavior $w^2(t) \sim t^{2\beta}$ is evident, with $\beta \simeq 0.63$ (solid line). The
lower branch crosses over to a constant, indicating a finite roughness in the pinned
phase. The upper branch turns into a line with a slope close to 2/3 (dashed line) at
large t, suggesting a crossover to the Kardar-Parisi-Zhang exponent $\beta = 1/3$.

The steady-state height-height correlation function $C(r) = [\overline{\langle (h_{i+r} - h_i)^2 \rangle}]^{1/2}$ exhibits an interesting behavior around the depinning transition. Figure 4 shows our data
for this quantity on both sides of the transition. The $q = 0.545$ data show a power-law
regime with a roughness exponent $\zeta \simeq 0.63$ as predicted by the directed percolation
mechanism. The $q = 0.530$ data show a similar scaling regime with a larger exponent
$\zeta \simeq 0.77$. In both cases the level off at larger distances is due to a combination of the
finite-size effect and a finite $\xi_\parallel$. On the moving side of the transition, a crossover to
$\zeta = 0.5$ is expected on length scales larger than $\xi_\parallel$. For $r < \xi_\parallel$, however, the interface consists of a mixture of pinned and moving segments which leads to an increase
of its roughness in this regime. Figure 5 shows a typical set of interface configurations for $q = 0.530$ at uniform time intervals. (Time increases from bottom to top.)
For the purpose of illustration each curve is shifted vertically upwards by an amount

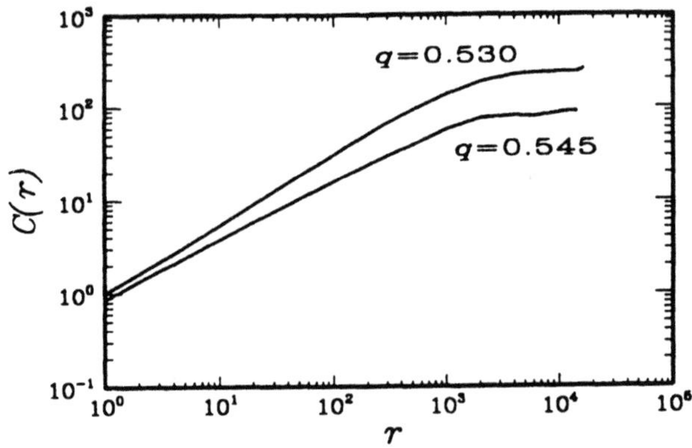

Figure 4. Steady-state height-height correlation function
above and below the pinning transition.

proportional to t. Dark regions in the figure are parts of the interface which remain stationary over each time interval, while the white regions are swept through by the moving part of the interface from an easy direction. Rather than a more or less uniform growth, the picture clearly indicates a separation of pinned and moving parts. Each pinned segment traces out a typical path on a directed percolation cluster, which has a power-law roughness with $\zeta = \nu_\perp/\nu_\parallel \simeq 0.63$. On the other hand, the moving segments tend to have a slope of order one or larger. On length scales less than $\xi_\parallel$ the interface can no longer be regarded as self-affine. We suspect that this type of inhomogeneous pattern may be the origin of the seemingly larger roughness exponent reported in the glass-bead experiment and simulations and may also explain the discrepancy among the quoted values of ζ.

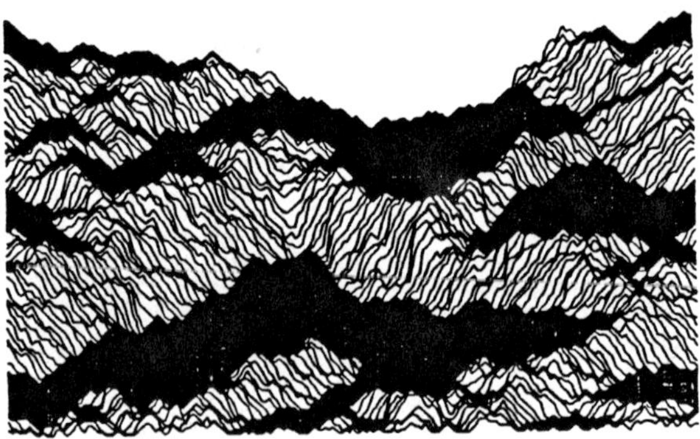

Figure 5. Snapshots of the moving interface at $q = 0.530$.

4. Summary

We discussed here a particular pinning mechanism—directed percolation of wide channels—to explain the anomalous roughening phenomenon observed in the glass-bead experiment. This mechanism may be relevant to a variety of other interface pinning problems in a two-dimensional disordered medium, where a collective growth stabilizes a planar interface. The roughness of a pinned interface close to the depinning transition satisfies self-affine scaling with an exponent $\zeta \simeq 0.63$. However, a moving interface close to the depinning transition may not be self-affine on small length scales. The effective roughness exponent in this regime may become larger than 0.63.

Directed percolation as a possible mechanism for interface pinning in $(1+1)$ dimensions has been independently proposed by Barabási et al.[19] This research is supported in part by the Deutche Forschungsgemeinschaft through SFB 166 and 237.

References

[1] M. Kardar, G. Parisi and Y.-C. Zhang, *Phys. Rev. Lett.* **56** 889 1986.

[2] F. Family and T. Vicsek, *Dynamics of Fractal Surfaces*, World Scientific, Singapore 1991.

[3] M. A. Rubio, C. A. Edwards, A. Dougherty and J. P. Gollub, *Phys. Rev. Lett.* **63** 1685 1989.

[4] V. K. Horváth, F. Family and T. Vicsek, *Phys. Rev. Lett.* **65** 1388 1990; *J. Phys.* **A 24** L25 1991.

[5] S. F. Edwards and D. R. Wilkinson, *Proc. R. Soc. London, Ser.* **A 381** 17 1982.

[0] J. S. Rowlinson and B. Widom, *Molecular Theory of Capillarity*, Oxford University Press, New York 1984.

[7] Y.-C. Zhang, *J. Phys. (France)* **51** 2129 1990.

[8] J. Koplik and H. Levine, *Phys. Rev.* **B 32** 280 1985.

[9] D. A. Kessler, H. Levine and Y. Tu, *Phys. Rev.* **A 43** 4551 1991.

[10] N. Martys, M. Cieplak and M. O. Robbins, *Phys. Rev. Lett.* **66** 1058 1991; N. Martys, M. O. Robbins and M. Cieplak, *Phys. Rev.* **B 44** 12294 1991.

[11] R. Bruinsma and G. Aeppli, *Phys. Rev. Lett.* **52** 1547 1984.

[12] M. V. Feĭgel'man, *Sov. Phys. JETP* **58** 1076 1983.

[13] D. Stauffer and A. Aharony, *Introduction to Percolation Theory* (2nd edition), Taylor & Francis, London 1992.

[14] W. Kinzel, in *Percolation Structures and Processes*, edited by G. Deutscher, R. Zallen and J. Adler, A. Hilger, Bristol 1983, p. 425.

[15] J. Kertész and D. E. Wolf, *Phys. Rev. Lett.* **62** 2571 1989.

[16] J. W. Essam, K. De'Bell, J. Adler and F. M. Bhatti, *Phys. Rev.* **B 33** 1982 1986; J. W. Essam, A. J. Guttmann and K. De'Bell, *J. Phys.* **A21** 3815 1988.

[17] S. Redner, *Phys. Rev.* **B 25** 5646 1982.

[18] L.-H. Tang and H. Leschhorn, *Phys. Rev.* **A** (in press).

[19] A.-L. Barabási, S. V. Buldyrev, F. Caserta, S. Havlin, H. E. Stanley and T. Vicsek, submitted to *Phys. Rev. Lett.*

KINETIC ROUGHENING IN THE PRESENCE OF QUENCHED RANDOM PHASES

Joachim Krug

*IBM Research Division, T.J. Watson Research Center,
Yorktown Heights, NY 10598*

Abstract. A driven restricted solid-on-solid (RSOS) model in which the fractional parts of the heights are quenched random variables is introduced. Several unusual features are found, such as (i) a first order depinning transition, (ii) logarithmic finite size corrections and (iii) invariance of the growth rate under small tilts.

1. Introduction

A moving interface can be kinetically roughened [1-3] by fluctuations in the microscopic displacement dynamics. The paradigmatic example is the growth of a solid from the vapor phase, where the fluctuations are due to the shot noise in the deposition flux. A rather different physical realization of kinetic roughening which has stimulated much recent experimental [4-6] and theoretical [7,8] work is the motion of a two-fluid interface through a random medium. This system displays a (presumably continuous) depinning transition at some critical value of the driving force [4]. While the roughening dynamics in the moving state is expected asymptotically to be governed by the conventional Kardar-Parisi-Zhang (KPZ) theory [1], there is some indication [7,8] of an intermediate scaling regime characterized by anomalous exponents.

Here I consider a situation in which the quenched disorder in the medium couples to the gradient of the interface position, rather than to the position itself. Specifically, let the height of the interface above some substrate lattice be given by real variables h_i with *quenched random fractional parts*, i.e. the allowed values of the height are

$$h_i = n_i + \eta_i \tag{1}$$

where n_i is an integer and η_i is, for a given sample, a fixed random number in the unit interval. This type of randomness can be thought to represent a bulk disordered crystal with short ranged translational but long ranged orientational order [9]. It also appears in other contexts, most notably in the sine-Gordon model of the phase dynamics of sliding charge density waves (CDW's) [10]. In fact, the model studied below can be viewed as a noisy, anharmonic variant of a cellular automaton introduced recently [11] in the context of mode locking in pulsed CDW's. Rather than attempting a realistic description of any particular physical system, however, the present paper aims at exploring some typical features of phase-disordered, driven interfaces through the study of a specific, simplified model. The discussion is restricted to the one-dimensional case, though several of the conclusions apply more generally.

2. The phase disordered RSOS model

The natural energy expression associated with the height variables (1) is, in one dimension,

$$H = \sum_{i=1}^{L-1} |h_i - h_{i+1}| = \sum_{i=1}^{L-1} |m_i + \delta_i|, \tag{2}$$

where I have introduced the bond variables

$$m_i = n_i - n_{i+1}, \quad \delta_i = \eta_i - \eta_{i+1}. \tag{3}$$

The ground state of (2) is obtained by setting $m_i = 1$ (-1) for $\delta_i < -1/2$ $(> 1/2)$, and $m_i = 0$ otherwise. The corresponding height configuration shows the conventional random walk type roughness of one-dimensional interfaces[1], $< (h_i - h_j)^2 >= \frac{1}{12}|i - j|$.

In the interest of simplicity, energetic considerations are ignored in the definition of the growth model. An effective interfacial tension is instead included through the restriction

$$|h_i - h_{i+1}| \leq \Delta \tag{4}$$

on the set of admissible configurations, where Δ is a continuous control variable. Admissible configurations always exist for $\Delta \geq 1/2$ (e.g., the ground state of (2) is admissible) and for $\Delta \geq 1$ the "flat" configuration $m_i \equiv 0$ is admissible. The boundary conditions are free for $\Delta < 1$ and periodic otherwise. The dynamics consists of sequential attempts to increase the height, $h_i \rightarrow h_i + 1$, at a randomly chosen site. An attempt is discarded if the resulting configuration violates (4), and is accepted otherwise. Growth models of this type (without disorder) were first introduced by Kim and Kosterlitz [12].

Many features of the model can be understood by considering the set of values of the integer slopes m_i which are allowed by a given value of the quenched phase difference δ_i and the control parameter Δ. Since deposition at a site i implies transport of one unit of slope across that site (i.e., $m_{i-1} \rightarrow m_{i-1} - 1$ and $m_i \rightarrow m_i + 1$), sites

---

[1] In contrast, the two dimensional equilibrium problem displays a nontrivial, superrough low temperature phase [9].

adjacent to bonds with many accessible values of m_i tend to have a higher acceptance rate of deposition moves and hence grow faster. This idea will be made precise in the following.

2.1. $1/2 < \Delta < 1$

In this region a finite fraction $P_0(\Delta) = 2(1 - \Delta)$ of bonds have only one allowed value of the slope m_i, which can be 0, 1 or -1 depending on δ_i. These bonds block the growth of the two sites adjacent to them, and can therefore be thought of as pinning centers. We conclude that the asymptotic growth velocity vanishes for $\Delta < 1$. The typical size of moving regions is $1/P_0(\Delta)$ and diverges as $(1 - \Delta)^{-1}$ for $\Delta \to 1$.

In a system of size L there is a finite probability[2] of the order $(1 - P_0)^L$ of not having any pinning center. This implies that the threshold is shifted below $\Delta = 1$ by an amount of order $1/L$, and the disorder averaged growth rate is $\mathcal{O}(e^{-L(1-\Delta)})$. It is also worth noting that the relaxation time T, defined as the number of MC steps (= deposition attempts per site) required to reach a completely frozen configuration, *diverges* with system size for all $\Delta < 1$. This is due to the fact that T is determined by the largest moving region in the sample. Since the probability of finding n adjacent bonds which are not pinning centers is proportional to $(1 - P_0)^n$, the typical size of the largest moving region (and therefore the typical relaxation time) in a system of size L is of the order of $\ln(L)/\ln(1/(1 - P_0))$. More of this type of reasoning will be encountered below.

2.2. $\Delta = 1$

Now every bond has two allowed states for m_i, which are -1 and 0 (1 and 0) for $\delta_i > 0$ ($\delta_i < 0$). Rather surprisingly, this implies an exact mapping to a model without disorder. For a given configuration of disorder, we define the spin variables

$$\sigma_i = 2m_i + \text{sign}(\delta_i). \tag{5}$$

It is then a straightforward matter to verify that the dynamics of the σ_i induced by (5) is precisely the asymmetric spin exchange dynamics of the single step model [13]: Exchanges of neighboring spins σ_i, σ_{i+1} occur at unit rate if $(\sigma_i, \sigma_{i+1}) = (-1, 1)$, and are forbidden otherwise. In a system with periodic boundary conditions the total magnetization is conserved. It is determined by the initial configurations of both the m_i and the δ_i, e.g. for an initially flat state, $m_i \equiv 0$, the quenched disorder produces a random magnetization of order $L^{1/2}$. Note that a flat initial state of the disordered model corresponds to a rough single step interface, and *vice versa*.

The steady state properties of the single step model are known exactly [13,14]. In particular, the asymptotic growth rate, defined here as the acceptance rate for deposition trials in the large system, long time limit, is equal to 1/4. This proves

---

[2] For the exact evaluation of this probability the short ranged correlations between the δ_i would have to be taken into account.

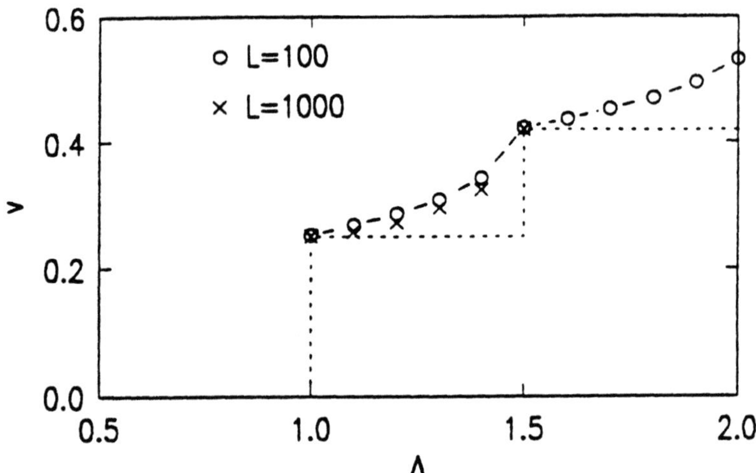

Figure 1 Disorder averaged growth rate for system size 100 (o) and 1000 (×). The predicted asymptotic growth rate is indicated by the dotted line.

that the depinning transition at $\Delta = 1$ is of first order. In the presence of an average inclination the growth rate is reduced to[3]

$$v(< m_i >) = \tfrac{1}{4} - < m_i >^2 \tag{6}$$

(Figure 2). Moreover it is well established [13] that the single step model belongs to the universality class of the the KPZ equation, with interface fluctuations growing as

$$\xi(t)^2 = < (h_i - < h_i >)^2 > = a_2 t^{2/3} \tag{7}$$

in the large system limit, where the amplitude a_2 can be computed from steady state properties [14]. This is confirmed by simulations of the disordered system at $\Delta = 1$ (Figure 3). At nonzero tilt the rough initial state of the corresponding single step model implies that (7) is replaced by a *linear* growth law [14].

Similar mappings to pure models can be found for all half-integer values of Δ. For $\Delta = N/2$ all bonds have N allowed states for m_i. For odd values of N this implies that the disordered model becomes equivalent to the pure RSOS model [12] with $\Delta = (N-1)/2$. Hence we have found a discrete set of parameter values where the disorder is irrelevant and the model is in the KPZ universality class. Nevertheless, the *generic* behavior is quite different.

2.3. $1 < \Delta < 3/2$

Typical disorder configurations consist of a mixture of two types of bonds, with either two (A bonds) or three (B bonds) allowed states for m_i. The average density of

---

[3] The largest possible inclination in the disordered model, $< m_i >= 1/2$, is obtained by setting $m_i = (1 - \text{sign}(\delta_i))/2$. Then $\sigma_i \equiv 1$ and no exchanges can occur.

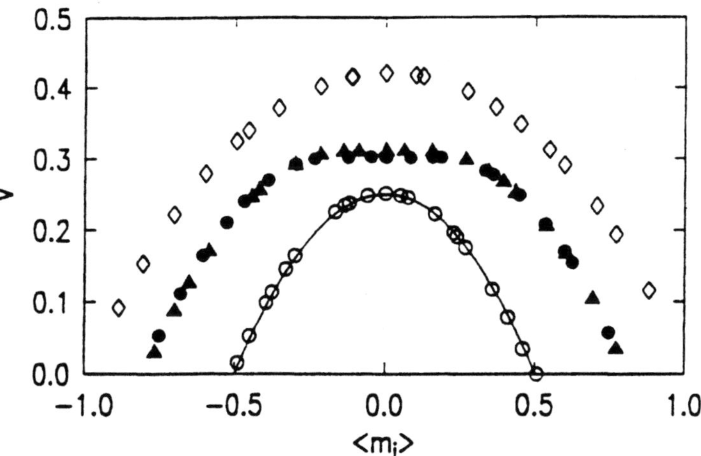

Figure 2 Inclination dependent growth rate for $\Delta = 1$ ($\circ$), $\Delta = 1.2$ (full symbols) and $\Delta = 1.5$ ($\diamond$). Each data point corresponds to a single long run (5×10^5 time steps) of a system of size $L = 200$, and the runs for each set of symbols were carried out with the same disorder configuration. The full line is the prediction (6).

A bonds is $P_A(\Delta) = (2 - \Delta)^2 - (\Delta - 1)^2$. A straightforward numerical measurement of the disorder averaged growth rate $v(\Delta)$ suggests that it interpolates smoothly between the limiting pure cases $\Delta = 1$, where $v = 1/4$, and $\Delta = 3/2$, where [14] $v \approx 0.419$, with a cusplike singularity at $\Delta = 3/2$ (Figure 1). I will now argue that the smooth variation is in fact a finite size effect, and that the asymptotic growth rate is $1/4$ in the whole interval $1 \leq \Delta < 3/2$.

A given disorder configuration can be decomposed into domains of consecutive bonds of a single type. According to the mapping described above, a domain of n consecutive A bonds can be viewed as a single step interface of length n. The neighboring B domains have a larger intrinsic growth rate, which forces the A domain to remain at zero inclination, where the growth rate is maximal. In contrast, the B domains can lower their growth rate by attaining a nonzero inclination (note that the inclination dependent growth rate is *convex*, as in (6), for the pure RSOS models (Figure 2)). The growth rate of a horizontal single step interface of n sites is $1/4 + c/n$ where c is a positive constant which depends on the boundary conditions [15]. Since the height differences are bounded by (4), in a stationary situation all domains have to grow at the same rate. The overall growth rate is then determined by the *slowest* domain, *i.e.* the largest domain of consecutive A bonds. An estimate of the type performed in Section 2.1 shows the typical size of the largest domain in a system of L sites to be of the order of $\ln(L)/\ln(1/P_A)$, which corresponds to a growth rate

$$v(\Delta) - 1/4 \approx c\ln(1/P_A(\Delta))/\ln(L). \tag{8}$$

This is consistent with the system size dependence of $v(\Delta)$ depicted in Figure 1.

A surprising consequence of the domain picture is the invariance of the growth rate under small inclinations away from the horizontal. Indeed, since the B domains

181

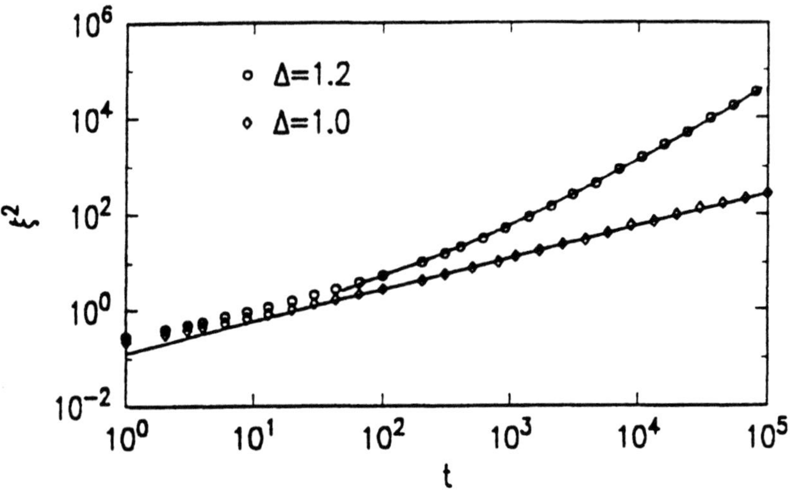

Figure 3 Interface width for $L = 10^5$ and $\Delta = 1.2$ (o) and $\Delta = 1$ ($\diamond$). The full lines indicate the analytic predictions.

can lower their growth rate by adopting either a positive or a negative inclination, the system can absorb a finite amount of net tilt by redistributing the inclination among and within the B domains. This is illustrated in Figure 2. It should be emphasized that the argument applies to *individual* (typical) disorder configurations and does not require a disorder average[4]. The disorder averaged growth rate will presumably pick up an exponentially small inclination dependence from rare configurations, such as pure A or pure B.

The logarithmic correction (8) also appears in the temporal relaxation of a large system. According to the scaling theory of kinetic roughening [2] the correlations in a large system at short times t (given a particular, say flat, initial state at $t = 0$) correspond to the stationary correlations in system of size $\xi_\parallel(t) \sim t^{1/z}$, where $z \geq 1$ is a dynamic exponent. Replacing L by $\xi_\parallel$ in (8) then predicts a logarithmic correction in *time*, which is in fact found in the simulation.

We now assume that this logarithmic correction sets the scale for velocity fluctuations between different domains in the transient regime. The velocity fluctuations drive the increase of the interface width ξ, so

$$d\xi/dt \sim 1/\ln(\xi_\parallel). \tag{9}$$

Together with the scaling relation [2] $\xi \sim \xi_\parallel^\zeta$, where ζ is a roughness exponent, this yields the long time behavior[5] $t = a\xi(\ln \xi - 1) + b$, which compares favorably with the

---

[4] This is in contrast to the disordered driven sine-Gordon model [10], where the tilt independence is a consequence of disorder averaging [16].

[5] Linear growth with logarithmic corrections has also been predicted for the interface width of the KPZ equation with static disorder [17,18].

numerical data for $\Delta = 1.2$ (Figure 3). Note that the *true* asymptotics $\xi \sim t/\ln(t)$ is not attained on the time scales studied here. Since the dynamic correlation length $\xi_\parallel$ cannot grow faster than t, it also follows that $\zeta = 1$, which is the maximal value of the roughness exponent allowed by (4).

3. Conclusion

In this paper, I have shown how the presence of quenched random phases in a simple model strongly enhances the kinetic roughening of a moving interface. Consequences for the related problems mentioned above [10,11,17,18], as well as extensions to higher dimensions and more realistic dynamics, will be discussed elsewhere.

Acknowledgement

I would like to thank Matthew Fisher for discussions which initiated this work.

REFERENCES

[1] M. Kardar, G. Parisi and Y.C.Zhang, *Phys. Rev. Lett.* **56** 889 1986.

[2] J. Krug and H. Spohn, *Solids far from Equilibrium* (C. Godrèche, ed.) Cambridge University Press, Cambridge 1991.

[3] F. Family and T. Vicsek (eds.), *Dynamics of Fractal Surfaces* World Scientific, Singapore 1991.

[4] J.P. Stokes, A.P. Kushnick and M.O. Robbins, *Phys. Rev. Lett.* **60** 1386 1988.

[5] M.A. Rubio, C.A. Edwards, A. Dougherty and J.P. Gollub, *Phys. Rev. Lett.* **63** 1685 1989.

[6] V.K. Horváth, F. Family and T. Vicsek, *J. Phys.* A **24** L25 1991.

[7] D.A. Kessler, H. Levine and Y. Tu, *Phys. Rev. A* **43** 4551 1991.

[8] G. Parisi, *Europhys. Lett.* **17** 673 1992.

[9] J. Toner and D.P. DiVincenzo, *Phys. Rev. B* **41** 632 1990.

[10] P.B. Littlewood, *Charge Density Waves in Solids* (L.P. Gorkov and G. Grüner, eds.) North Holland, Amsterdam 1989.

[11] A.A. Middleton, O. Biham, P.B. Littlewood and P. Sibani, *Phys. Rev. Lett.* **68** 1586 1992.

[12] J.M. Kim and J.M. Kosterlitz, *Phys. Rev. Lett.* **62** 2289 1989.

[13] P. Meakin, P. Ramanlal, L.M. Sander and R.C. Ball, *Phys. Rev. A* **34** 5091 1986.

[14] J. Krug, P. Meakin and T. Halpin-Healy, *Phys. Rev. A* **45** 638 1992.

[15] J. Krug and P. Meakin, *J. Phys. A* **23** L987 1990.

[16] D.S. Fisher, *private communication*

[17] M.E. Cates and R.C. Ball, *J. Physique* **49** 2009 1988.

[18] T. Nattermann and W. Renz, *Phys. Rev. A* **40** 4675 1989.

Surface Disordering:
Growth, Roughening and
Phase Transitions

CRITICAL PHENOMENA IN FLUID INVASION: TRANSITIONS IN GROWTH MORPHOLOGY

Marek Cieplak[1] and Mark O. Robbins
*Department of Physics and Astronomy, The Johns Hopkins University,
Baltimore, MD 21218*

Abstract. Results of simulations of capillary invasion in model 2-D porous media
as a function of the contact angle are reviewed. Various critical phenomena involved
are described. Randomness in the contact angle is shown to enhance the regime of
parameters in which fractal growth takes place.

1. Introduction

The displacement of one fluid by an immiscible fluid in a porous medium displays
a wealth of pattern-formation processes, depending, e.g., on velocity, viscosity, and
pore geometry [1-11]. In the limit of very low capillary numbers, pattern formation is
controlled by the invading fluid's wetting properties (contact angle θ) and the applied
pressure P. Recent two dimensional ($2D$) simulations [8-10,12] in the quasistatic limit
have revealed that for each θ there is a critical pressure, P_c, at which the invading fluid
spans the entire system. When θ is large, i.e. the invasion is non-wetting, the large
scale structure of the invaded pattern is a fractal characteristic of percolation. However,
below a typical fingerwidth, $\bar{w}$, the patterns are compact. As θ approaches a critical
angle θ_c from above, $\bar{w}$ is found to have a power-law divergence with a non-universal
exponent.

What happens below θ_c is governed by collective growth mechanisms and depends
on the degree of disorder which is present in the system and on the type of geometry
which defines the porous medium. The self-similar patterns are replaced either by self-

---

[1]   Permanent address: Institute of Physics, Polish Academy of Sciences, 02-668 Warsaw,
Poland.

185

affine fractals or by faceted growth. The former structure arises in media constructed by placing disks of random radii on the sites of a square or hexagonal lattice, except when the disorder was weak [8]. It is also observed experimentally in glass bead packs [5,6]. The faceted morphology is seen in computer and experimental models consisting of ducts of random width placed along the bonds of a square network [2,12]. In general, the sequence of morphologies as the degree of disorder decreases can be : self-similar, self-affine, and faceted, with one of the compact phases possibly being suppressed due to the nature of randomness.

Understanding of the nature of transitions between various growth patterns has been aided by studies of random-field Ising models [13,14] in an external magnetic field, which acts like the forcing pressure. In these models, a given spin orientation corresponds to presence of a given liquid. The random fields mimick the effect of randomly sized throats in a porous medium. The Ising spin analogs of invasion can easily be studied in $3\text{-}D$ and it is in the $3\text{-}D$ models that all of the three growth morphologies are observed [13].

The transitions between various growth morphologies are critical phenomena characterized by diverging coherence lengths [9,10]. The onset of interface motion at a given θ is also a critical transition in the self-similar and self-affine regimes. Increasing the pressure towards P_c produces larger and larger incremental advances of the interface. Their mean size diverges at P_c and they follow a power law distribution. The exponents describing these and other quantities can be related through scaling laws [10]. Only the fractal dimensions and a single additional exponent are needed to determine all other exponents.

Here, we focus on interface growth in the random disk models and begin by describing the growth rules used in our studies. In Section 3 we compare the collective growth processes which occur below θ_c to the independent growth characteristic of invasion percolation. Finally, in Section 4, we focus on the role of random wettability. Random wettability is common in rocks and it is usually due to their chemically diversified grain structure. This structure causes pieces of equilibrium fluid interfaces to intersect the medium at a contact angle which varies from grain to grain. Variations in θ act like an extra disorder imposed on the tortuous geometry of the medium. Our results suggest that this additional disorder may significantly affect the interface morphology, usually by extending the range of parameters in which self-similar growth takes place.

2. Growth model

Porous media in our simulations were modelled by assigning disks of randomly varying radii to sites on a $2\text{-}D$ lattice. The effective degree of disorder in the random-disk porous media depends on two factors. The first is the geometrical disorder in the pore space. It can be measured by specifying the ratio between the maximum and minimum widths of the throats between disks. The second factor affecting the growth morphology is θ.

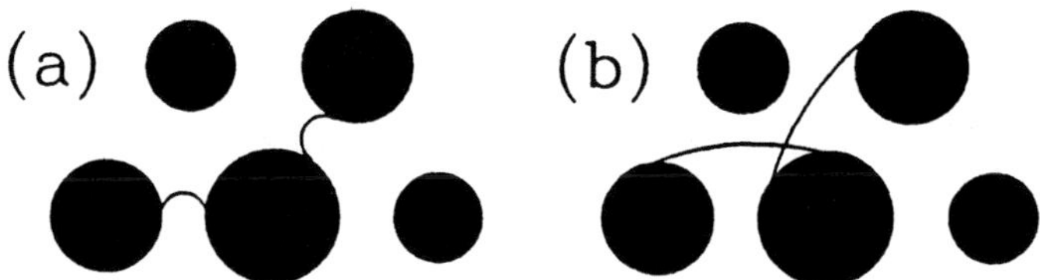

Figure 1. Arcs connecting successive beads along the interface for (a) $\theta = 180°$ and (b) $\theta = 30°$. The invading fluid is advancing from below, and the bond angle between three successive disks along the interface is 120°.

Figure 1 illustrates why θ controls effective interactions between neighboring pores. The invasion is driven by an external pressure. In quasi-static invasion this pressure drop appears everywhere along the interface. In 2D, the interface consists of circular arcs with curvature P/γ, where γ is the surface tension and the boundary of the invading fluid is convex for positive P's. These arcs must also intersect the disks at θ.

When the invading fluid is non-wetting, the narrow throats between disks are the hardest to penetrate because the required pressure is largest (Fig. 1(a)). Arcs are confined within their respective throats, and neighboring arcs advance independently by *bursts*. When the invading fluid is more wetting, arcs are sucked rapidly through the narrow throats and the pore spaces are the hardest regions to pass. Since arcs in neighboring throats share a common pore space, interactions between them are enhanced. In Fig. 1(b) each arc would be stable individually. However, when both are present in the pore they *overlap* and the kink in their combined interface is unstable. The probability of an overlap depends on the bond angle between successive disks along the interface. In the figure the bond angle is 120°. Overlaps preferentially remove sharp bends in the interface. Since overlaps become more likely as θ decreases, the effective degree of coupling increases and the relative degree of disorder decreases.

Growth was initiated from a flat interface or central ring. P was initially set to P_0, the pressure at which the first segment of the interface became unstable. This segment of the interface was advanced. Any resulting instabilities were advanced in turn until a new stable interface was reached. P was then increased until a segment of this interface became unstable, and the process was repeated. When advancing a segment causes the interface to surround a region of the defending fluid, the region becomes trapped and can not shrink further since the fluid is assumed to be incompressible.

3. Percolation versus depinning

Far above θ_c, the interface advances along the path of least resistance to single-arc instabilities (bursts). As P increases, more of the throats between disks are passable and P_c corresponds to the pressure at which the set of passable throats first spans the system, ie. percolates. As θ decreases, overlap instabilities become important and adjacent segments of the interface advance coherently. However, for $\theta > \theta_c$ the range of correlations is finite and comparable to the finger width.

Below θ_c the entire interface is coupled. This coherence is evidenced in two ways. The first is that the fingerwidth diverges. As P increases to P_c, the characteristic width of interface segments which advance after each increase in pressure diverges. Thus the entire interface advances coherently at P_c. The second indication of coherence is the emergence of a non-zero order parameter, much like the magnetization found below an equilibrium magnetic transition. One can define a surface-normal correlation function: $S(l) = \langle \hat{n}(l' + l) \cdot \hat{n}(l') \rangle$ where l is the arc length along the interface and $\hat{n}$ is the local surface normal. In the self-affine regime the long-distance limit of this correlation function is a constant, $S(\infty)$. The value of $S(\infty)$ rises continuously from zero at the critical transition and reaches unity in the faceted regime. An alternative order parameter is an effective macroscopic surface tension which also becomes non-zero below θ_c [9].

The difference in growth morphology above and below θ_c leads to different universality classes for the critical transition as P increases to P_c. This critical behavior is fairly well characterized in both regimes and scaling relations between critical exponents have been derived and verified numerically [9,10,13].

The compact patterns grown in the self-affine regime can be characterized by the roughness exponent, α, of the interface. The roughness exponent calculated in our simulations is 0.81 ± 0.04 for the disk model [9] and 0.67 ± 0.03 for the 3D Ising model [13]. The latter value is consistent with the prediction of scaling arguments: $\alpha = 2/3$. The result for the random disk model is not yet fully understood, but agrees well with experimental values for invasion of glass bead packs by a wetting fluid [5,6]. Indeed, our simulations suggest an explanation for the difference between reported values of the roughness exponent. We find that the size and frequency of overhangs on the interface depend on θ. Excluding overhangs tends to increase the apparent value of α [9,14,15]. Rubio et al. [5] analyzed the full interface and found $\alpha = 0.73 \pm 0.03$, while Horvath et al. [6] removed overhangs and found a larger value, $\alpha = 0.81$.

The observation of interfaces with $\alpha > 0.5$ was unexpected and sparked substantial interest. Most growth models map into the continuum growth equation of Kardar, Parisi and Zhang [16]. This yields $\alpha = 0.5$ in 2D and $\alpha < 0.5$ in 3D for uncorrelated noise. Larger values of α can only be produced within this model by assuming power-law correlated noise [17]. It is possible that the power law distribution of growth events at P_c acts like power-law correlated noise in producing a larger value of α.

4. Effects of random wettability

We now discuss what happens when the contact angle varies randomly. We consider a porous system in which the disk radii are uniformly distributed between $0.05a$ and $0.49a$ (system A in [8]) where a is the lattice constant of the hexagonal lattice on which the disks are placed. In this system, θ_c was approximately 49° when θ was uniform.

We now repeat our simulations for situations in which θ is allowed to be a random variable: each disk has its own specific contact angle. The distribution of the contact angles is not correlated to the distribution of the radii and it can be characterized by a mean value, $\bar{\theta}$, and a dispersion. We have considered the following three models of randomness in θ:

   [I:] A bimodal mixture of disks, half with contact angle $25°$ and half with a variable contact angle θ which is larger than $25°$. The properties of the patterns are then studied as function of θ.

   [II:] Contact angles distributed uniformly in a range from $-\sigma$ to $+\sigma$ around a specified mean value $\bar{\theta}$. We took $\sigma=5°$ and $20°$ and in the uniform case $\sigma=0$.

   [III:] Disks with a given contact angle θ (greater than θ_c in the uniform case) mixed with wetting disks with a contact angle of $25°$. The fraction x of wetting disks was varied, and the values of θ were $62°$, $70°$, and $90°$.

Fig. 2 shows the invasion patterns for three realizations of model I: $\theta=179°$, $60°$, and $50°$ from left to right in the figure, respectively. The corresponding values of $\bar{\theta}$ are $102°$, $42.5°$, and $37.5°$. The last two values are substantially below the critical angle of $49°$ for the uniform θ system, yet the growth patterns are fractal instead of compact. More surprisingly, one gets fractal patterns even when one mixes $48°$ grains with $25°$ grains.

As for uniform θ systems, we identify a critical angle θ_c at which the finger width $\bar{w}$ diverges. To measure $\bar{w}$, we make parallel slices through the pattern along lines of nearest-neighbor disks. The slices have segments which lie entirely within the invaded region, and $\bar{w}$ is determined by calculating the average length of these segments as explained in [8]. System sizes between 300 x 300 and 900 x 900 disks were studied.

Fig. 3 shows $\bar{w}$ in model I for θ down to $46°$ at which value we get $\bar{w}$ of order $67.6a$. We conclude that the critical $\bar{\theta}_c$ at which the finger width diverges (signalling a transition to self-affine growth) has moved down substantially relative to the homogeneous θ situation in the same geometry.

The results for the finger widths in model II are shown in Fig. 4. Here the contact angles are uniformly dispersed about $\bar{\theta}$. When the dispersion is zero, we get the homogenous θ result. Increasing the dispersion sharply decreases $\bar{\theta}$ and thus θ_c. The lowest $\bar{\theta}$ we have considered in this model is $30°$ and the largest dispersion is $\pm 20°$. In this case almost all of the contact angles are below the θ_c of the uniform system but the pattern is fractal and the finger width is close to $86.5a$. It appears that $\bar{\theta}_c$ is close to zero for this degree of dispersion.

Similar effects are seen in the model III systems, as shown in Fig.5. Mixing non-wetting grains with wetting ones does not result in any substantial increase in $\bar{w}$ until one reaches x in the range of 80-90%. The critical fraction at which the finger width diverges is difficult to determine with the system sizes we could study, but it is close to 100%. The value of $\bar{\theta}_c$ is thus close to the contact angle of the wetting disks ($25°$).

We conclude that randomness in θ shifts $\bar{\theta}_c$ to smaller values, and may even entirely supress the transition to compact growth. Note that even if all of the beads have random θ's that are less than θ_c for the homogenous system, the growth dynamics may be still determined by the physics of invasion percolation. The reason for the

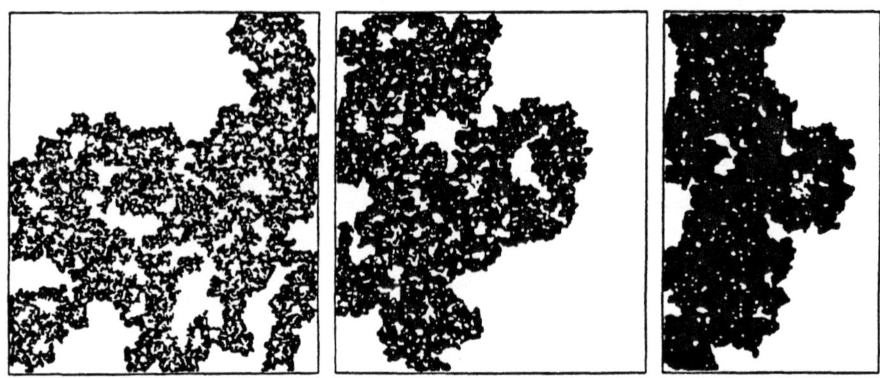

Figure 2. The growth patterns obtained for model I of random wettability. In this model one has 50% of disks with contact angle of 25° and 50% of disks with another value of θ. The left, middle, and right figures are for θ of 179°, 60°, and 50° respectively. The system size is 300 x 300.

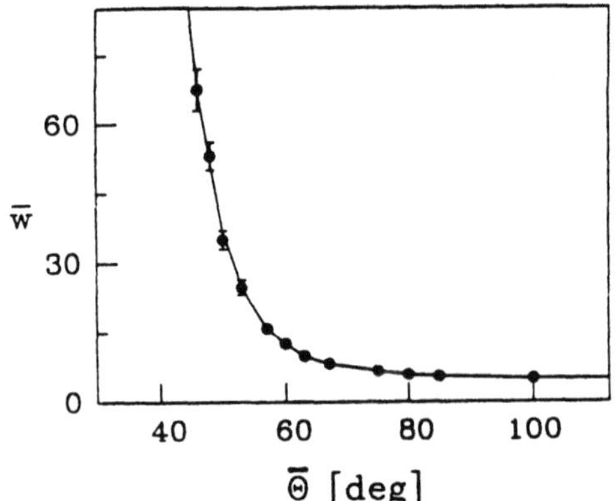

Figure 3. The finger width as a function of $\bar{\theta}$ in model I. The finger width is in units of the lattice constant a. The line is a guide to the eye.

enhancement of percolative behavior can be explained by studying properties of individual meniscii. Equations that enable one to determine the geometry of the meniscus at a given pressure are given in ref. [8]. It turns out that the stable arcs connecting a wetting disk to an non-unwetting one are situated essentially diagonally across the throat between disks. They are thus well screened from interactions with neighboring meniscii and occurrence of a cooperative overlap is unlikely. In this sense arcs in random wettability are similar to arcs in uniform θ systems at large contact angles. It is

190

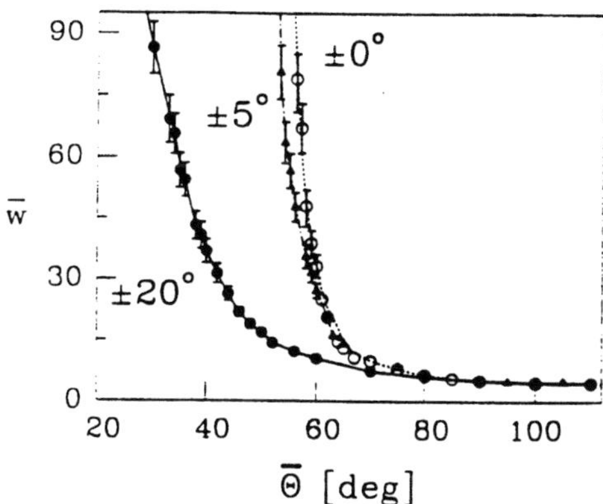

Figure 4. The finger width for model II as a function of the average contact angle in the system. The numbers inside the figure indicate the dispersion in θ

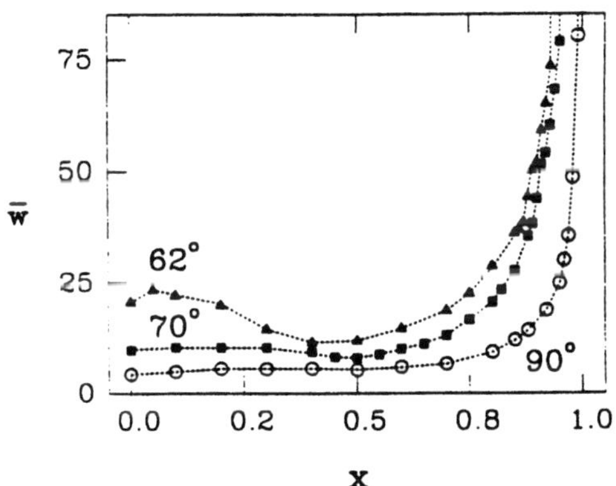

Figure 5. The finger width for model III as a function of the percentage of disks with $\theta=25°$. The other disks have the contact angle indicated in the figure.

only for low contact angles on both disks that the arcs can vacate the throat and allow for an overlap.

In summary, we have illustrated the rich variety of behavior which is found when disorder influences interface growth in porous media. The growth can be compact when the disorder is small, otherwise it is fractal. The physics of growth at finite capillary numbers remains to be elucidated.

191

Acknowlegdements

Support from the National Science Foundation through Grant no. DMR-9110004, from the Donors of the Petroleum Research Foundation, administered by the American Chemical Society, and from the Polish KBN is gratefully acknowledged.

REFERENCES

[1] See eg., J. Feder, *Fractals* Plenum Press, New York, 1988; P.-Z. Wong, *Phys. Today* 41 No 12, 24 1988; G. M. Homsy, *Annu. Rev. Fluid Mech.* 19 271 1987; D. Bensimon, L. P. Kadanoff, S. Liang, B. I. Shraiman, and C. Tang, *Rev. Mod. Phys.* 58 977 1986.

[2] R. Lenormand and C. Zarcone, *Phys. Rev. Lett.* 54 2226 1985; R. Lenormand, C. Zarcone, and A. Zorr, *J. Fluid. Mech.* 135 337 1985; R. Lenormand, *J. Phys.: Condens. Matter* 2 SA79 1990.

[3] R. Lenormand and S. Bories, *C. R. Acad. Sci. Ser. B*, 291 279 1980. R. Chandler, J. Koplik, K. Lerman and J. F. Willemsen, *J. Fluid Mech.* 119 249 1982.

[4] L. Patterson, *Phys. Rev. Lett.* 52 1621 1984; K. Maloy, J. Feder, and T. Jossang, *Phys. Rev. Lett.* 55 2688 1985.

[5] M. A. Rubio, C. Edwards, A. Dougherty and J. P. Gollub, *Phys. Rev. Lett.* 63 1685 1989; *ibid.* 65 1389 1990.

[6] V. K. Horváth, F. Family, and T. Vicsek, *Phys. Rev. Lett.* 65 1388 1990.

[7] M. O. Robbins, P. M. Chaikin, and H. M. Lindsay, *Phys. Rev. Lett.* 57 1718 1986.

[8] M. Cieplak and M. O. Robbins, *Phys. Rev. Lett.* 60 2042 1988; *Phys. Rev. B* 41 11508 1990.

[9] N. Martys, M. Cieplak, and M. O. Robbins, *Phys. Rev. Lett.* 66 1058 1991.

[10] N. Martys, M. O. Robbins, and M. Cieplak, *Phys. Rev. B* 44 12294 1991.

[11] J. P. Stokes, A. P. Kushnick and M. O. Robbins, *Phys. Rev. Lett.* 60 1386 1988.

[12] B. Koiller, H. Ji and M. O. Robbins, *Phys. Rev. B* 45 7762 1992.

[13] H. Ji and M. O. Robbins, *Phys. Rev. A* 44 2538 1991, and to be published.

[14] B. Koiller, H. Ji, and M. O. Robbins, *Phys. Rev. B* in press.

[15] M. O. Robbins, M. Cieplak, H. Ji, B. Koiller, and N. Martys, in *Proc. of the NATO ARW on Growth Patterns in Physical Sciences and Biology* Granada, Spain, 7-11 October, 1991. L. Sander and P. Meakin eds., Plenum Press, New York 1992.

[16] M. Kardar, G. Parisi, and Y. Zhang, *Phys. Rev. Lett.* 64 543 1990.

[17] E. Medina, T. Hwa, M. Kardar and Y.-C. Zhang, *Phys. Rev. A39* 3053 1989.

Surface Disordering:
Growth, Roughening and
Phase Transitions

IMBIBITION IN POROUS MEDIA: EXPERIMENT AND THEORY

Albert-László Barabási,[1] Sergey V. Buldyrev,[1] Shlomo Havlin,[1,2] Greg Huber,[1] H. Eugene Stanley[1] and Tamás Vicsek[3]

[1]*Center for Polymer Studies & Department of Physics, Boston University, Boston, MA 02215*
[2]*Department of Physics, Bar-Ilan University, Ramat Gan, Israel*
[3]*Department of Atomic Physics, Eötvös University, Budapest, H-1445, Hungary*

Abstract. We review our recent imbibition experiments in both $1 + 1$ and $2 + 1$ dimensions—using simple materials as the random media and various aqueous suspensions as wetting fluids. We measure the width $w(\ell, t)$ of the resulting interface and find it scales with length ℓ as $w(\ell, \infty) \sim l^\alpha$ with $\alpha = 0.63 \pm 0.04$. We develop a new imbibition model that describes quantitatively our experiments as well as a wide range of interface-roughening phenomena. For $d = 1 + 1$, the model can be mapped to directed percolation; for $d = 2 + 1$, it corresponds to a new *anisotropic* surface percolation problem. Our model leads to exponents $\alpha = 0.633 \pm 0.001$ for a pinned (static) interface and $\alpha_{\text{dyn}} = 0.70 \pm 0.05$ for a moving (dynamic) interface in $d = 1 + 1$; we find $\alpha \simeq \alpha_{\text{dyn}} \simeq 0.50 \pm 0.05$ in $d = 2 + 1$. Also, we find $\beta \approx \alpha$, where β, the dynamical exponent, is defined through the relation $w \sim t^\beta$.

1. Introduction

The growth of rough interfaces in random media is a topic of current interdisciplinary interest [1-4]. For the most part, two types of models have been applied to interface-roughening phenomena: (a) nonlinear Langevin equations—principally the KPZ equation [5,6]—resulting in self-affine interfaces, and (b) spreading and invasion percolation models [7], generating self-similar interfaces (Fig. 1).

193

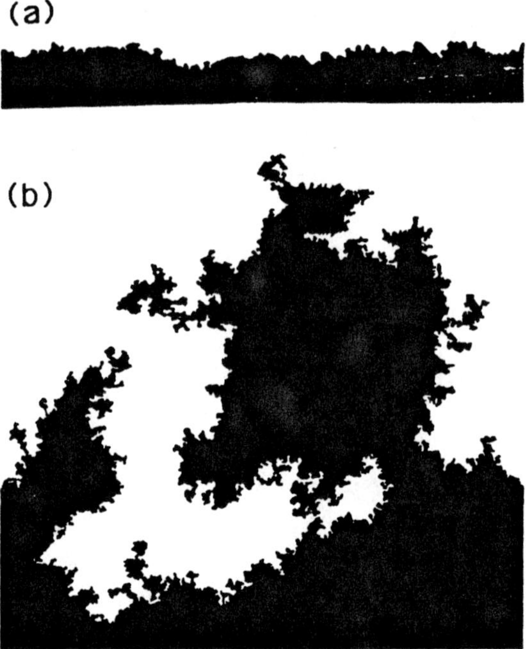

Figure 1. (a) A typical example of the KPZ-type interface: Eden growth in a strip geometry. Note that the overhangs are present but are very small in comparison with the overall width of the interface. (b) An example of a self-similar interface for critical spreading percolation in a strip geometry with periodic boundary conditions.

The self-affine interface (a) can be characterized by the rms surface width

$$w(\ell, t) \equiv \langle (h(x,t) - < h(x,t) >)^2 \rangle^{1/2}. \tag{1}$$

Here $h(x,t)$ is the surface height at time t, and the angular brackets denote the average over x belonging to an interval of size ℓ.

An alternative and equivalent quantity is the height-height correlation function $c(\ell, t)$. The scaling exponents obtained from w and c are believed to be identical, so we use them interchangeably. Analysis of the KPZ equation implies the scaling law [4]

$$w(\ell, t) \sim c(\ell, t) \sim l^\alpha f\left(\frac{t}{l^{\alpha/\beta}}\right), \tag{2a}$$

where

$$f(u) \sim \begin{cases} u^\beta & u \ll 1 \\ \text{const} & u \gg 1. \end{cases} \tag{2b}$$

The roughness exponent $\alpha = 1/2$ and the dynamical exponent $\beta = 1/3$ for $d = 1 + 1$ [5,6]. Numerical studies of the KPZ equation give $\alpha \simeq 0.4$ in $d = 2 + 1$ [8-10].

Table 1: Our Simulation Results for the Imbibition Model

$d = 1 + 1$	$d = 2 + 1$

PINNED INTERFACE

$\alpha = 0.63 \pm 0.01$	$\alpha = 0.49 \pm 0.05$

MOVING INTERFACE

$\alpha = 0.70 \pm 0.05$	$\alpha = 0.5 \pm 0.05$
$\beta = 0.70 \pm 0.05$	$\beta = 0.5 \pm 0.05$

FRACTAL DUST BELOW p_c

$p_c = 0.470 \pm 0.005$	$p_c = 0.74 \pm 0.01$
$d_f = 0.55 \pm 0.05$	$d_f = 1.25 \pm 0.01$
$\xi_\parallel \sim \lvert p - p_c \rvert^{-1.73 \pm 0.05}$	$\xi_\parallel \sim \lvert p - p_c \rvert^{-1.1 \pm 0.2}$

AVALANCHE DISTRIBUTION ABOVE p_c

$\tau_a = 1.245 \pm 0.05$	$\tau_a = 1.46 \pm 0.1$
$\xi_\parallel \sim \lvert p - p_c \rvert^{-1.73 \pm 0.02}$	$\xi_\parallel \sim \lvert p - p_c \rvert^{-1.06 \pm 0.1}$
$\xi_\perp \sim \lvert p - p_c \rvert^{-1.1 \pm 0.02}$	$\xi_\perp \sim \lvert p - p_c \rvert^{-0.47 \pm 0.1}$

On the other hand, approach (b) (percolation-type models) produces self-similar (fractal) interfaces [7], for which $w(\ell, \infty) \sim \ell^\alpha$ with $\alpha = 1$ and $\beta = 1$ (Fig. 1b). For many phenomena—from bacterial growth [11] to viscous flows [12,13] to the tearing [14] and burning [15] of paper—self-affine surfaces are found with anomalous exponents α and β significantly larger than the KPZ values but less than 1.

For the past two years, we have been addressing the following question: Do these experiments, in fact, represent a crossover from self-affine to self-similar behavior (Fig. 1), or is there a new universality class of growth models that produces self-affine interfaces with an anomalous $\alpha > 1/2$?

As a step in answering this question, we designed some imbibition experiments using simple materials as the random media and various aqueous suspensions as wetting fluids. We find for the case $d = 1 + 1$ a roughness exponent $\alpha \simeq 0.63$ (after the surface has stopped propagating), and for $d = 2 + 1$ a corresponding value $\alpha \simeq 0.5$. We develop a model of spreading percolation with anisotropy in the growth direction which belongs to a new universality class characterized by a self-affine interface with the roughness exponents given in Table 1 [16-18]. The $(1+1)$-dimensional model is mapped to directed percolation and the $(2+1)$-dimensional model can be mapped to a novel type of surface percolation. Alternative explanations for the anomalous roughening exponents can be found in Refs. 19-25.

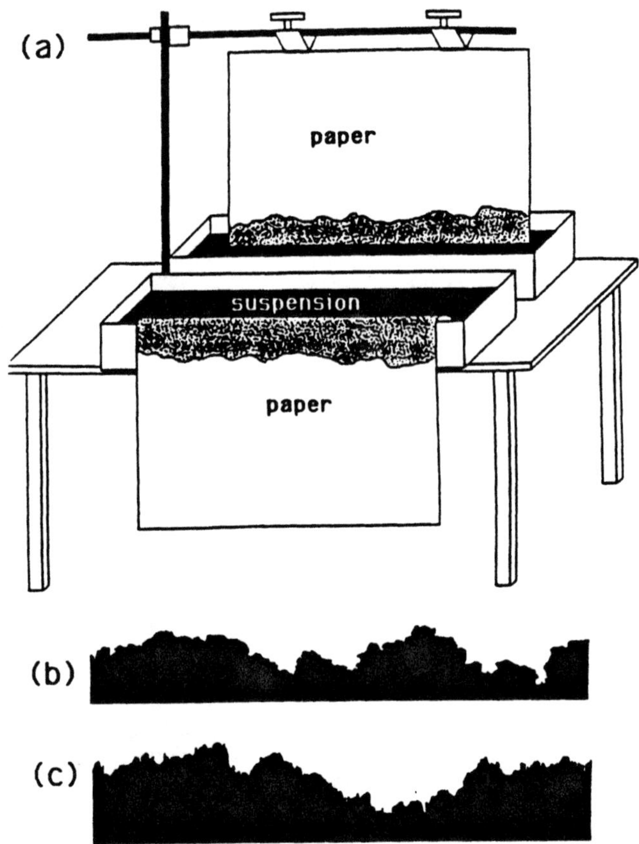

Figure 2. (a) Schematic illustration of the experimental setup. Aqueous suspensions of coffee, ink and various food colorings were used. Parameters such as type of paper, temperature, humidity, direction of growth and concentration of coffee were varied systematically. These changes affect the area, the speed of wetting, and the global width of the rough surface, but they do not affect the scaling properties of the surface. (b) Digitized *experimental* interface, using an Apple Scanner^{TM} with resolution 300 pixels per inch; the horizontal size of the paper was 20 cm. The function $h(x, t \to \infty)$ was obtained as the highest dark pixel in column x. (c) Typical result of the *model* with width $L = 400$ and $p \simeq p_c = 0.47$.

2. Dimension $1 + 1$: Experiment

Our imbibition experiment is performed by dipping a 20 cm strip of paper ("Industrial Roll Towel," Scott Co.) into a basin filled with suspensions of ink or coffee and clipping it to a ring stand, or allowing the paper to hang down over the basin edge (Fig. 2a). The suspension is absorbed into the paper, creating a rough interface between wet and dry regions. We allow the interface to rise up (or move down) the paper until it stops, so that no change in either the height or shape of the interface is observed. The stopping can be attributed to the evaporation of the fluid in the wet regions. After drying, we digitize this rough interface (Fig. 2b). We then calculate the width on different length scales ℓ, averaging over 10 different samples hanging up and also hanging down. Presenting the data in a log-log plot, we find a scaling of the form $w(\ell, \infty) \sim \ell^\alpha$ with

$$\alpha = 0.63 \pm 0.04 \qquad \text{[EXPERIMENT]}. \tag{3a}$$

No difference in α is observed between fronts propagating up and fronts propagating down—i.e., we find that *gravity plays a negligible role* relative to other driving forces.

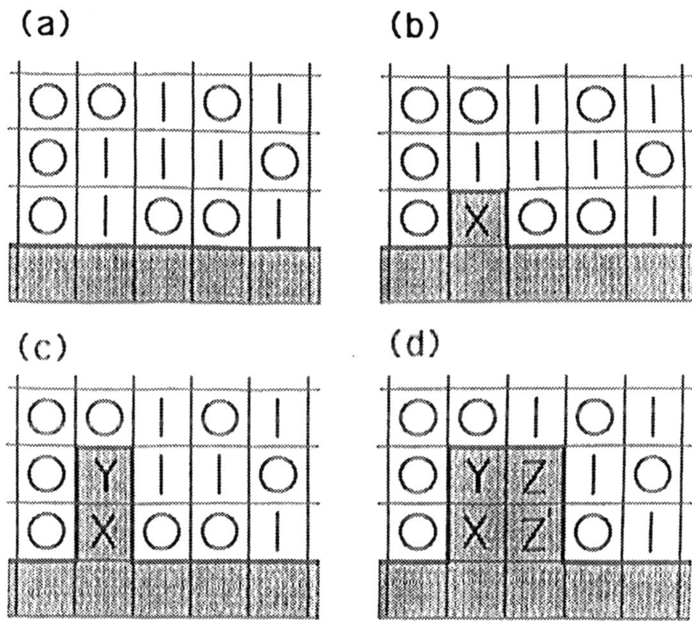

Figure 3. Explanation of our model of imbibition with erosion of overhangs. Wet cells are indicated by shaded cells. Dry cells are randomly blocked with probability p (indicated by the symbol "$\bigcirc$") or unblocked with probability $1 - p$ (indicated by a vertical line). The interface between wet and dry cells is shown as a heavy line. (a) $t = 0$, (b) $t = 1$, (c) $t = 2$ and (d) $t = 3$.

3. Dimension $1 + 1$: Model

We define our model as follows: on a square lattice of edge L (with periodic boundary conditions) we block a fraction p of the cells to correspond to the inhomogeneous nature

197

of the paper. At $t = 0$, the "interface" is the bold horizontal line shown in Fig. 3a. At $t = 1$ we randomly choose a cell (labeled X in Fig. 3b) which is one of the unblocked dry cells that are nearest neighbors to the interface. We wet cell X and *all cells below it in the same column*. This process is then iterated. For example, Fig. 3c shows that at $t = 2$ we choose cell Y, a second unblocked cell, to wet; Fig. 3d shows that at $t = 3$ we wet cell Z *and also cell Z' below it*.

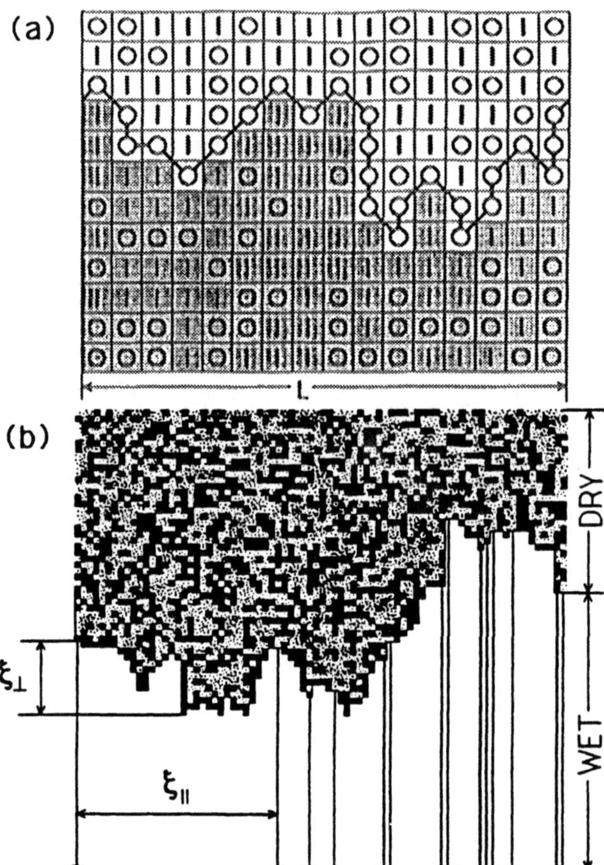

Figure 4. (a) The bold line is a *spanning* path formed by the connected nearest-neighbor and next-nearest neighbor blocked cells that pin the interface. Note that the various *non-spanning* clusters of blocked cells are insufficient to pin the interface. (b) A larger system still evolving near the critical point ($p = 0.44 < p_c$). Black squares are dry blocked cells, shaded squares are unblocked dry cells. Wet cells are not shown. Large portions of the interface are blocked by directed paths of blocked cells, similar to those shown in (a). Correlation lengths $\xi_\parallel$ and $\xi_\perp$ are the typical sizes of blocked regions. Columns which still have unblocked cells on the interface form a fractal dust, shown by the rectangles at the bottom of each "live column."

198

We find that for p below a critical threshold $p_c = p_c(L)$ [16-18], the interface propagates without stopping, while for p above p_c, the interface does not propagate. Figure 2c displays the interface of the model at criticality after it has stopped propagating. Averaging over 10^4 samples for systems with $L = 2^{15}$ we find that

$$\alpha = 0.63 \pm 0.02 \qquad \text{[SIMULATION]}, \qquad (3b)$$

in excellent agreement with the experimental value of Eq. (3a).

Figure 4a displays a typical model interface, after the growth has stopped; this occurs just when the surface meets a spanning path of blocked cells connected through nearest and next-nearest neighbors. This path cannot have overhangs due to the definition of the model, and thus it is equivalent to a spanning path on the square lattice which goes from left to right, may turn up or down, but can never turn left. Such a directed path is, in fact, a path on a directed percolation cluster [26,27].

When the probability of blocked cells p is close to p_c, the growth is halted in many places by the paths of a directed percolation cluster (Fig. 4b). Each path can be characterized by two correlation lengths $\xi_\perp$ and $\xi_\parallel$. When the path is spanning, i.e., when the growth is stopped completely, $\xi_\parallel$ is equal to the system size L, and $\xi_\perp$ is proportional to the width of the interface w. It is known from the theory of directed percolation [26,27] that the correlation lengths diverge in the vicinity of p_c,

$$\xi_\perp \sim |p - p_c|^{-\nu_\perp}, \qquad (4a)$$

and

$$\xi_\parallel \sim |p - p_c|^{-\nu_\parallel}, \qquad (4b)$$

where $\nu_\parallel \simeq 1.733$, $\nu_\perp \simeq 1.097$ [27]. Thus

$$w \sim \xi_\perp \sim L^{\nu_\perp/\nu_\parallel} \equiv L^\alpha, \qquad (4c)$$

where

$$\alpha = \nu_\perp/\nu_\parallel \simeq 0.633 \pm 0.001 \qquad \text{[THEORY]}. \qquad (5)$$

This imbibition model describes the final state of a wetting front completely pinned by inhomogeneities in the paper. Our theoretical value of α [Eq. (5)] is in excellent agreement with both our simulations [Eq. (3a)] and our experiments [Eq. (3b)].

4. Dimension $2 + 1$: Experiments and Model

We also performed *experiments* for the case $d = 2 + 1$ [18]. Preliminary results suggest that the roughness exponent is $\alpha \approx 0.5$. We also generalized our *model* to the case $d = 2 + 1$ [17,18]. We find that the probability of blocked cells equals a critical value $p = p_c \simeq 0.74$ and that $\alpha = 0.50 \pm 0.05$. Note that the analogy with directed percolation is transparent only for $d = 1 + 1$; for $d = 2 + 1$ our model corresponds to a new problem in the directed percolation of surfaces. (See Ref. 28 for an isotropic model of percolating hypersurfaces.) We find the correlation-length and roughness exponents for this $(2 + 1)$-dimensional surface to be

$$\nu_\| = 1.06 \pm 0.1, \quad \nu_\perp = 0.47 \pm 0.1, \quad \alpha = \nu_\perp/\nu_\| = 0.44 \pm 0.1 \qquad \text{[SIMULATION]}. \qquad (6)$$

5. Dynamics

A natural question is what are the *dynamics* of imbibition. To answer this, we study the dynamical behavior of the model below and above p_c. Figure 4b shows a snapshot of the wetting front as it continues to propagate in the $(1+1)$-dimensional media when $p < p_c$. Large sections of the interface are already pinned and the growth is occurring only in columns that contain unblocked cells on the wet boundary.

The effective roughness exponent of this *moving* interface is slightly larger: we find that in $d = 1 + 1$,

$$\alpha_{\mathrm{dyn}} = 0.70 \pm 0.05 \qquad \text{[SIMULATION]}. \qquad (7)$$

This might be due to the unblocked regions which generally have larger slopes than the blocked regions. The growth is now mostly a fast erosion of steep slopes, propagating horizontally with constant speed. This observation implies that the dynamical exponent $z_{\mathrm{dyn}} \equiv \alpha_{\mathrm{dyn}}/\beta$ has a value close to 1. Thus $\beta \simeq \alpha_{\mathrm{dyn}}$, in good agreement with our numerical results [16-18]. This large value of α_{dyn} may explain the large values $0.7 - 0.8$ seen in dynamical experiments [11-15].

6. Fractal Dust and Avalanches

The projection of the live columns forms a *fractal dust* (see Figs. 4b and 5). It is known that in isotropic spreading percolation, live cells also form a fractal dust with a fractal dimension approximately equal to the fractal dimension of red bonds [3,29,30]. In order to check this result in the anisotropic case, we have studied the density-density correlation function $G(r)$ of the live columns projected onto the $(d - 1)$-dimensional hyperplane perpendicular to the growth direction. We found that $G(r)$ has a scaling form

$$G(r) \sim r^{d_f - d + 1} f\left(\frac{r}{\xi_\|}\right), \qquad (8a)$$

where d_f is the fractal dimension of the dust and $f(x)$ is a scaling function such that

$$f(u) \sim \begin{cases} \text{const} & u \ll 1 \\ u^{d - d_f - 1} & u \gg 1. \end{cases} \qquad (8b)$$

In $d = 1 + 1$, we find $d_f = 0.55 \pm 0.03$ and $\xi_\| \sim |p - p_c|^{-1.73 \pm 0.05}$, which are in good agreement with the longitudinal fractal dimension of the red bonds $d_\|^{\mathrm{red}} = 1/\nu_\|$ and

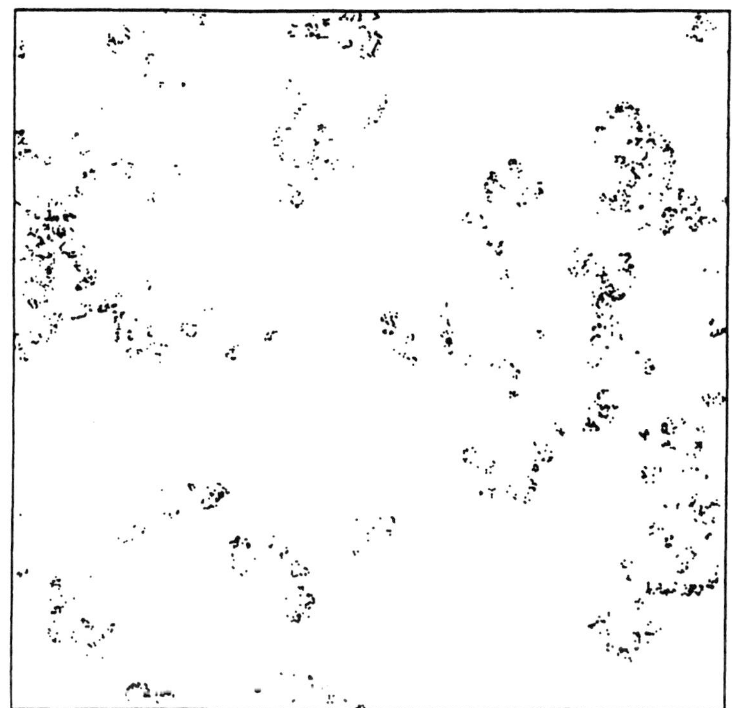

Figure 5. Fractal dust of live columns in dimension $2 + 1$, $p = 0.739$, $L = 512$; view from top.

the correlation exponent $\nu_\parallel$ in the directed percolation problem [30]. For the results of the $(2 + 1)$-dimensional model, see Table 1.

Above p_c, the growth is stopped by the spanning path of a directed percolation cluster in $d = 1 + 1$, or by a directed surface in $d = 2 + 1$. However, we can modify our model and assume that even when the growth is completely stopped, the blocked cells on the interface may still erode—but at an infinitesimal rate. With this assumption, we can remove blocked cells at random when the interface is completely stopped. Each removal will produce an avalanche of growth which eventually will die out when the front reaches another directed spanning path, or directed surface of blocked cells (see Fig. 6).

We also study the distribution of avalanche sizes $P(V)$, and find [30]

$$P(V) \sim V^{-\tau_s} F\left(\frac{V}{V_0}\right), \qquad (9a)$$

where V is the number of sites removed in an avalanche, and $V_0 \sim \xi_\parallel^{d-1}\xi_\perp$ is the characteristic volume. The probability $P(V)$ is estimated to be the ratio of the number of avalanches of size V to the total number of avalanches. In $d = 1 + 1$, the maximum

201

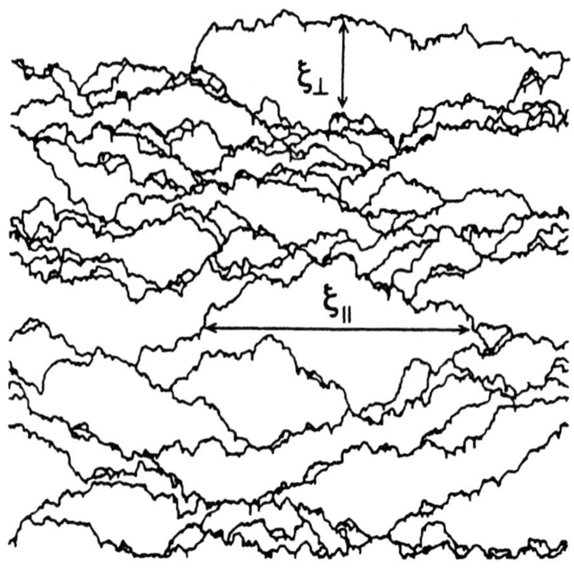

Figure 6. Successive series of pinned interfaces, showing the boundaries of avalanches, produced by removing a randomly-chosen blocked cell from the previously-pinned interface. $L = 400$, $p = 0.5 > p_c$. Correlation lengthes $\xi_\parallel$ and $\xi_\perp$ are the typical sizes of the avalanches.

linear extent of the avalanches (Fig. 6), in the longitudinal and transverse directions, is found to scale with exponents

$$\nu_\parallel^{\text{aval}} = 1.73 \pm 0.02, \quad \nu_\perp^{\text{aval}} = 1.10 \pm 0.02 \qquad \text{[SIMULATION]}, \tag{9b}$$

in excellent agreement with the correlation-length exponents of directed percolation. Moreover, we find

$$\tau_a = 1.245 \pm 0.02 \qquad \text{[SIMULATION]}. \tag{9c}$$

The corresponding simulation results for $d = 2 + 1$ are presented in Table 1.

7. Average Front Height

Another quantity of dynamical significance is the propagation of the average front height $\bar{h}(t)$. Simulations based on our model in $d = 1 + 1$ indicate that $\bar{h}(t) \sim t^\delta$, with

$$\delta \simeq 0.70 \pm 0.05 \qquad \text{[SIMULATION]}. \tag{10}$$

A simple scaling argument can be used to calculate $\bar{h}(t)$ near p_c. The height at a live cell jumps by distances proportional to $\xi_\perp$, since this is the typical size of an eroded column of cells (Fig. 4b). On the other hand, erosion only occurs at weak cells which are separated by the length $\xi_\parallel$ (Fig. 4b). Thus the height is inversely proportional to $\xi_\parallel$, and we predict that $\bar{h}(t) \sim \left(\frac{\xi_\perp}{\xi_\parallel}\right) t$. Since $\xi_\parallel \sim t^{1/z}$ ($z \equiv \alpha/\beta$) and $\xi_\perp \sim \xi_\parallel^{\nu_\perp/\nu_\parallel} \sim t^{\nu_\perp/z\nu_\parallel}$, we obtain $\delta = 1 + (\nu_\perp - \nu_\parallel)/z\nu_\parallel = 1 + (\alpha - 1)/z$. We assume that this relation holds for the dynamical exponents; since our simulations yield $z \simeq 1$, we then predict that $\delta \simeq \alpha_{\mathrm{dyn}}$—in agreement with our simulations. This result in our model is obtained using the assumption that the driving force is constant . In real imbibition experiments the driving force decreases with time, therefore may lead to a different δ.

8. Discussion

We measured the roughness exponent α in paper wetting [16]. We also developed a model—based on directed percolation—which is in excellent agreement with the experiment. The importance of quenched noise and pinning as a mechanism of surface roughening was suggested by several authors [7,31-34]. Directed percolation as a mechanism for interface pinning in $1+1$ dimensions has been independently proposed by Tang and Leschhorn [34]. Results on dynamical experiments are also presented in this conference [35]. Other experiments [12-14] produce rougher interfaces than displayed by our experiments and model; this may correspond to the moving phase of our model for $p < p_c$, where $\alpha_{\mathrm{dyn}} \approx 0.75$ was observed. In $d = 2 + 1$, there are no previous experiments—except for a study of certain mountainous regions [36], which gives $\alpha = 0.57$, a value 10% larger than our model predicts.

Our interest in the experimental aspects of this project benefited from seminal work of D. Wolf. We wish to thank K. Shaknovich for technical assistance, M. Araujo, M. Gyure, J. Kertesz, M. O. Robbins and S. Schwarzer for helpful discussions, and the Hungary-USA exchange program of the Hungarian Academy of Sciences and the NSF for financial support.

REFERENCES

[1] J. Krug and H. Spohn in *Solids Far From Equilibrium: Growth, Morphology and Defects* edited by C. Godréche, Cambridge Univ. Press, Cambridge, England 1991.

[2] D. E. Wolf in *Kinetics of Ordering and Growth at Surfaces* edited by M. Lagally, Plenum, NY 1990.

[3] T. Vicsek *Fractal Growth Phenomena (2nd edition)* World Scientific, Singapore 1992.

[4] *Dynamics of Fractal Surfaces* eds. F. Family and T. Vicsek, World Scientific, Singapore 1991.

[5] M. Kardar, G. Parisi and Y.-C. Zhang, *Phys. Rev. Lett.* **56** 889 1986.

[6] E. Medina, T. Hwa, M. Kardar and Y.-C. Zhang, *Phys. Rev. A* **39** 3053 1989.

[7] N. Martys, M. Cieplak and M. O. Robbins, *Phys. Rev. Lett.* **66** 1058 1991.

[8] J. M. Kim and J. M. Kosterlitz, *Phys. Rev. Lett.* **62** 2289 1989.

[9] B. M. Forrest and L.-H. Tang, *Phys. Rev. Lett.* **64** 1405 1990.

[10] K. Moser, J. Kertész and D. E. Wolf, *Physica A* **178** 215 1991.

[11] T. Vicsek, M. Cserzö and V. K. Horváth, *Physica A* **167** 315 1990.

[12] M. A. Rubio, C. A. Edwards, A. Dougherty and J. P. Gollub, *Phys. Rev. Lett.* **63** 1685 1990.

[13] V. K. Horváth, F. Family and T. Vicsek, *J. Phys. A* **24** L25 1991.

[14] J. Kertész, V. K. Horváth and F. Weber, *(this volume)* 1992.

[15] J. Zhang, Y.-C. Zhang, P. Alstrøm and M. T. Levinsen, *(this volume)* 1992.

[16] S. V. Buldyrev, A.-L. Barabási, F. Caserta, S. Havlin, H. E. Stanley and T. Vicsek, *Phys. Rev. A* **45** R-8313 1992.

[17] S. Havlin, A.-L. Barabási, S. V. Buldyrev, C. K. Peng, M. Schwartz, H. E. Stanley and T. Vicsek in *Growth Patterns in Physical Sciences and Biology* eds. E. Louis, L. Sander and P. Meakin, Plenum, NY 1992.

[18] S. V. Buldyrev, G. Huber, A.-L. Barabási, S. Havlin, J. Kertész and H. E. Stanley, *(unpublished)* 1992.

[19] Y.-C. Zhang, *J. de Physique* **51** 2113 1990.

[20] J. Krug, *J. de Physique (1)* **1** 9 1991.

[21] T. Vicsek, *(this volume)* 1992.

[22] S. V. Buldyrev, S. Havlin, J. Kertész, H. E. Stanley and T. Vicsek, *Phys. Rev. A* **43** 7113 1991.

[23] S. Havlin, S. V. Buldyrev, H. E. Stanley and G. H. Weiss, *J. Phys. A* **24** L925 1991.

[24] V. K. Horváth, F. Family and T. Vicsek, *Phys. Rev. Lett.* **67** 3207 1991.

[25] C. K. Peng, S. Havlin, M. Schwartz and H. E. Stanley, *Phys. Rev. A* **44** 2239 1991.

[26] W. Kinzel in *Percolation Structures and Processes* eds. G. Deutscher, R. Zallen and J. Adler, A. Hilger, Bristol 1983.

[27] *Fractals and Disordered Systems* eds. A. Bunde and S. Havlin, Springer Verlag, Heidelberg 1991.

[28] J. Kertész and H. J. Herrmann, *J. Phys. A* **18** L1109 1985.

[29] S. Havlin and R. Nossal, *J. Phys. A* **17** L427 1984.

[30] G. Huber, S. V. Buldyrev, S. Havlin and H. E. Stanley, *(preprint)* 1992.

[31] D. A. Kessler, H. Levine and Y. Tu, *Phys. Rev. A* **43** 4551 1991.

[32] G. Parisi, *Europhys. Lett.* **17** 673 1992.

[33] M. H. Jensen and I. Procaccia, *(preprint)* 1991.

[34] L.-H. Tang and H. Leschhorn, *Phys. Rev. A* **45** R-8309 1992.

[35] F. Family, K. C. B. Chan and J. G. Amar, *(this volume)* 1992.

[36] M. Matsushita and S. Ouchi, *Physica D* **38** 246 1989.

Surface Disordering:
Growth, Roughening and
Phase Transitions

DYNAMICS OF INTERFACE ROUGHENING IN IMBIBITION

Fereydoon Family, K.C.B. Chan, and Jacques G. Amar
Department of Physics, Emory University, Atlanta, GA 30322

Abstract. The results of a simple imbibition experiment with paper as the porous medium and water as the wetting fluid are presented. The average height of the water-air interface $\bar{h}$ is found to scale with time as $\bar{h} \sim t^{\delta}$ with $\delta \simeq 0.7$ rather than linearly as in forced-flow experiments. The interface width $w(L,t)$ (r.m.s. fluctuation of the height) is found to scale as $w(t) \sim t^{\beta}$ with $\beta = 0.3 - 0.4$ and $w(\bar{h}) \sim \bar{h}^{\beta'}$ with $\beta' = 0.41 - 0.55$. The roughness exponent α has also been measured, and found to be in the range $0.62 - 0.78$, consistent with previous experiments.

1. Introduction

Recently, there has been a great deal of interest in the dynamics of growing surfaces and interfaces [1]. Much of this interest is based on the observation that surface fluctuations exhibit scaling behavior in both time and space. In particular, assuming an initially flat interface, the scaling of the interface width $w(L,t)$ on length scale L at time t, is expected to be of the form [2], $w(L,t) = L^{\alpha} f\left(t/L^{\alpha/\beta}\right)$, where $f(x) \sim x^{\beta}$ for $x \ll 1$, and $f(x) \to const$ for $x \gg 1$. Simulations of a wide variety of surface growth models [2-6] as well as experiments have been conducted and found to satisfy this scaling form.

Recent experiments [7,8] on the forced-flow of a wetting fluid through a porous medium in two dimensions (one-dimensional interface) have obtained exponents which are significantly higher ($\alpha \simeq 0.81$, $\beta \simeq 0.65$) than predicted by standard models [2-4] of nonlinear interface growth ($\alpha = 1/2, \beta = 1/3$). A possible explanation of this behavior has been obtained by introducing a power-law, rather than a Gaussian noise distribution in these models which gives rise to continuously varying exponents [9,10]. In fact, the existence of such a power-law noise distribution has recently been observed in experiments [11]. We also note that a quasi-static numerical simulation [12] also finds $\alpha \simeq 0.81$ in good agreement with forced flow experiments.

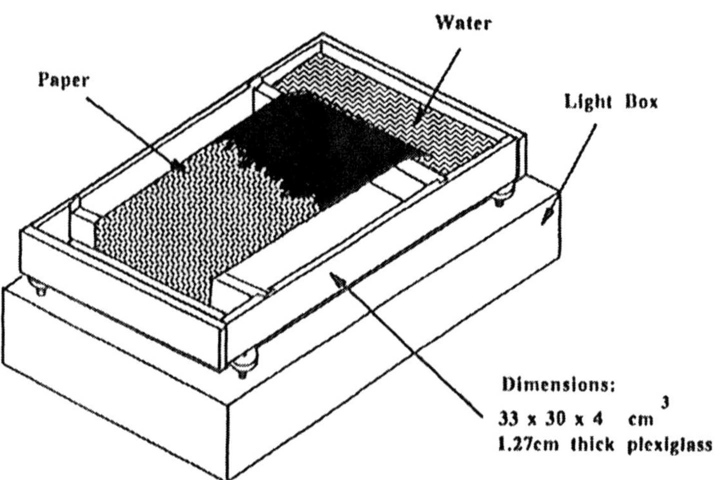

Figure 1: Experimental setup for imbibition experiment.

Many important processes of technological and environmental interest, such as seepage, ground water flow, and spreading of wetting fluids in porous media occur via imbibition rather than forced flow. Thus, it would be interesting to study the interfacial dynamics in this case. In the experiments to be described we rely on the natural capillary pressure (imbibition) to draw the wetting fluid through the medium. One question of particular interest, is whether the same scaling behavior as in the forced-flow experiments will be observed.

2. Experimental Setup

The experimental setup [13, 14] is shown in Fig. 1, with paper as the porous medium. As is shown in Fig. 1 the paper was suspended horizontally between two Plexiglass supports with one end of the paper in contact with a water reservoir. The water was then absorbed by the paper from the reservoir, so that the water-air interface moved horizontally over the distance (approximately 8-10 centimeters) between the supports. Depending on the type of paper used, the time for the interface to travel from one end to the other varied from 1/2 hour to 2 hours. A reservoir of water was kept underneath the paper in order to try to reduce the effects of evaporation.

The motion of the interface was recorded on videotape using a video camera. The data were then digitized (640 x 480 pixel resolution) and the interface was analysed using image processing software. Fig. 2 shows typical pictures of the interface at a sequence of different times. The size of the paper used was typically 20 cm x 26 cm (20 centimeter wide interface).

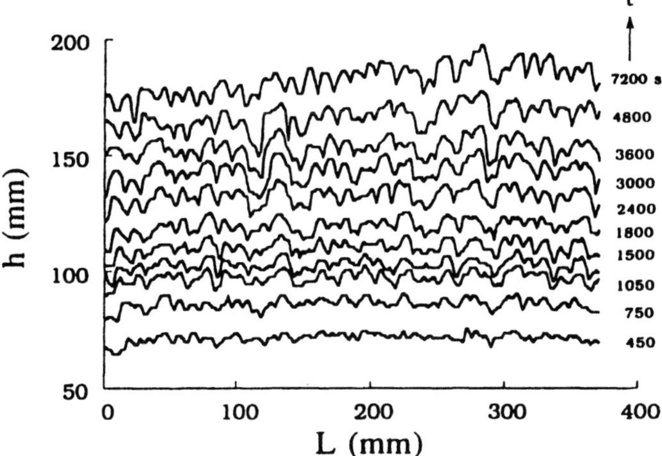

Figure 2: Picture of sequence of interfaces for newsprint paper at different times.

Using the digitized data at each time t the average height $\bar{h}$ across the interface was calculated, using as a reference $(t = 0, \bar{h} = 0)$ a fixed point near the edge of the first plexiglass wall. The interface width $w(L, t)$ at each time was also calculated from the rms fluctuation of the interface. At late times the interface width as a function of window size L less than the system size was also calculated, in order to determine the roughness exponent α.

Three types of paper were used: newsprint, Chinese painting paper, and Bounty brand paper towel. For all papers used, the interface developed a very small amount of overhangs. These were typically eliminated during the analysis of the interface width, and the highest value of h for each x determined. However, given the small amount of overhangs observed we expect this should not affect most of our results, except possibly at the latest times. For the case of the Chinese painting paper, it was found that at very late times strong overhangs were present, however this very-late time regime was not studied in our scaling analysis.

3. Results

Figures 3 and 4, respectively, show log-log plots of the average surface height $\bar{h}$, and width $w(L, t)$ as a function of time for the Chinese painting paper. Similar results for the other two types of paper were also obtained and the results are summarised in Table I. We note that unlike the case of driven fluid-flow [7-8,11], the average interface height scales as $\bar{h}(t) \sim t^{\delta}$ with $\delta \simeq 0.70 - 0.74$ rather than linearly. This result appears to be fairly robust and independent of paper type. However, there does appear to be a very slight crossover to a lower exponent at later times, possibly due to evaporation .

207

effects. For the exponent characterising the growth of the interface width with time, we find $\beta \simeq 0.3 - 0.4$, again significantly lower than in the forced flow experiments. In addition, the interface width scales as $w(\bar{h}) \sim \bar{h}^{\beta'}$ with $\beta' \simeq 0.4 - 0.5$. We note that β and β' are trivially related via the relation $\beta' = \beta\delta$.

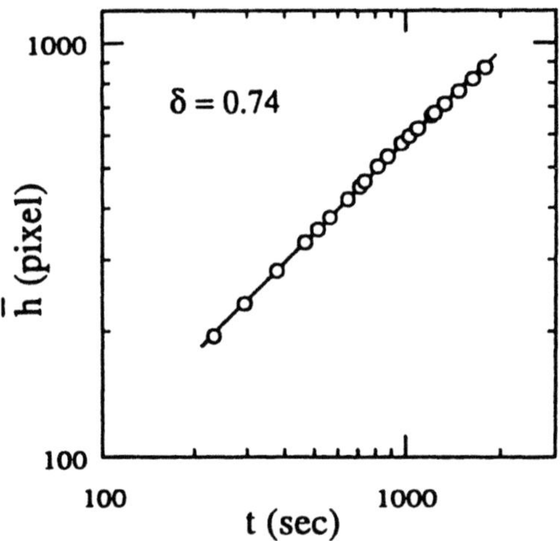

Figure 3: Log-log plot of average interface height $\bar{h}$ versus time t for Chinese painting paper. Slope is $\delta = 0.74$.

Figure 5 shows a typical plot of the saturation width as a function of window sizer. For all three types of paper, power-law scaling of the form $w(L, \infty) \sim L^{\alpha}$ with $\alpha = 0.62 - 0.78$ is observed, with α close to the values observed in previous forced-flow experiments. However, there appears to be a relatively rapid crossover to a different behavior at larger length scales, possibly due to the existence of a characteristic length scale at late times (see Fig. 2).

Table I. Summary of scaling results for various kinds of absorbing paper.

Paper	δ	α	β'	β
Chinese	0.74	0.76	0.52	0.38
Newsprint 1	0.72	0.62	0.45	0.33
Newsprint 2	0.74	0.69	0.41	0.29
Newsprint 3	0.73	0.75	0.55	0.40
Bounty	-	0.78	-	-

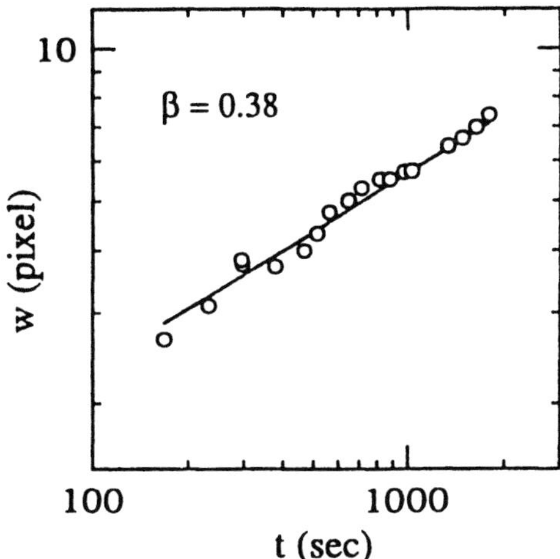

Figure 4: Log-log plot of interface width as function of time t for Chinese painting paper. Slope is $\beta = 0.38$.

4. Discussion

In previous experiments [7-8,11], the wetting fluid was forced to flow at a uniform rate (velocity) through the porous medium, so that the average height $\bar{h}$ scaled linearly with time t. One of the main results of our imbibition experiments is the existence of a nonlinear growth law for the average height ($\bar{h} \sim t^{0.7}$). As already mentioned, this result appears to be fairly robust and independent of the type of paper. Intuitively, the existence of a sublinear growth law for the interface height may be explained by the fact that the capillary pressure driving the interface is pushing larger amounts of fluid through the porous medium as time goes on. In fact a simple argument which neglects the pressure drop across the moving interface, but equates the pressure gradient within the wetting fluid ($\nabla P = P/h$) to the flow velocity using Darcy's law ($v = \kappa \nabla P$) gives (assuming power-law scaling) $\bar{h}(t) \sim t^{1/2}$ i.e. a value for δ which is smaller than observed [15]. One important question is to what extent the interface roughness (due to random variation in pore sizes) is responsible for the larger value observed for this exponent. We note that a number of additional effects such as the dependence of the interface contact angle on velocity may also be important in determining the growth behavior.

The scaling of the interface width as a function of time t or average height $\bar{h}$ is also of interest, although not at present understood. Since the velocity of the interface is not constant, we do not expect KPZ behavior to be observed nor does the scaling relation [5] $\alpha + \alpha/\beta = 2$ appear to hold. We also note that our results for the scaling behavior of the interface width do not agree with the Laplacian growth model previously studied by Krug and Meakin [16]. This is not surprising since although capillary effects are taken into account, the interface in this model is assumed to travel with a constant velocity. It is possible that a modified version of the Laplacian roughening model, in

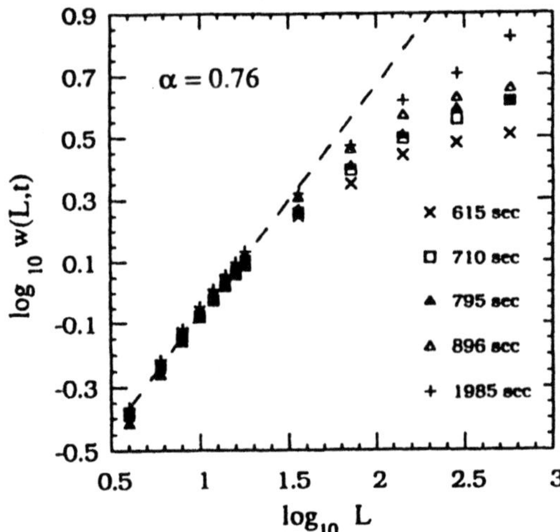

Figure 5: Log-log plot of saturation width as a function of window size L at 4 different times for the Chinese paper. Slope of fit is $\alpha = 0.76$.

which the existence of quenched randomness is taken into account, may be helpful in explaining the observed dynamics.

As already noted, on short length scales we see scaling of the interface width with system size similar to the forced flow experiments although there is a much more rapid crossover. At late times, it appears that overhangs become more important, and as already noted, for the Chinese paper, a regime that looks like the beginning of invasion percolation appears to be approached. This may be due to chemical dissolution of paper binding material which occurs at late time. We are currently looking at theoretical models to try to understand this late-time transition and behavior.

Recently, similar results for the exponent α were obtained in an imbibition experiment in a gravitational field which is reported elsewhere in this volume [17]. However, the dynamics of the interface was not studied and it is not clear how the gravitational field affects the dynamical scaling behavior. In addition, the theoretical models presented elsewhere in this volume, [18-19] for describing the evolution of an interface in a flow in porous media do not appear to agree with our dynamical imbibition results for the exponents β and δ.

Acknowledgements

This research was supported by the Office of Naval Research.

REFERENCES

[1] F. Family and T. Vicsek, eds., *Dynamics of Fractal Surfaces* World-Scientific, Singapore 1991.

[2] F. Family and T. Vicsek, *J. Phys. A: Math. Gen.* **18** L75 1985.

[3] R. Jullien and R. Botet, *J. Phys. A: Math. Gen.* **18** 2279 1985.

[4] M. Kardar, G. Parisi and Y.-C. Zhang, *Phys. Rev. Lett.* **56** 889 1986.

[5] P. Meakin, P. Ramanlal, L. M. Sander and R. C. Ball, *Phys. Rev. A* **34** 5091 1986.

[6] J. M. Kim and J. M. Kosterlitz, *Phys. Rev. Lett.* **62** 2289 1989.

[7] M. A. Rubio, C. A. Edwards, A. Dougherty and J. P. Gollub, *Phys. Rev. Lett.* **63** 1685 1989.

[8] V. K. Horváth, F. Family and T. Vicsek, *J. Phys. A: Math. Gen.* **24** L25 1991.

[9] Y.-C. Zhang, *J. de Physique* **51** 2129 1990.

[10] J. G. Amar and F. Family, *J. Phys A: Math. Gen* **24** L79 1991.

[11] V. K. Horváth, F. Family and T. Vicsek, *Phys. Rev. Lett.* **67** 3207 1991.

[12] N. Martys, M. Cieplak and M.O. Robbins, *Phys. Rev. Lett.* **66** 1058 1991.

[13] K.C.B. Chan, J.G. Amar, and F. Family, *Bull. Am. Phys. Soc.* **37** 184 1992.

[14] K.C.B. Chan, J.G. Amar, and F. Family, *Preprint.*

[15] We thank Mark Robbins for useful discussions about this point.

[16] J. Krug and P. Meakin, *Phys. Rev. Lett.* **66** 703 1990.

[17] A.-L. Barbási, S. V. Buldyrev, F. Caserta, S. Havlin, H. E. Stanley and T. Vicsek, *This volume.*

[18] T. Vicsek *et al* , *This volume.*

[19] L. H. Tang and H. Leschhorn, *This volume.*

FLUID FRONTS IN HETEROGENEOUS POROUS MEDIUM

O. Sudre and R. Lenormand

Institut Français du Pétrole, BP 311, 92506 Rueil-Malmaison, Cedex, France

Abstract.   Strongly heterogeneous porous media are generated by a multiplicative process similar to the construction of a multifractal set. Various fluid injections are studied in such structures and the scaling laws for tracer injection are obtained analytically and by computer simulations.

1. Introduction

Geologic porous media such as petroleum reservoirs or aquifers are strongly heterogeneous and existing models for fluid transport are not directly suitable. This paper describes fluid transport through a strongly correlated permeability field, generated with a multiplicative process.

Recently, fractal geometry has appeared as a complement of standard geostatistics. Some authors represent the geometry of a porous medium as a fractal percolation cluster near the threshold [1]. This approach is suitable for consolidated media with very low porosity, but does not seem relevant for well connected media.

Another way to use fractal concepts is through long range permeability correlations. Hewett and coworkers have proposed a model of reservoirs derived from the construction of fractal landscapes [2] in which the elevation is replaced by the porosity. This approach is based on a generalization of random noise called fractional noise, which introduces correlations at all length scales. The level of correlations depends on a parameter H, called the Hurst exponent. Weatcraft and Tyler [3] have presented an other approach that is more suitable for tracer injection. It consists in modeling the trajectory of the particles as a fractional random walk. The main result is the ability to model dispersion with variance in t^{2H} (anomalous dispersion).

These techniques based on fractional random walks are successful because they introduce correlations at all length scales, which seems to be a property of geological structures. However, these methods are quite complicated and we prefer to use a multifractal approach which is a more direct way of characterizing non-uniform distributions.

In this paper we presents some results of fluid injection in a porous medium with a multifractal permeability field. The first part presents the construction of a 2-dimensional permeability field with correlations at all length scales and simulations of immiscible fluid injections. The results are compared to the case of random porous media without correlation. The second part present analytical calculations of the dispersion of tracer in a 1-D multifractal medium. Finally, some numerical simulations are presented.

2. Two-Dimensional Flow Simulations

In this part, we construct a permeability field with correlations at all the scales and perform computer simulations of fluid injection with viscous and capillary effects.

2.1. Multifractal Permeability Field

The construction is based on an iterative process [4,5]. Let us consider a probability p which can take equally likely values in the interval $[1 - \lambda, 1 + \lambda]$ (fig. 1). First, a value $p1$ is chosen at random and given to all the pixels of the square grid which represents the medium. The square is divided in four and four different values $p2$ are attributed to each of the new squares. The same process is repeated till the size of the pixel is reached. We define now the permeability K of each pixel to be proportional to the product of the $n + 1$ values of p corresponding to this pixel. The result of such a construction with $n = 6$ and a grey level for K is shown in fig. 2a.

The first property of this construction process is to introduce spatial correlations at all scales. The second is to lead to a log-normal distribution of the permeability, a property experimentally verified for real rocks.

2.2. Flow Simulations

Simulations of two-phase flow have been performed on a network simulator [6]. Three kinds of displacements have been simulated for different values of the viscosity ratio (viscosity of the displaced fluid over the viscosity of the injected fluid) and the capillary number Nca (viscosity forces over capillary forces). The fluid is injected on the left hand side (in black) and the simulation is stopped when it reaches the opposite side.

For the random pattern (fig. 2-b) the three simulations show standard results of dispersion, viscous fingering (DLA) and capillary fingering (percolation). For the multifractal pattern (fig. 2a), the fluid distribution is essentially controlled by the heterogeneities whatever the flow conditions.

In order to study the evolution of interface between the fluids in a correlated medium, and especially the scaling laws for the front, we have developed a 1-dimensional version of this model, which permits analytical calculations.

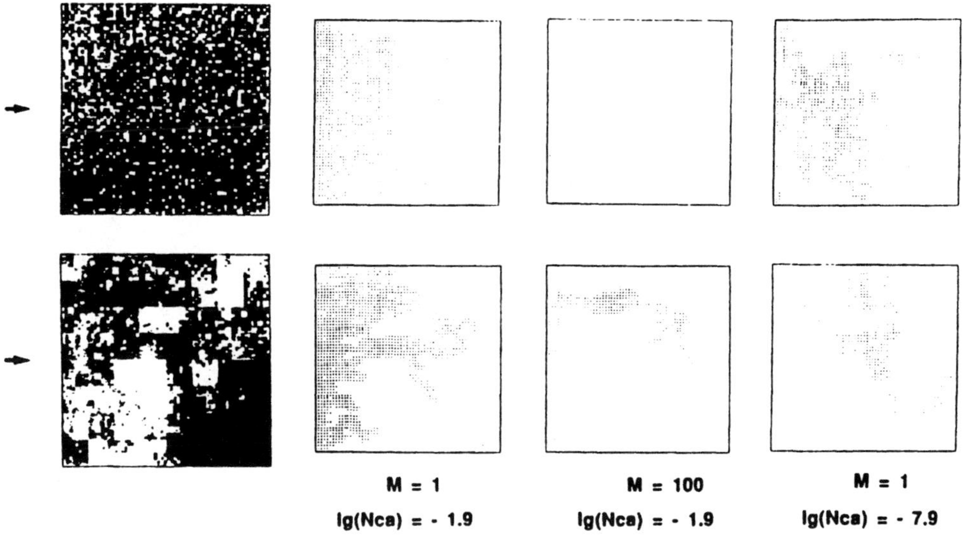

Figure 1: Construction rule of a multifractal network

3. Dispersion in a 1-D Multifractal Porous Medium

3.1. Flow Model

The porous medium is modelled by a bundle of a large number of stream tubes. Each tube contains a fraction of the fluid flowing at a constant volumic flow rate q. Each tube is made of $N = 2^n$ porous elements with a constant length a. The permeability takes a stochastic value k and the section area A can vary like in a real stream tube. We assume that there is no cross flow between the tubes. Consequently, the pressure must be the same at any point of a cross-section of the medium. One solution is to assume a constant pressure gradient ΔP over each element. The section is related to the permeability and fluid viscosity μ by using Darcy's law:

$$\frac{\Delta P}{a} = \frac{\mu q}{Ak}.$$

With the constant flow rate through the tube, each element is filled during a different time τ, proportional to the volume aA of the element. Consequently τ is proportional to $1/k$. This random filling time is the origin of the dispersion of the tracer. If a spike is injected at the entrance of the bundle of tubes, the travel time will fluctuate from one tube to another, leading to macroscopic dispersion when averaging

215

over all the tubes, even if there is no molecular diffusion nor microscopic dispersion in each tube (fig. 3). If there is no correlation between the permeabilities of the different elements, these model leads to standard macrodispersion.

3.2. Permeability Distribution

The permeability K is the product of $n + 1$ uncorrelated random variables p. Consequently, the average of the product is the product of the average: $< K >= 1$. The time τ of presence of the tracer in each element is proportional to $1/K$ and its probability distribution function (pdf) is also log-normal. Its average $< \tau >$ and variance can be calculated as functions of λ.

The permeability correlation function is defined as the average

$$< K(y)K(y + x) > .$$

We study the special case where $y = 1$ and $x = 2^m$. The permeabilities $K(O)$ and $K(m)$ are both a product of $(n + 1)$ different values of p. From the construction rule, $(n - m - 1)$ values are the same and $(m + 2)$ values are different. The average can be written as a product of uncorrelated and corraleted variables. It can be shown that:

$$< K(y)K(y + x) > \propto \left(\frac{x}{L}\right)^{\beta}$$

with:

$$\beta = log_2(1 + \frac{\lambda^2}{3}).$$

3.3. Arrival Times

The process is analogous to a Continuous Time Random Walk with spatial correlations. By using the recursive properties of the medium, the different moments of the arrival time t at distance $x = 2^m$, are calculated.

$$< t >= \frac{x}{a} < \tau >$$

$$< t^2 >= A < t >^2 \left(\frac{x}{L}\right)^{-\alpha}$$

with:

$$\alpha = log_2(1 + \frac{\sigma_\tau^2}{< \tau >^2})$$

3.4. Spatial Dispersion

We have calculated the moments of the probability distribution $f(x, t)$ of arrival times at a distance x. Now we want to calculate the moments of the spatial distribution

$C(x,t)$ at time t. when a spike of tracer is injected. Starting with the general relationship obtained by mass balance:

$$\frac{\partial f(x,t)}{\partial x} = -\frac{\partial C(x,t)}{\partial t}$$

and using the Laplace Transform in time which generates the moments, we obtain:

$$< x > = a\frac{A}{2}(2-\alpha)\left(\frac{L}{a}\right)^{\alpha}\left(\frac{t}{<\tau>}\right)^{1-\alpha}$$

and, for small values of dispersion:

$$< x^2 > = a^2\left(\frac{L\tau}{at}\right)^{3\alpha}\left(\frac{t}{<\tau>}\right)^{2}$$

The variance is no longer a power law:

$$< \sigma_x^2 > = 3a^2 A\alpha\left(\frac{t}{<\tau>}\right)^{2} Log\left(\frac{at}{L\tau}\right)$$

4. Numerical Simulations

We have studied the spatial dispersion of a spike of tracer injected through a bundle of 1000 tubes. Each tube contains 2048 elements. The injected particle (or tracer) spends a random time in each element, and its position x is recorded at different times t. Then, x is averaged over all the tubes and the moments are calculated.

The results are in agreement with the analytical calculation. The first and second moments of the travelled distance varies as a power law of time (fig. 4).

For the second moment, the exponent varies as: $(2 - 1.14\lambda^2)$, where I is the width of the distribution used to construct the permeability field (fig. 5).

5. Conclusion

We have used a simple geometrical construction to generate a porous medium with strong correlations at all length scales.

Two dimensional immiscible flow simulations show that the invasion front is mainly governed by the heterogeneities.

A one-dimensional model of tracer injection leads to the different moments of the spatial dispersion. Contrarily to the standard dispersion, and anomalous dispersion, the variance is no longer a power law of time, but varies as the square of time multiplied by $log(t)$. However, the second moment is a power law of time with an exponent which is a function of the width of the distribution used to construct the permeability field.

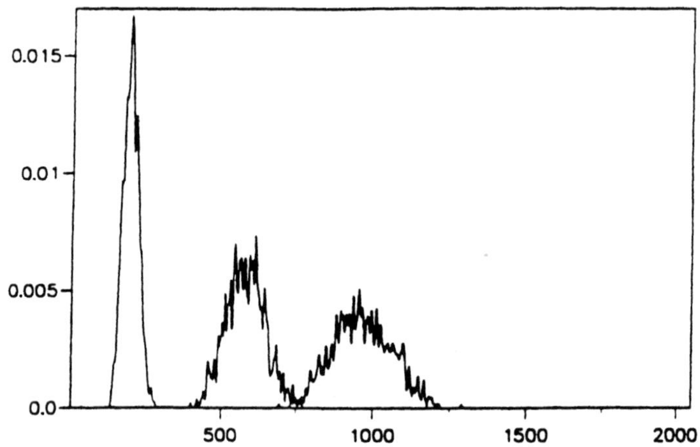

Fig. 2: Concentration $C(x,t)$ of a spike of tracer in the bundle of tubes at three different times.

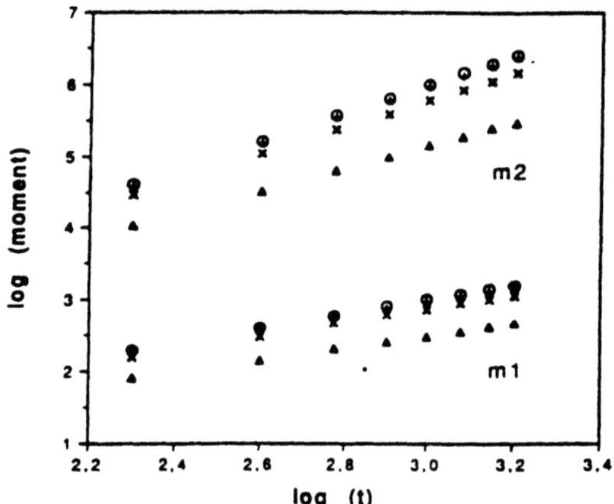
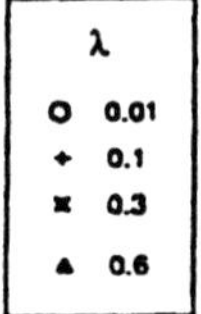

Fig. 3: First (m1) and second moment (m2) as functions of time for different values of λ.

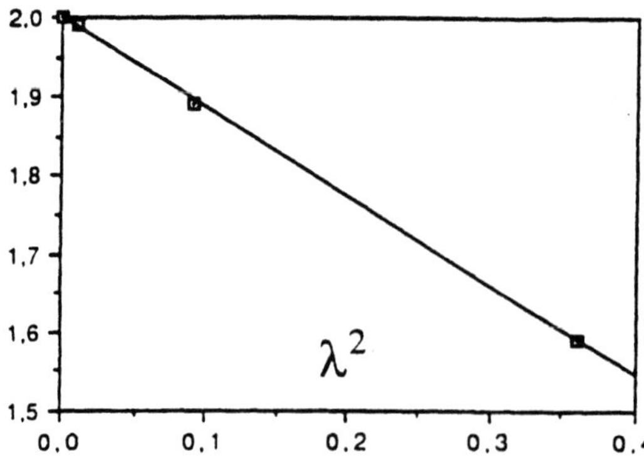

Fig. 4 Evolution of the exponent of the 2nd moment with λ.

Acknowledgement

We would like to thank B. Noetinger and F. Kalaydijian for helpful discussions, and
B. Mandelbrot for suggesting the 1-D model for analytical calculations.

REFERENCES

[1] U. Oxaal,M. Murat, F. Boger, A. Aharony, J. Feder and T. Jossang, *Nature* **329**
32 1987.

[2] R.F. Voss, in: *Fundamentals Algorithms for Computer Graphics* (R.A. Earnshaw,
ed.) Springler, Berlin 1985.

[3] W. Weatcraft and S.W. Tyler, *Water Resources Research* **24** 566 1988.

[4] P. Meakin, *PHys. Rev. A* **36** 2833 1987.

[5] R. Lenormand, F. Kalaydijian, M.T. Bieber and J.M. Lombard,   Paper 20475,
presented at the 65th annual conference of the Soc. of Petroleum Eng., New-Orleans
1990.

[6] R. Lenormand, E. Touboul and C. Zarcone, *J. Fluid Mech.* **189** 165 1988.

SELF-AFFINE FIRE FRONTS FROM PAPER BURNING

Jun Zhang*, Y.-C. Zhang†, P. Alstrøm*, and M.T. Levinsen*

*Physics Laboratory H.C. Ørsted Inst. Universitetsparken 5, DK-2100, Copenhagen, Denmark
†Institut de Physique Théorique, Pérolles, Univ.de Fribourg, CH-1700, Switzerland

Abstract. We study the propagation of a flameless fire on a thin piece of paper. We find that fire fronts on a piece of paper follow quite well self-affine scaling statistics, with χ around 0.70, well above the value 1/2 of theoretical analysis using gaussian noise.

Fire propagation on a two-dimensional substance is often encountered in daily life. Here we report a paper burning experiment under rather idealized conditions. An initially straight fire front will develop into a rather rough shape, as a result of local inhomogeneities. We are interested in the fluctuation statistics of fire fronts as they propagate.

Fire spread is just an example of the much general mechanism of interface growth, which is one of the current focuses of statistical physics [1]. Fire fronts can be modelled by a non-linear stochastic differential equation. The theoretical prediction based on analytical calculations and computer simulations shows that fire fronts should be a self-affine fractal with a roughening exponent $\chi = 1/2$. Recent experiments however, show significant departure from the theoretical prediction.

We choose an optical paper (lens tissue) made by Whatman Paper Ltd., UK. In fact, almost any paper may serve the purpose, even this page of the proceedings. This special choice is due to its extreme lightness, $9.1 g/m^2$ (or thickness $\sim 45\mu$m), about one ninth of that of Xerox copy paper. This helps keeping the heat production by fire to a minimum, and this in turn reduces complication to fire propagation by a strong air circulation. For the same reason, we want flameless, relatively slow burning. Paper usually does not sustain a flameless fire consistently. To ensure a uniform fire propagation, we treat the paper with a solution of KNO_3 which is a common oxidization

221

aid, used in many ordinary explosives. We adjust KNO$_3$ concentration as a means of controlling fire propagation speed. There appears to be a lower threshold concentration below which fire would not be sustained continuously; an upper threshold above which the flameless fire would burst into flame. The above concentration range is 0.87 $\sim$ 1.6g/m^2. The corresponding mean fire propagation speed lies within 5.5 $\sim$ 8.2mm/sec. Our experiment is performed mostly for concentrations close to the lower threshold value, in line with our minimal heat consideration.

We carefully dry the treated paper while keeping it flat and smooth. The format is 46cm wide and 110cm long (high). We put the paper in an upright frame and apply a slight tension on its two lateral sides, making it immobile during burning. When fire approaches the metallic holders on the sides it stops upon touching. To avoid boundary effects we discard a region of a few centimeters on both sides in the analysis. The experiment takes place under normal room conditions with no forced ventilation. Fire is ignited from the bottom side by a straight electric heating wire. The initially straight fire front gradually develops into irregular shapes, on ever larger length scales.

We use the two-step digitization method: We take high resolution photographic pictures to record the fire front propagation. The pictures are then captured by a video camera into a (512x750) pixel buffer with multi-level grey scale. To obtain better resolution than that offered by ordinary video equipments, we digitize only sub-regions of a picture and they are then patched up into the whole in the computer. This indirect method allows in principle arbitrary resolution. We typically divide pictures into four patches and the effective pixel grid is thus about 512x3000. We identify the fire front as the burning region which is 0.5 $\sim$ 3mm wide in the vertical direction. We define fire-line as the sharp boundary between the burning part and the intact part. From original two-dimensional data we single out (selecting desired grey scales) the one-dimensional fire-line, represented by $h(x,t)$ its position at point x (discrete after digitization) and time t, as shown in Fig. 1.

The quantity of interest is the mean square deviation, or width $W(L,t)$:

$$W^2(L,t) = <\frac{1}{L}\sum_{x=1}^{L}[h(x,t)-\bar{h}(t)]^2>, \quad \text{where} \quad \bar{h}(t) = \frac{1}{L}\sum_{x=1}^{L}h(x,t). \tag{1}$$

L (0.05$cm \leq L \leq$ 18.3cm) is the length of a segment of a fire-line, large compared to microscopic scales but it is still smaller than the paper's width (46cm), to avoid boundary effects. $< \cdot >$ denotes the average over all the segments, obtained from a single fire-line by shifting the starting point ($x = 1$) each time.

Fig. 2 shows the data of $W(L,t)$ against L for a large and fixed t, on a double-log scale. One sees that in a region about two decades in the x-direction the data follow rather well a straight line, with the estimated slope $\sim 0.70 \pm 0.03$.

It is remarkable that even a single line already show reasonable scaling. The sample-averaged χ value we estimate to be 0.71 ± 0.05.

No matter how uniform a fuel-bed, for example every effort is made in the preparation of our paper, there are unavoidable local irregularities in the forms of random fibre network, porosity, chemical composition, etc. The noise contribution is actually

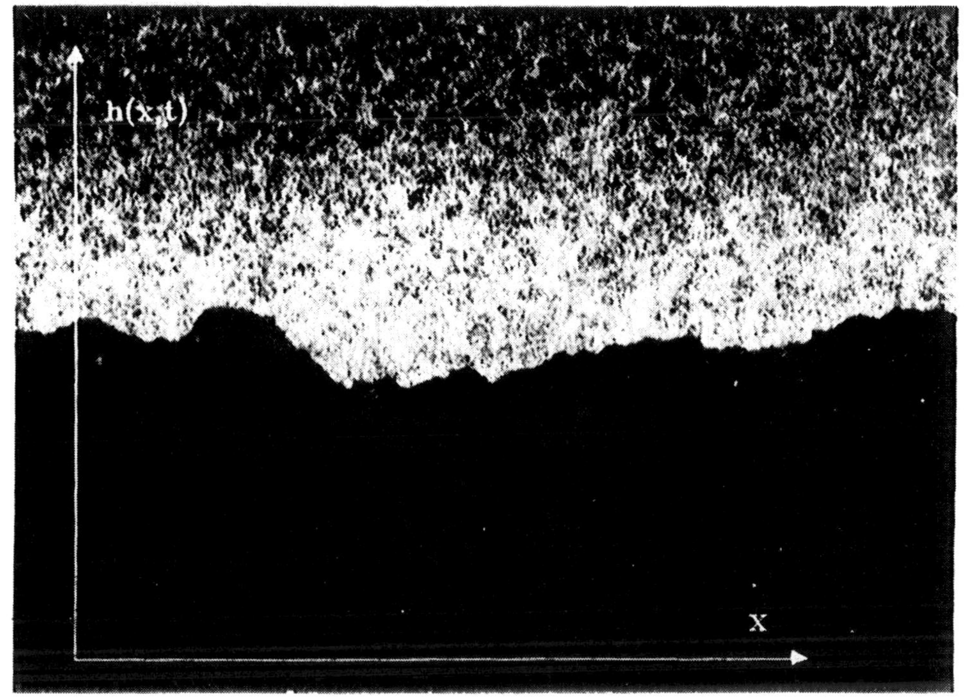

Figure 1. This photo shows a segment of a piece of burning paper. The transverse size is about 8.5cm, which is only a small part of the total ($\sim$ 46cm). Fire is propagating upwards, the smoke indicates slow and laminar air circulation. The actuel fire front is a few mm wide, but by using a strong background light the sharp fire-line can be identified.

fixed on the paper, not dependent on time. However in our experiment the fire front is moving everywhere, from its own point of view the irregularities are effectively a function of time. Fig.3 is a microscope picture of magnification 20, where we see that the paper is actually a network of intertwined fibres. All sorts of random factors influence fire propagation, making it difficult to determine exact origins of noise.

The noise for our case can be regarded practically uncorrelated in space and time, this is also the most studied theoretical case. For uncorrelated gaussian or truncated noise, it has been shown analytically that the roughening exponent is $\chi = 1/2$. Many numerical simulations, not based on eq.(3) but directly on various growth rules, also yield $\chi \simeq 0.5$.

It was pointed out [2] that in real experiments χ maybe non-universal and always larger than the theoretical value. This is because the theoretical assumption of gaussian

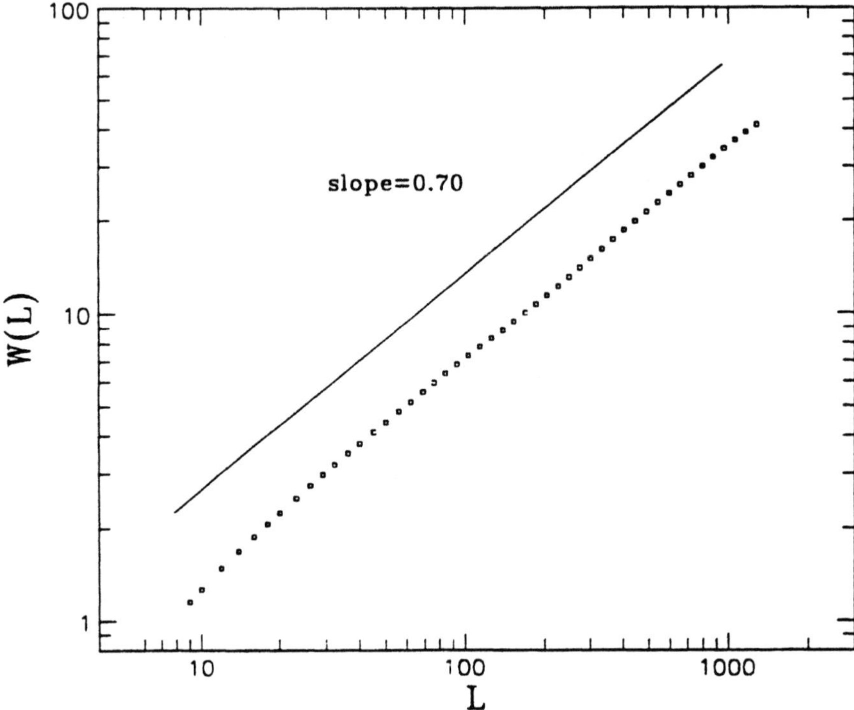

Figure 2. The mean fire-line width $W(L)$ plotted versus the transverse l ength L. The data is from a single instantaneous fire-line including the segment shown in Fig.1. The average is over different segments by shifting their starting point on the fire-line. The data appear to follow a straight line of a slope 0.70, over about two decades.

or truncated noise may not materialize in reality. For uncorrelated noise with a power-law distribution one can show that χ can vary from 1/2 to 1, in numerical simulations [2]. The above result of $\chi > 1/2$ is consistent with this suggestion.

There are previous experiments on interface growth, which are thought to share the same mechanism of fire propagation. Notably, one [3] is on the water-air immiscible displacement in a random medium; the other [4] is on bacterium-colony expansion on an agar nutrient plate. They report also non-universal, larger values of χ. In fact, power-law noise is reported being observed in one case [5]. However, these experiments suffer the drawback that bulk material on one or both sides of an interface still participate in the growth dynamics. This may cast doubts on the adequacy of an interface-only description. In contrast, the present burning experiment is a strict

Figure 3. The apparent uniform paper reveals much of its inhomogeneity at microscopic scales, in this photo at magnification 20. The intrinsic fibre network and uneven KNO_3 concentration on it all contribute to the noise affecting the fire propagation.

interface phenomenon: burned paper becomes ash thus no further participation; intact paper does not change until reached by fire.

We have reported a simple paper burning experiment. The fire fronts show interesting self-affine scaling behavior. Besides its implications in practice our experiment can be regarded as a true realization of the interface growth equation, thus we can benefit much from recent theoretical progresses, and *vice versa*. In the near future we plan to perform the experiment in a quasi-vacuum condition, with oxygen provided chemically by the burning process itself. This will help fend off the eventual criticism that hot air might have influenced the scaling behavior.

REFERENCES

[1] *Dynamics of Fractal Surfaces* F. Family and T. Vicsek eds., World Scientific, Singapore 1991 and references therein.

[2] Y.-C. Zhang,, *J. de Physique* **50** 2129 1990.
  Physica **170** 1 1990▷

[3] M.A. Rubio, C.A. Edwards, A. Dougherty, and J.P. Gollub, *Phys. Rev. Lett.* **63** 1685 1989.
  V.K. Horváth, F. Family, and T. Vicsek, *J.Phys.* *A* **24** L24 1991▷

[4] T. Vicsek, M. Cserzö, and V.K. Horváth, *Physica* **167** 315 1990.

[5] V.K. Horváth, F. Family, and T. Vicsek, *Phys. Rev. Lett.* **67** 3207 1991.

NON-UNIVERSALITY IN PINNING, ROUGHENING, TEARING, CRUMPLING, CRINKLING AND CRUSHING EXPERIMENTS

Xin Yun Zhang Li*, Fulbrecht-László Hárthanásy*,
Dieter-Wolfgang Achtung* and J. Dupont JeGenne[†‡]

*Interplanetary Institute for Virtual Supercalculations, Dustbin Road, Saturn, IL 69696

[†]Royal Academy of Blood, Sweat and Tears, Marks and Spencer's Road, London, LE1 3PI, United Kingdom

Abstract. We report measurements of the penetration of various fluids in Physical Review Letters papersheets. Systematic experiments were performed using 10 complete collections of Vol. 17 and Vol. 66 ($\sim 2.10^5$ pages, excluding covers) immersed in several laboratory fluids : coffee, Ginger-Ale, Peking duck coating, Bourgogne "Aloxe Corton" 1966, and graduate students sweat. We measured the roughness exponent of the interface using STM, TGV, TEM, UHV, MBE, HIP and investigated the pinning threshold using FBI, CIA (and the older KGB) techniques. We find : $w(\ell) \sim \ell^\zeta$ with $\zeta = .7189423212 \pm .01$. Conformal invariance suggests the value $\zeta = 219$ (per square inch) and we attribute the discrepancy to lack of luck. Experiments with "grenadine" were also performed, confirming a bold conjecture of D. Berriba that in this case ζ satisfies $K_0(\zeta^\zeta) = 1$. This actually coincides with the fractal dimension of dry dogbiscuits (between 350 and 400°C). Crushing experiments of the Connection Machine® are found to be in a different universality class. A fishy theoretical explanation of our results is given, which also explains any experiments in which a linear combination of the above values of ζ is found.

The full text of this contribution will appear in crumpled Phys. Rev. Lett. (Les Houches special issue).

[‡]To whom correspondence should be sent.

DIRECTED POLYMERS

INTERFERENCE OF DIRECTED PATHS IN STRONG LOCALIZATION

Mehran Kardar* and Ernesto Medina[†]

*Department of Physics, Massachusetts Institute of Technology, Cambridge, MA 02139 USA

[†]Intevep SA, Apdo 76343, Caracas 1070A, Venezuela

Abstract. Quantum interference phenomena for electrons *localized* by disorder are dominated by virtual directed paths in the random potential. The sum over such *forward scattering paths* can be computed efficiently by the transfer matrix algorithm, and studied analytically by replica analysis. By such methods we characterize the probability distribution for tunneling between sites separated a distance t. We find that the scaling properties of the distribution are intimately related to those of directed polymers in random media, and hence related to the scaling of fluctuations of growing interfaces. In particular, the distribution is approximately log–normal; its mean proportional to t, and its variance growing as $t^{2\omega}$, with ω equal to the exponent β for the growth of interface fluctuations! From the point of view of localization, an important question is how the tunneling rate is modified by a magnetic field. Without spin–orbit scattering (SO) we find that a magnetic field leads to an increase in the localization length, initially scaling as $B^{1/2}$. With SO, there is still a positive magnetoconductance (initially growing as $B^2 t^3$), but no change in the localization length.

1. Introduction

At first glance the topic addressed here appears quite unrelated to the theme of this workshop. After all what correspondence is there between quantum mechanical phenomena in mesoscopic insulators, and the non–equilibrium roughening of interfaces. There is in fact an intimate connection between the two problems that is more mathematical in nature. The modern era of critical phenomena has established that *scaling* laws are quite fundamental, and grouped in a few *universality* classes. We will show that the universality classes that govern the dynamic roughening of interfaces are related to those that control the scaling of the tunneling probability distribution for strongly localized electrons. This discussion starts with a brief introduction to the

231

phenomena associated with quantum interference, and proceeds to elucidate the above connection.

The influence of quantum interference phenomena on magnetoconductance (MC) and conductivity fluctuations has been extensively studied for *weakly localized* electrons (for recent reviews see[1]). In the absence of spin–orbit scattering (SO), a magnetic field causes an increase in the localization length (a positive MC), and a factor of 2 decrease in the conductance fluctuations. These results are attributed to suppression of *backscattering* loops by a magnetic field. With SO, the magnetic field has the opposite effect of decreasing the localization length (a negative MC), but still reduces the conductance fluctuations[2]. Symmetries of the underlying Hamiltonian, and their modification by a magnetic field, are also be invoked to support these conclusions[3].

By contrast, the behavior of conductivity and its fluctuations for *strongly localized* electrons is more controversial and less well understood. The main mechanism for conductivity in this regime is by thermally activated, variable range, electron tunneling[4]. Nguyen, Spivak, and Shklovskii (NSS)[5] have emphasized the importance of quantum interference of *forward scattering* paths to the tunneling probability. Treatments that ignore the correlations between such paths conclude a positive magnetoconductance, but no change in the localization length, whether in the absence[5][6] or presence[7] of SO. On the other hand, a random–matrix approach[8] predicts that a magnetic field leads to a doubling of the localization length ξ (big, positive MC) without SO, but a halving of ξ (big, negative MC) in the presence of SO. Over the past three years we undertook extensive numerical studies of the NSS model. We used a transfer matrix procedure that allows exact summation of forward scattering paths for very large systems ($t \approx 10^3 - 10^4$), and obtained statistics by averaging over many realizations ($\approx 10^3$) of randomness. These results, combined with some analytical insights, provide a coherent description of the behavior of strongly localized electrons[9][10].

2. Model

The starting point is the Anderson Hamiltonian

$$\mathcal{H} = \sum_{i,\sigma} \epsilon_i a_{i,\sigma}^{\dagger} a_{i,\sigma} + \sum_{<ij>,\sigma\sigma'} V_{ij,\sigma\sigma'} a_{i,\sigma}^{\dagger} a_{j,\sigma'} \quad , \tag{1}$$

where ϵ_i are the random site energies. The *nearest–neighbor* hopping elements are set to $V_{ij} = VU_{ij}e^{i\mathcal{A}_{ij}}$, where V is a constant, U_{ij} is a randomly chosen $SU(2)$ matrix describing the spin rotation due to a strong SO scatterer[3][7], and $\mathcal{A}_{ij}$ is the magnetic vector potential. The tunneling probability between two sites is related to their overlap, which using a "locator" expansion[11] can be written as

$$< i\sigma|G(E)|f\sigma' > = \sum_{\Gamma} \prod_{i_r} \frac{Ve^{i\mathcal{A}}U}{E - \epsilon_{i_r}}. \tag{2}$$

The above expression is a Feynman sum over all possible trajectories Γ between the initial (i) and final (f) sites. Each bond along the path contributes a random spin rotation U, and a phase factor from the magnetic vector potential $\mathcal{A}$.

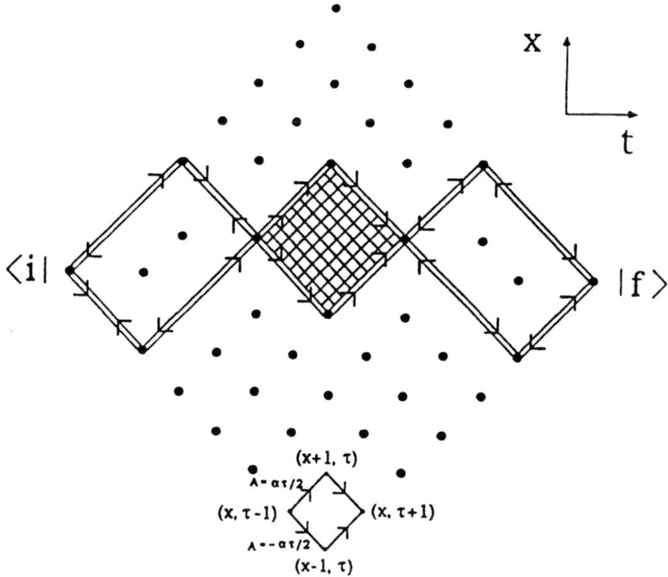

Figure 1. Directed paths contributing to the tunneling between diagonally separated endpoints. The averaging over SO impurities pairs forward and time–reversed paths, and their spins, as indicated.

To simplify the energy denominators, we use the NSS[5] model in which the energy of the initial and final sites is set to zero, while the intermediate ϵ_i take on values of $+W$ or $-W$ with equal probability. A path of length ℓ now contributes an amplitude $W(V/W)^\ell$ to the sum, as well as an overall sign and rotation matrix. For $V/W \ll 1$, corresponding to strongly localized electrons, the sum is convergent[11], and eq.(2) is dominated by the shortest paths connecting the end points. Following NSS, we maximize the interference between such *forward scattering* paths by choosing i and f to lie along the diagonal of a hypercubic lattice, as depicted for two dimensions in Figure 1. All shortest paths Γ' now have the same length t, and the tunneling amplitude simplifies to

$$A = <i\sigma|G(0)|f\sigma'> = W(V/W)^t J(t); \qquad J(t) = \sum_{\Gamma'} \prod_{i_{\Gamma'}} \text{sign}(\epsilon_{i_{\Gamma'}})e^{iA}U. \qquad (3)$$

After averaging over the initial spin, and summing over the final spin, the tunneling probability is

$$T = \frac{1}{2}\text{Tr}(A^\dagger A) = W^2(V/W)^{2t}I(t); \qquad I(t) = \frac{1}{2}\text{Tr}(J^\dagger J). \qquad (4)$$

All interference information is now contained in $I(t)$.

3. The Tunneling Probability Distribution

For each realization of randomness, the contribution of forward scattering paths to $J(t)$ in eq.(3) can be computed exactly using a *transfer matrix method*[9][10]. We numerically studied the statistical properties of $I(t)$ for t of up to 2000, and for over 2000 realizations of the random Hamiltonian. In all cases, with and without a magnetic field, in the presence or absence of SO, we find a probability distribution for $I(t)$ that is broad (almost log–normal). The quantity $\ln I(t)$ is similar to a quenched average free energy, and we confirmed that, as any extensive quantity, it grows linearly with t. Since the Green's function in eq.(3) is expected to typically decay as $\exp(-t/\xi)$, the localization length ξ has a contribution from $\ln I(t)$. In fact, we define *local* and *global* contributions to ξ by setting

$$\frac{\langle \ln | < i|G|f > | \rangle}{t} \equiv \xi^{-1} \equiv \xi_0^{-1} + \xi_g^{-1},\qquad (5)$$

with

$$\xi_0^{-1} \equiv \ln \left(\frac{W}{\sqrt{2}V} \right) \quad , \text{ and } \quad \xi_g^{-1} \equiv \ln \sqrt{2} - \frac{\langle \ln I(t) \rangle}{2t}.\qquad (6)$$

Thus, an important feature of the sum over forward scattering paths in eq.(4) is the appearance of *two independent parameters*. The local factors V and W in the starting Hamiltonian only contribute to ξ_0. They do not appear in ξ_g or $I(t)$, and hence the fluctuations are completely independent of the Anderson parameter[11] (V/W). This is in agreement with previous results by Cohen *et al* [12] who argue that a one parameter scaling arises only in the limit of weak disorder. They also argue in favor of two parameter scaling in strong localization.

Fluctuations of $\ln I(t)$ also increase with t, but with a slower power law: Figure 2 shows the variance of $\ln I(t)$ fitted to a scaling form $t^{2\omega}$ with $\omega = 0.33 \pm 0.05$. Log–normal distributions arise quite naturally for the conductance of one dimensional systems[13]. Indeed, in a one–dimensional description of forward scattering, $I(t)$ is simply the product of the random contributions of t consecutive bonds, and hence manifestly log–normal. However, for such cases the variance of $\ln I(t)$ is proportional to t, i.e. $\omega = 1/2$. We shall argue later on that the exponent ω is a universal characteristic of the distribution which depends on dimensionality ($\omega = 1/3$ in $d = 2$). In fact it should equal the universal exponent $\beta = \chi/z$ which describes the short time growth of interface width in ballistic deposition[14].

4. Magnetic Field Response

The response of the system to a magnetic field B (measured in units of quantum flux per plaquette), with and without SO, is plotted in Figure 3. To emphasize the magnetoconductance (MC) we have subtracted the zero field averages (without SO on the bottom, and with SO on the top portion). The asymptotic slopes on the bottom portion indicate the increase in the global contribution ξ_g^{-1} to the (inverse) localization length. We first note that the introduction of SO (indicated by +) is accompanied by

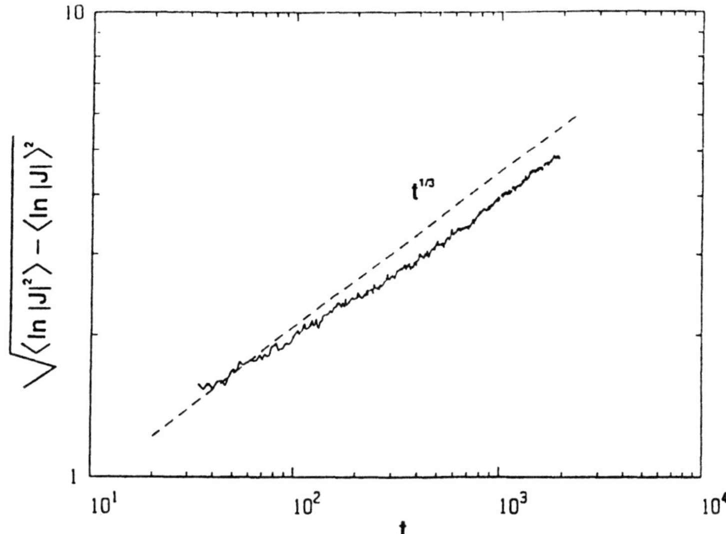

Figure 2. Standard deviation of the logarithm of the sum over directed paths versus the path length t. The dashed line has a slope $\omega = 1/3$.

a significant increase in tunneling, i.e an increase in ξ. Secondly, the addition of a magnetic field leads to an increase in $\langle \ln I(t) \rangle$, but in qualitatively different manners in the absence or presence of SO. Without SO, there is a change in slope, i.e. the most important effect of the field is to increase the localization length. (The increase in SO seems to saturate to a limit corresponding to addition of random phases (indicated by *) on bonds, mimicking random magnetic impurities.) This an enhancement of tunneling that grows exponentially in t. By contrast, with SO, we observe that the slopes in Figure 2 remain unchanged. Thus the magnetic field enhances the tunneling rate by a t–independent constant, i.e. there is no change in the localization length.

The appropriate scaling laws in a small magnetic field can be obtained by collapsing the data in Figure 3 for different values of B and t. Such a collapse in the absence of SO is obtained as a function of Bt^2 and the slopes are well fitted to

$$\xi_g^{-1} = (0.053 \pm 0.02) - (0.15 \pm 0.03)\,(\phi/\phi_0)^{1/2} . \tag{7}$$

By contrast, the appropriate scaling combination with SO is $Bt^{3/2}$, and the scaling function takes the form

$$\langle \ln I(t,B)_{SO} \rangle - \langle \ln I(t,0)_{SO} \rangle = \begin{cases} cB^2 t^3 & \text{if } B^2 t^3 < 1 \\ C & \text{if } B^2 t^3 > 1 \end{cases} . \tag{8}$$

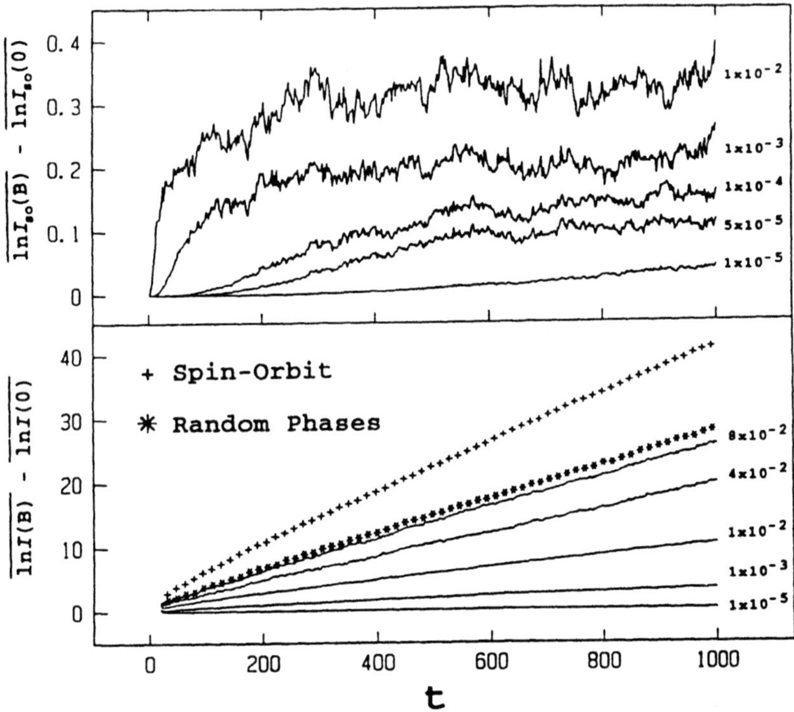

Figure 3. Bottom: Increase in tunneling (MC) without SO (solid curves) from the zero field value I_0. A large MC is also obtained by introducing random phases (*), or SO (+), on the bonds. Top: The analogous MC with SO is much smaller and plotted at enlarged scale.

5. Moment Analysis

We can gain some analytic understanding of the distribution function for $I(t,B)$ by examining the moments $\langle I(t)^n \rangle$. From eqs.(3) and (4) we see that each $I(t)$ represents a forward path from i to f, and a time reversed path from f to i. For $\langle I(t)^n \rangle$, we have to average over the contributions of n such pairs of paths. Averaging over the random signs of the site energies forces a *pairing* of the $2n$ paths (since any site crossed by an odd number of paths leads to a zero contribution)[9]. To understand the MC it is useful to distinguish two classes of pairings: **(1)** *Neutral paths* in which one member is selected from J and the other from $J^\dagger$. Such pairs do not feel the field since the phase factors of e^{iA} picked up by one member on each bond are canceled by the conjugate factors e^{-iA} collected by its partner. **(2)** *Charged paths* in which both elements are taken from J or from $J^\dagger$. Such pairs couple to the magnetic field like particles of charge $\pm 2e$. In the presence of SO, averaging over the random $SU(2)$ matrices further forces neutral paths to carry parallel spins, while the spins on the two partners of charged paths must be antiparallel. These constraints are indicated in Figure 1 which shows a possible contribution to $\langle I^2 \rangle$ with SO.

Since at each site any path has a choice of two directions, we may naively expect $\langle I(t)^n \rangle$ to asymptotically scale as 2^{nt}. This ignores the correlations between paths which manifest themselves when two paired paths intersect[15]. Since at each intersection the pairs may exchange partners there is an additional statistical attraction for such crossings. The attraction factor depends crucially on the symmetries of the Hamiltonian in eq.(1): For $B = 0$ and without SO, the Hamiltonian has *orthogonal* symmetry. All pairings are allowed and the attraction factor is 3 since an incoming (12)(34) can go out as (12)(34), (13)(24), or (14)(23). Note that even if both incoming paths are neutral, one of the exchanged configurations is charged (see Figure 1). A magnetic field breaks time reversal symmetry, discourages charged configurations, and reduces the exchange attraction. The limiting case of a 'large' magnetic field is mimicked by replacing the gauge factors with random phases. In this extreme, the Hamiltonian has *unitary* symmetry, and only neutral paths are allowed. The exchange factor is now reduced to 2; from $(11^*)(22^*) \rightarrow (11^*)(22^*)$, or $(12^*)(21^*)$. With SO, we must also take into account the allowed spin exchanges, and we find that the intersection of two paired paths results[10] in an exchange attraction of 3/2 (*symplectic* symmetry).

6. Equivalence to Directed Polymers

Calculating $\langle I(t)^n \rangle$ is now reduced to finding the sum over n paths with the above exchange attractions. This is precisely the same problem as calculating the moments of the partition function of a directed polymer (DP) in a random medium[16],

$$W(\mathbf{x}, t) = \int_{(0,0)}^{(\mathbf{x},t)} \mathcal{D}\mathbf{x}(\tau) \exp\left\{ - \int d\tau \left[\frac{\dot{\mathbf{x}}^2}{2\nu} + \mu(\mathbf{x}, \tau) \right] \right\}. \tag{9}$$

Assuming that the random energies are Gaussian distributed with $\langle \mu(x,t) \rangle = 0$ and $\langle \mu(x,t)\mu(x',t') \rangle = \sigma^2 \delta(x - x')\delta(t - t')$, the averaged n^{th} moment is given by

$$\langle W^n(t) \rangle = \int \mathcal{D}\mathbf{x}_1 \cdots \mathcal{D}\mathbf{x}_n \exp\left\{ - \int d\tau \left[\sum_\alpha \frac{\dot{\mathbf{x}}_\alpha^2}{2\nu} + \sigma^2 \sum_{\alpha,\beta} \delta(\mathbf{x}_\alpha - \mathbf{x}_\beta) \right] \right\}. \tag{10}$$

The attraction in this case depends on the variance of randomness, whereas for $\langle I(t)^n \rangle$ it reflects the symmetry group of the Anderson Hamiltonian. However, whenever the attraction (or randomness) is relevant, the resulting scaling behaviors are independent of its magnitude. In this sense the two problems belong to the same universality class. (For a general introduction to the literature on the directed polymer and interface growth problems, see references[17][18] and [14].)

Moments of the DP partition function in 1+1 dimensions have been calculated[19] by a transfer matrix method, and the final result is

$$\langle I(t)^n \rangle = A(n)2^{nt} \exp[2\rho n(n^2 - 1)t]. \tag{11}$$

The parameter $\rho \propto \sigma^4/\nu$ is an increasing function of the strength of the attraction between paths. We have also included an overall amplitude $A(n)$.

Cumulants C_i of the characteristic function for $\ln I(t)$ are defined by the powers of n in the expression

$$\langle I(t)^n \rangle = \langle \exp(n \ln I(t)) \rangle \equiv \exp\left(\sum_i \frac{n^i}{i!} C_i [\ln I(t)] \right). \tag{12}$$

C_1 is the average, $\langle \ln I(t) \rangle$, while the absence of an n^2 term in eq.(11) indicates no second cumulant at order of t. The n^3 term reflects a third cumulant scaling as t, i.e.

$$\begin{aligned}
\langle \ln I(t) \rangle &= [\ln 2 - 2\rho]\, t, \\
C_2(\ln I(t)) &= 0 \quad \text{(to order } t\text{)}, \\
C_3(\ln I(t)) &= 12\rho t.
\end{aligned} \tag{13}$$

Comparing the above with eqs.(5) and (6), we note that the global contribution to the localization length is $\xi_g = 1/\rho$, i.e. directly related to the strength of the bound state. Also it follows that $\ln I(t)$ is approximately normal, with fluctuations given by

$$\delta \ln I(t) \sim |\frac{t}{\xi_g}|^{1/3}. \tag{14}$$

We can now appreciate the trends in Figure 3, as the slopes are indicative of the statistical attraction factors. Without SO, the magnetic field gradually reduces the attraction factor from 3 to 2 leading to the increase in slope. Addition of SO to the Hamiltonian has the similar effect of suddenly decreasing the attraction to 3/2. Why does the addition of the magnetic field lead to no further change in ρ in the presence of SO? Without SO, the origin of the continuous change in the attraction factor is the type of exchange indicated in Figure 1 whereby a charged bubble appears from the intersection of two neutral paths[9]. In the presence of SO, it can be shown[10] that the contribution of such configurations is zero. Thus the neutral paths traverse the system without being effected by the magnetic field; their attraction factor, and hence ρ and ξ_g remain unchanged. The smaller positive MC observed in the simulations is due to the quenching of the charged paths by a magnetic field[10]. The resulting change is thus only in the amplitude $A(n)$ of eq.(11).

The exchange attraction between neutral paths can also be computed for $SU(n)$ impurities, and equals $1 + 1/n$, which reproduces 2 for $U(1)$ or random phases, and 3/2 for $SU(2)$ or SO scattering. The attraction vanishes in the $n \to \infty$ limit, where the paths become independent. The statistical exchange factors are thus universal numbers, simply related to the symmetries of the underlying Hamiltonian. The attractions in turn are responsible for the formation of bound states in replica space, and the universal scaling of the moments in eq.(11). In fact since the single parameter ρ completely characterizes the distribution, the variations in the mean and variance of $\ln I(t)$ should be perfectly correlated. This can be tested numerically by examining respectively coefficients of the mean and the variance from eq.(13) for the different cases

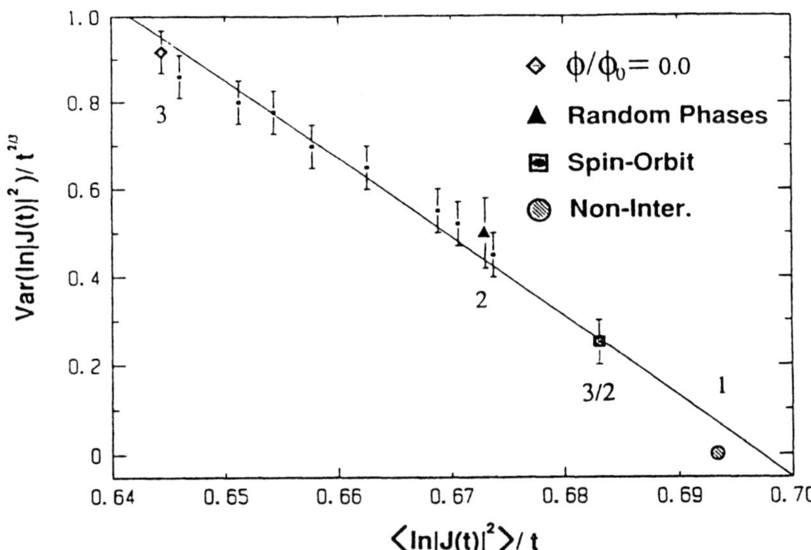

Figure 4. Complete hierarchy of the exchange attractions between paired paths, reflecting the various Hamiltonian symmetries; Diamond: orthogonal; triangle: unitary; square: symplectic; circle: the non-interacting limit corresponding to independent paths, and also to $SU(n)$ with $n \to \infty$. Small squares correspond to finite magnetic fields but no SO.

studied. The results plotted in Figure 4 do indeed fall on a single line, parameterized by ρ. The largest value corresponds to the NSS model for $B = 0$ and no SO (orthogonal symmetry, exchange attraction 3). Introduction of a field gradually reduces ρ until saturated at the limit of random phases (unitary symmetry, exchange attraction 2). SO scattering reduces ρ further (symplectic symmetry, exchange attraction 3/2). The final point corresponds to independent paths with $\rho = 0$.

7. Relationship to the KPZ equation

The calculated exchange attractions between the paths representing $\langle I(t)^n \rangle$ are in fact valid in all dimensions. But the sum over the attracting paths in eqs.(9) and (10) is only calculable in dimensions $d = 0 + 1$ and 1+1. Fortunately in higher dimensions we can take advantage of a mapping to the interface growth problem[14] where extensive numerical results are available. First note that the path integral in eq.(9) is the formal solution to the equation

$$\partial_t W = \nu \nabla^2 W + \mu(\mathbf{x}, t)W; \tag{15}$$

which through the Cole–Hopf transformation $W = \exp\left(\lambda h/2\nu\right)$ can be mapped onto the KPZ equation[20]

$$\partial_t h = \nu \nabla^2 h + \frac{\lambda}{2}(\nabla h)^2 + \eta(\mathbf{x}, t). \tag{16}$$

Since eq.(16) should describe the growth of self–affine interfaces, the accumulated literature on this topic[14] is also relevant to the tunneling probability distribution. For example, the Kim and Kosterlitz conjecture[21] implies that fluctuations in $\ln I(t)$ grow as $t^{1/d+1}$, when randomness is relevant in d dimensions. Actually this conjecture has quite an appealing form in terms of moments, as it generalizes the small n behavior in eq.(11) to

$$\langle |I(t)|^n \rangle = A(n)\langle |I(t)| \rangle^n \exp[\rho n(n^d - 1)t]. \tag{17}$$

Note that this also reproduces the log–normal distribution in one dimensions[13][12], with $\omega = 1/2$.

Zhang[22][23] has disputed the equivalence between the tunneling problem (sometimes called CDP for complex directed polymers) and the DP and KPZ universality classes. He points out that weights for intersections of 3 or more paths are different in the two cases. If relevant, such higher order interactions will lead to different universality classes. Actually his most compelling evidence is from numerical results for the transverse fluctuations of the paths which grow as $t^{2/3}$ for DP, and are reasonably well fitted to $t^{3/4}$ for CDP[22]. Zhang does not provide any numerical results for the variance of $\ln I$, as in Figure 2. Two recent, rather extensive, numerical studies[24][25] shed more light on this problem. Both simulations seem to equivocally point to the importance of including corrections to scaling in the fits. In 1+1 dimensions they indeed find $\omega = 1/3$ for the variance, and $\nu = 2/3$ (with a large correction to scaling term) for transverse fluctuations. We close by pointing out that the numerical transfer matrix computations for DP and CDP are computationally straightforward as they can be carried out in polynomial time. Hence these, and future, extensive simulations should easily be capable of establishing, or disproving, the equivalence between DP, KPZ, and tunneling problems.

Acknowledgements. We have benefited from discussions with B. Altshuler, Y. Meir, Y. Shapir, N. Wingreen, and X.R. Wang. This research was supported by the NSF through grant number DMR–90–01519, the PYI program (MK).

References

[1] P.A. Lee and B.L. Altshuler, *Physics Today* **41** 36 1988; P.A. Lee and T.B. Ramakrishnan, *Rev. Mod. Phys.* **57** 287 1985.

[2] B.L. Altshuler and B.I. Shklovskii, *Zh. Eksp. Teor. Fiz.* **91** 220 1986; P.A. Lee, A.D. Stone and H. Fukuyama, *Phys. Rev. B* **35** 1039 1987.

[3] N. Zannon and J.L. Pichard, *J. Phys. (Paris)* **49** 907 1988.

[4] O. Faran and Z. Ovadyahu, *Phys. Rev.* **B38** 5457 1988.

[5] V.L. Nguyen, B.Z. Spivak and B.I. Shklovskii, *Pis'ma Zh. Eksp. Teor. Fiz.* **41** 35 1985 [*JETP Lett.* **41** 42 1985]; *Zh. Eksp. Teor. Fiz.* **89** 11 1985 [*JETP Sov. Phys.* **62** 1021 1985].

[6] U. Sivan, O. Entin–Wohlman and Y. Imry, *Phys. Rev. Lett.* **60** 1566 1988; O. Entin-Wohlman, Y. Imry and U. Sivan, *Phys. Rev. B* **40** 8342 1988.

[7] Y. Meir, N.S. Wingreen, O. Entin–Wohlman and B.L. Altshuler, *Phys. Rev. Lett.* **66** 1517 1991.

[8] J.L. Pichard, M. Sanquer, K. Slevin and P. Debray, *Phys. Rev. Lett.* **65** 1812 1990.

[9] E. Medina, M. Kardar, Y. Shapir and X. R. Wang, *Phys. Rev. Lett.* **62** 941 1989; *ibid* **64** 1816 1990.

[10] E. Medina, and M. Kardar, *Phys. Rev. Lett.* **66** 3187 1991.

[11] P.W. Anderson, *Phys. Rev.* **109** 1492 1958.

[12] A. Cohen, Y. Roth and B. Shapiro, *Phys. Rev. B* **38** 12125 1988.

[13] P.W. Anderson, D.J. Thouless, E. Abrahams and D.S. Fisher, *Phys. Rev. B* **22** 3519 1980.

[14] See, e.g. *Dynamics of Fractal Surfaces*, (F. Family and T. Vicsek eds.) World Scientific, Singapore 1991.

[15] Y. Shapir and X.R. Wang, *Europhys. Lett.* **4** 1165 1987.

[16] M. Kardar and Y.-C. Zhang, *Phys. Rev. Lett.* **58** 2087 1987.

[17] M. Kardar, *New Trends in Magnetism* (M.D. Coutinho–Filho and S.M. Rezende, eds.) World Scientific, Singapore 1990.

[18] M. Kardar, *Disorder and Fracture*, (J.C. Charmet, S. Roux and E. Guyon, eds.) Plenum Press, New York 1990.

[19] M. Kardar *Nucl. Phys.* **B290** [FS20] 582 1987.

[20] M. Kardar, G. Parisi and Y. Zhang, *Phys. Rev. Lett.* **56** 889 1986.

[21] J.M. Kim and J.M. Kosterlitz, *Phys. Rev. Lett.* **62** 2289 1989.

[22] Y.-C. Zhang, *Phys. Rev. Lett.* **62** 979 1989.

[23] Y.-C. Zhang, *Europhys. Lett.* **9** 113 1989.

[24] M.P. Gelfand, *Physica A* **177** xxx 1991.

[25] Y. Y. Goldschmidt and T. Blum, preprint 1992.

LOW TEMPERATURE PROPERTIES OF DIRECTED WALKS WITH RANDOM SELF INTERACTIONS

B. Derrida and P.G. Higgs

Service de Physique Théorique[1] CE-Saclay, F-91191 Gif-sur-Yvette Cedex, France

Abstract. We consider a simple model for a chain with random self interactions. Each monomer on the chain carries a charge which is a random variable and the configurations of the chain are directed walks. We show that at zero temperature the end to end distance R and the energy E of chains with total charge $Q \sim N^x$ scale as $R \sim N^{\nu(x)}$ and $E \sim N^{\theta(x)}$. The exponents thus depend on the total charge through the parameter x. These exponents are measured numerically and compared to the predictions of some simple arguments. At finite temperature R is extensive and appears to be independent of x.

1. Introduction

Many models have been proposed to understand the problem of the folding of heteropolymers like proteins or DNA molecules[1, 5]. For many of these models, ideas coming from the mean field theory of spin glasses have proved to be very useful. Recently, it was shown that even very simplified one dimensional models of heteropolymers[6, 7] could present interesting properties. Kantor and Kardar[6] considered a simple model where the heteropolymer consists of a sequence of N random charges $q_i = \pm 1$, the allowed configurations of which are the 1d random walks $\{r_i\}$ with $r_{i+1} - r_i = +1$ or -1. The energy $\tilde{E}(\{r_i\})$ of a configuration $\{r_i\}$ of the polymer is defined by

$$\tilde{E}(\{r_i\}) = \sum_{1 \leq i < j \leq N} q_i q_j \delta_{r_i, r_j} \tag{1}$$

---

[1] Laboratoire de la Direction des Sciences de la Matière du Commissariat à l'Energie Atomique,

Despite its simplicity, this model is not easy to attack either by analytical or by numerical approaches. At zero temperature, using an exact enumeration procedure to find the ground state configurations, Kantor and Kardar predicted power laws for the end to end distance $R \sim N^\nu$ and the fluctuation of the ground state energy $\Delta \tilde{E} \sim N^\theta$ (with $\nu \simeq .574$ and $\theta \simeq .58$). Because there is so far no alternative to the exact enumeration procedure, it is difficult to check the validity of these results for sizes $N \geq 20$.

More recently, an even simpler model was considered in a joint work with R.B. Griffiths[7]. The model is very similar to the one studied by Kantor and Kardar, the only difference being that the configurations $\{r_i\}$ are directed walks ($r_{i+1} - r_i = 0$ or 1). It is possible to obtain much more information when the walk is directed because transfer matrix techniques can be used allowing for the numerical study of much longer chains ($10^4 - 10^5$).

When the charges q_i are $+1$ or -1, the results[7] revealed the existence of a weak freezing transition temperature below which some degrees of freedom get frozen. At zero temperature, however, the problem looks simple since the ground state energy of a given chain is always related to the absolute value of the total charge of the chain. This can be seen by rewriting the energy $\tilde{E}(\{r_i\})$ defined by (1) as

$$2\tilde{E}(\{r_i\}) + \sum_i q_i^2 = E(\{r_i\}) = \sum_r [Q(r)]^2 \tag{2}$$

where

$$Q(r) = \sum_i q_i \delta_{r_i, r} \tag{3}$$

is the total charge on site r. In what follows, for convenience we will always call $E(\{r_i\})$ defined by (2) the energy of the system. Obviously, when the $q_i = \pm 1$, the ground state energy is always equal to the absolute value of the total charge $|\sum_i q_i|$ since one can easily build ground state configurations where each occupied site r has a total charge $Q(r)$ which is 0 or ± 1 so that $\sum_r Q^2(r) = |\sum_i q_i|$. However, the ground state being usually very degenerate, it is not easy to obtain the end to end distance R at zero temperature since one would need to average over all the ground state configurations. (Only when the total charge $Q = 0$, one can show that $R \sim N^{1/2}$ because the calculation of R can be formulated as the number of returns to the origin of a random walker[7]).

In this paper, we discuss the low temperature properties of this one dimensional directed model[7] when the charges q_i have a continuous distribution $\rho(q_i)$ that we will take to be a Gaussian of width 1

$$\rho(q_i) = \frac{1}{\sqrt{2\pi}} \exp\left(-q_i^2/2\right) \tag{4}$$

For continuous charges q_i, even the N dependence of the ground state energy of directed walks is a non trivial problem. We will see that an important parameter is the way the total charge Q of the chain scales with N. So in what follows, we study chains of N random Gaussian charges q_i with the constraint that the sum Q is given by

$$Q = \sum_{i=1}^{N} q_i = N^x \tag{5}$$

where x is a fixed parameter when we vary N. (It is easy to choose numerically N Gaussian numbers satisfying the constraint (5)).

Because the chain is directed, one can use transfer matrices[7]. The partition function Z_N of a chain of N charges satisfies the following recursion

$$Z_N = \sum_{i=0}^{N-1} Z_i \exp\left[-\left(\sum_{j=i+1}^{N} q_j\right)^2 / T\right] \tag{6}$$

where T is the temperature and $Z_0 = 1$. The calculation of the average size R_N (more precisely R_N is the average number of sites occupied by the chain) can also be done recursively

$$R_N = 1 + \frac{1}{Z_N} \sum_{i=0}^{N-1} R_i \, Z_i \, \exp\left[-\left(\sum_{j=i+1}^{N} q_j\right)^2 / T\right] \tag{7}$$

with $R_0 = 0$.

At zero temperature, (6) induces a very simple recursion for the ground state energy E_N :

$$E_N = \min_{0 \le i \le N-1} \left[E_i + \left(\sum_{j=i+1}^{N} q_j\right)^2\right] \tag{8}$$

with $E_0 = 0$. Note that in contrast to the case $q_i = \pm 1$, the ground state is non degenerate for continuous variables q_i.

It is clear from recursions (6), (7) and (8) that the time required to calculate the properties of a chain of length N increases at most as N^2.

Figure 1 shows the structure of the ground state for a random chain of $N = 2000$ monomers with various choices of x (see (5)). For each x, the plot shows the total charge Q_i up to the ith charge

$$Q_i = \sum_{j=1}^{i} q_j \tag{9}$$

versus i and the vertical bars are the break points in the ground state configuration (i.e. all the monomers between two successive vertical bars are on the same site). Clearly the number of vertical bars is equal to R.

We see in figure 1 that as x decreases, R decreases (since the number of bars decreases) and the fluctuations of the distances ℓ between successive bars become more and more important.

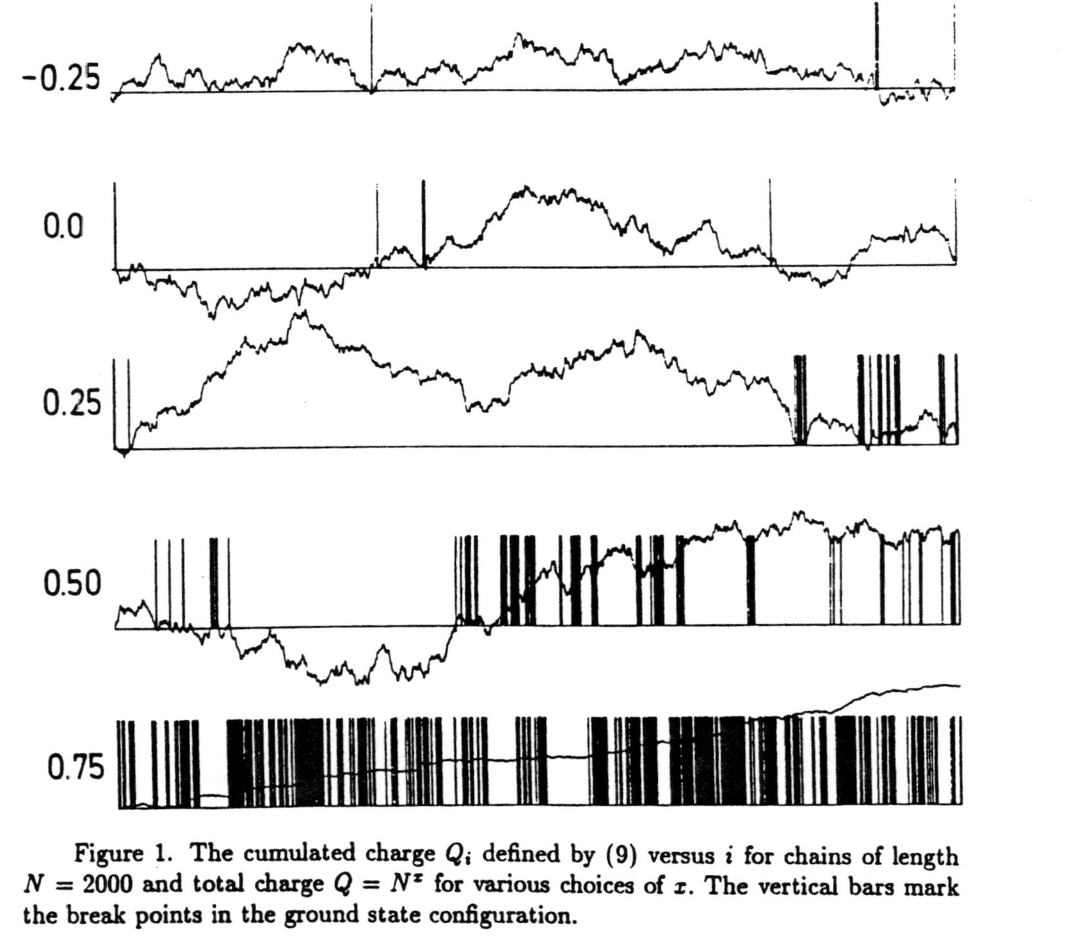

Figure 1. The cumulated charge Q_i defined by (9) versus i for chains of length $N = 2000$ and total charge $Q = N^x$ for various choices of x. The vertical bars mark the break points in the ground state configuration.

2. The scaling of the ground state energy E and of the number of occupied sites R

When one averages the ground state energy E or the number of occupied sites R (at $T = 0$) over many samples (here 5000 samples), one finds evidence that both E and R behave like power laws of N with exponents $\theta(x)$ and $\nu(x)$ which depend on x (see figure 2).

Because one can write the energy as

$$E = \sum_{r=1}^{R} [Q(r)]^2 \quad \text{and} \quad Q = N^x = \sum_{r=1}^{R} Q(r) \tag{10}$$

one always has

$$E \geq Q^2 / R \tag{11}$$

which implies that

$$\theta(x) \geq 2x - \nu(x) \tag{12}$$

In figure 2c, the ratios ER/Q^2 versus N is shown for various values of x. In all cases, the ratio seems to reach a limit as N increases, indicating that (12) is an equality rather than an inequality. The equality would mean that the total charge Q is evenly distributed among the R sites since a charge Q/R on each site would give an energy $E = R(Q/R)^2$.

As always, it is difficult to extract very reliable exponents from log-log plots. To estimate $\theta(x)$ and $\nu(x)$, we generated 2000 chains of sizes 50, 500 and 5000 and we estimated these exponents by comparing $[\theta = \log(E_{500}/E_{50})/\log 10]$ the sizes 50 and 500 (white circles) and the sizes 500 and 5000 (black circles). Both for $\theta(x)$ and $\nu(x)$, the results (figure 3) show little difference between the two sets of data, giving confidence that $N \geq 50$ is close to the asymptotic regime.

The problem, of course, is to understand the x dependence of these exponents. We have not been able to find an exact expression of $\theta(x)$ and $\nu(x)$ for all x. Instead, we can give an argument which leads to

$$\begin{aligned}
\nu(x) &= \frac{1 + 2x}{3} \quad \text{for } x \text{ close to} \quad 1 \\
\nu(x) &= \frac{1 + 2x}{4} \quad \text{for } x \text{ close to} \quad -\frac{1}{2}
\end{aligned} \tag{13}$$

These predictions together with the corresponding predictions for $\theta(x)$ using the equality in (12) are the straight lines shown on figure 3. They seem to be in good agreement with all data except for the neighborhood of $x = 1/2$ where there seems to be a crossover from one regime to the other.

Let us first present the argument which gives $\nu(x) = (1 + 2x)/3$. To obtain the ground state energy, the best is to divide the total charge $Q = N^x$ as much as possible. If the chain occupies R sites, the best would be to have on each site a charge Q/R and the resulting energy would be

$$E = Q^2 / R \tag{14}$$

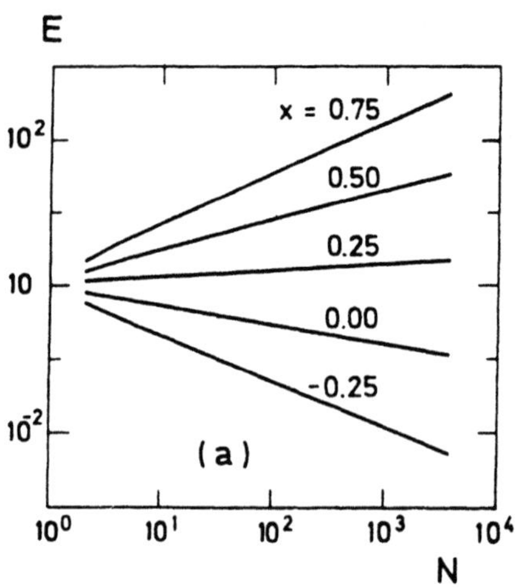

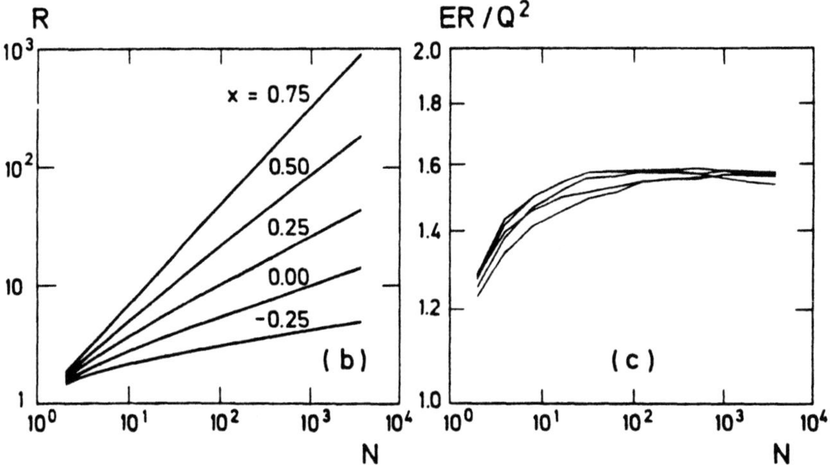

Figure 2. Log-log plot of the energy (Fig.2a) and R (Fig.2b) averaged over 5000 samples versus N for various choices of x. Fig.2c shows the ratio ER/Q^2 which seems to have a limit as N increases, indicating that (12) is an equality.

This expression tells us that to decrease E, R has to be as large as possible (and one would conclude that $R = N$ in the best). However, this is not possible because if $R \sim N$ there would be of order 1 monomer per site and hence a charge of order 1 per site, which is not consistent with $Q/R \sim N^{x-\nu(x)}$ as required to satisfy (14). It is therefore necessary to find the maximal value of R consistent with a charge of order Q/R on each site.

Here, it is useful to mention a simple fact on the sum S_ℓ of ℓ continuous random variables

$$S_\ell = \sum_{i=1}^{\ell} q_i \tag{15}$$

If one asks what is the minimum value $S_{\min}$ of $|S_k|$ for $1 \leq k \leq \ell$, one finds that

$$S_{\min} = \min_{1 \leq k \leq \ell} |S_k| \sim \ell^{-1/2} \tag{16}$$

This is because the average number of points S_k in an interval of size ΔS around the origin varies like $\ell\,\Delta S/\sqrt{2\pi\ell}$. When $\Delta S \simeq S_{\min}$, this number is of order 1.

So if the chain is divided into R occupied sites and **if one assumes that the number ℓ of monomers does not fluctuate too much from site to site**, then ℓ is given by $\ell \sim N/R$, and the charge on each occupied site is of order $\ell^{-1/2} \sim (N/R)^{-1/2}$. Then the energy E can be obtained using (14) and the fact that

$$Q/R \sim (N/R)^{-1/2} \tag{17}$$

i.e.

$$R \sim N^{(1+2x)/3} \quad \text{and} \quad E \sim N^{(4x-1)/3} \tag{18}$$

These predictions shown on figures 3 are in good agreement with the numerical simulations for x close to 1, but below $x = 1/2$, the agreement is clearly not satisfactory.

An important assumption to arrive at (18) was that the number ℓ of monomers on each occupied site does not fluctuate too much. This seems not to be true according to figure 1 (at least for the smaller values of x). Therefore one needs to know if the large fluctuations of ℓ are sufficient to invalidate the relation $\ell = N/R$. If one considers the total charge Q_i up to site i (9), Q_i performs a random walk and when $|Q_{i+\ell} - Q_i|$ is small enough, a break point (a vertical bar in figure 1) becomes possible. So the problem of the distance ℓ between successive vertical bars is similar to that of the first return close to the origin of a random walk. One knows from the theory of random walks[8, 9] that the probability of returning close to the origin for the first time after ℓ steps behaves like $\ell^{-3/2}$ for large ℓ. Therefore if there are R occupied sites and if the typical number of monomers is ℓ_0 on each site, one expects that the site with the largest number of monomers will have a number of monomers $\ell_{\max}$ given by

$$R \int_{(\ell_{\max}/\ell_0)}^{\infty} \frac{\mathrm{d}(\ell/\ell_0)}{(\ell/\ell_0)^{3/2}} \sim 1 \quad \text{i.e.} \quad \ell_{\max} \sim R^2 \ell_0 \tag{19}$$

This implies that most monomers are located on a very small number of sites, since $\ell_{\max} \gg \ell_0 R$. Therefore one arrives at

$$N \sim \ell_{\max} \sim R^2 \ell_0 \tag{20}$$

which together with the relation we had before that $Q/R \sim \ell_0^{-1/2}$ and (14) leads to

$$R \sim N^{(1+2x)/4} \quad \text{and} \quad E \sim N^{(6x-1)/4} \tag{21}$$

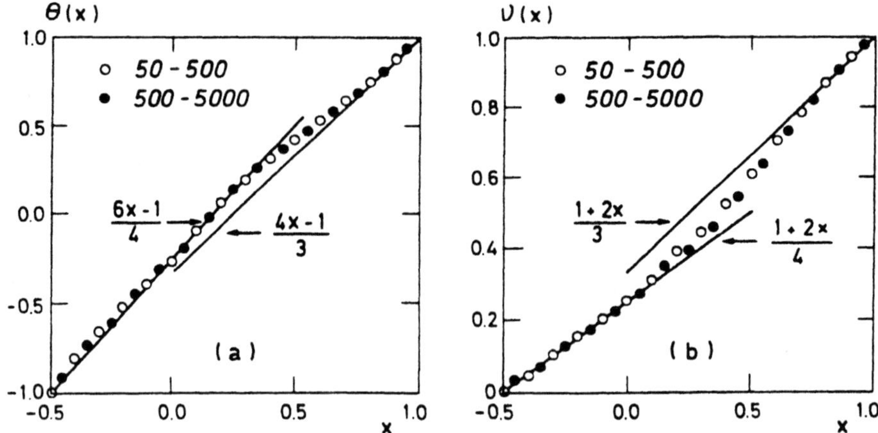

Figure 3. Estimates of the exponents $\theta(x)$ (Fig.3a) and $\nu(x)$ (Fig.3b) obtained by comparing results for $N = 50$ and 500 (white circles) and $N = 500$ and 5000 (black circles). The statistics were on 2000 chains.

These predictions are in good agreement with the simulations at least for $x \leq 0$ (see figures 3).

It is of course clear that $\nu(-1/2) = 0$ since if $Q = N^{-1/2}$ there is at most of order one other point with Q_i as small as $N^{-1/2}$.

Of course, one can wonder why (18) is more likely to be valid for x close to 1 and (21) for x close to $-1/2$. There is no very strong argument for that except for one more comparison with the statistics of the returns of a random walker close to the origin.

The main difference between the discussion which leads to (21) and the case $x > 1/2$ is that for a random walker, a strong enough bias (when $x > 1/2$) makes the probability of first return close to the origin after ℓ steps decay much faster than $\ell^{-3/2}$ for large ℓ. So the fluctuations of the number ℓ of monomers from site to site are expected to be much less important for $x > 1/2$ and considering that the ℓ are all roughly the same is not too bad an approximation.

Even if the numerical results of figures 3 are well fitted for $x > 1/2$ and $x < 0$ by expressions (18) and (21), the intermediate range does not seem to agree with either of these formulae and it is not clear to us whether another expression of the exponents should hold in this intermediate regime or whether (18) and (21) could be valid respectively for $\frac{1}{2} < x < 1$ and $-\frac{1}{2} < x < \frac{1}{2}$, the discrepancy in figure 3 being due to a slow convergence problem.

3. Conclusion

We have seen that for the problem of directed random heteropolymers with a continuous distribution of charges, the energy E and the end to end distance R at $T = 0$ seem to scale like power laws $E \sim N^{\theta(x)}$ and $R \sim N^{\nu(x)}$ with exponents $\theta(x)$ and $\nu(x)$ which vary with the total charge $Q = N^x$ of the chain.

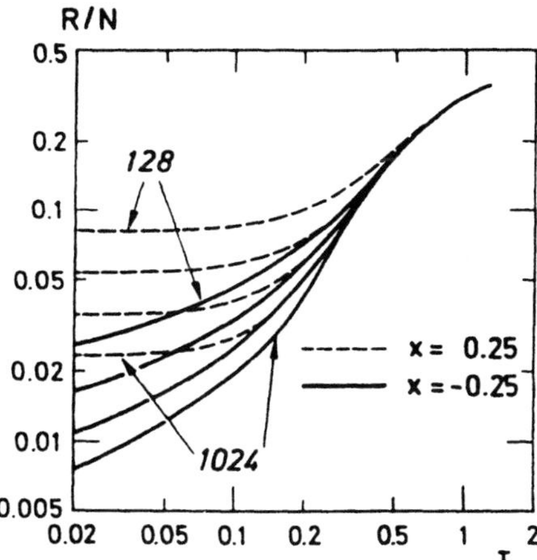

Figure 4. R averaged over 1000 samples for sizes $N = 128, 256, 512$ and 1024 and with $x = 0.25$ and $x = -0.25$. As N increases, the range of temperatures where finite size effects are visible decreases.

We have obtained approximate expressions of these exponents (18) and (21) which agree rather well with the results of simulations at least in some ranges of value of x. For unconstrained chains (i.e. when Q is not fixed), the exponent ν is equal to $\nu(1/2) \simeq .6$ according to Fig.3. It is worth noting that this is rather close to the prediction of Kantor and Kardar for the undirected case.

One can wonder how the x dependences of the zero temperature properties can manifest themselves at non zero temperature. Figure 4 shows the temperature dependence of R for various sizes N and various x. The results indicate that the finite size effects and the x dependence seem to occur in a range of temperature which shrinks as $N \longrightarrow \infty$.

So we think that a picture consistent with the results of figure 4 is that when $N \longrightarrow \infty$ at fixed temperature, the number of occupied sites R is extensive

$$\frac{R}{N} \longrightarrow r(T) \tag{22}$$

and $r(T)$ is independent of x. The range of temperature where the x dependence is noticeable varies probably like a power law $N^{-\lambda(x)}$ with an exponent $\lambda(x)$ which should vary with x and so for $T \sim N^{-\lambda(x)}$, one would expect

$$R \sim N^{\nu(x)} h\left(T\, N^{\lambda(x)}, x\right) \tag{23}$$

We did not find a good way of extracting the exponent $\lambda(x)$ from our data of figure 4 and so we do not present here any prediction for $\lambda(x)$. However if the picture of the two regimes (22) and (23) is right, then for R/N to be independent of x for $T\ N^{\lambda(x)}$ large, one would need that $(1 - \nu(x))/\lambda(x)$ is independent of x.

A similar low temperature range where the exponents are x dependent and which shrinks as N increases seems to be present in several other models and we hope to discuss this in more details in a future work.

If this picture is right, it would be the existence of these low temperature regimes which are responsible for the non trivial exponents $\theta(x)$ and $\nu(x)$. In these low temperature regimes, the sample to sample fluctuations would become important. It is worth noting that in these low temperature regimes, the exponents $\nu(x)$ are due to non extensive changes of the energy. This is rather reminiscent of the fact that for homopolymers with excluded volume, the Flory approximation predicts an exponent ν by calculating contributions to the free energy which are non extensive.

Acknowledgements

We thank J. des Cloizeaux and E.R. Speer for interesting discussions.

REFERENCES

[1] S.P. Obukhov, *J. Phys.* **A19**, 3655 1986.
[2] J.D. Bryngelson and P.G. Wolynes, *Proc. Natl. Acad. Sci.* **84**, 7524 1987; *Biopolymers* **30** 177 1990.
[3] T. Garel and H. Orland, *Europhysics Letters* **6** 307 1988 and **6** 597 1988.
[4] E.I. Shakhnovich and A.M. Gutin, *Europhysics Letters* **8** 327 1989 and *J. Phys.* **A22** 1647 1989.
[5] P.G. Higgs and J.F. Joanny, *J. Chem. Phys.* **94** 1543 1991.
[6] Y. Kantor and M. Kardar, *Europhysics Letters* **14** 421 1991.
[7] B. Derrida, R.B. Griffiths and P.G. Higgs, *Europhysics Letters* **18** 361 1992.
[8] E.W. Montroll, B.J. West, in *Fluctuation Phenomena* E. Montroll and J. Lebowitz eds., North Holland 1979.
[9] J.W. Haus and K.W. Kehr, *Phys. Rep.* **150** 265 1987.

DISTURBING THE RANDOM ENERGY LANDSCAPE

Tim Halpin-Healy and Devorah Herbert
*Physics Department, Barnard College, Columbia University, New York, NY
10027-6598 USA*

Abstract. We examine the effects of correlated perturbations upon globally optimal paths through a random energy landscape. Motivated by Zhang's early numerical investigations into ground state instabilities of disordered systems as well as the work of Shapir on random perturbations of roughened manifolds, we have studied the specific case of random bond interfaces unsettled by small random fields, confirming recent predictions for the instability exponents. Implications for disordered magnets and growing surfaces are discussed.

1. Introduction

Much of the present interest in kinetic roughening phenomena [1] can be traced to the tremendous outpouring of research on the dynamic scaling properties of Eden clusters and ballistic deposits [2], following the introduction of a noisy Burgers' equation by Kardar, Parisi and Zhang [3]. In the past five years, we have discovered the extraordinary richness of this intriguing equation; within the realm of KPZ can be found a wonderfully varied collection of apparently unrelated physical problems which, beyond the stochastic growth models mentioned above, include long time tails of randomly stirred fluids, the asymptotics of flame front propagation, as well as the roughening of vortex flux lines in ceramic superconductors. This last system [4], which in a different guise concerns the meandering of a directed polymer in a random medium (DPRM) [5-6], is essentially a baby version of the very difficult spin-glass problem [7] that has plagued the statistical physics community for the last decade, replete with an ultrametric free energy landscape and potentially severe replica symmetry breaking [8], though blessed by a number of simplifying features that make it one of the few tractable problems of ill-condensed matter. Controlled by a strong disorder zero temperature fixed point, the DPRM is a classic global optimization problem in which one seeks to minimize the total energy of a directed path through a random environment. Performing averages of many realizations of the random energy landscape yields highly nontrivial geometric and thermodynamic properties that characterize the ensemble of optimal paths. The present paper, motivated initially by the work of Zhang [9], and later influenced by the efforts of Shapir [10] and Mezard [11], investigates the resistance of these optimal paths to random disturbances of the disordered landscape. Our results address specifically the role of correlated drifts in the DPRM random energy landscape and confirms important predictions for random field (RF) perturbations of random bond (RB) domain walls in disordered two-dimensional magnets.

2. The Model

Our starting point is Zhang's formulation of the zero-temperature DPRM, which is most amenable to extensive numerical simulation. One considers a directed walker who, starting at the origin of a square lattice, has the option of making an immediate step diagonally left or right to $(x,t) = (\pm 1, 1)$. These and succeeding diagonal bonds have random energies drawn uniformly between 0 and 1. Neighboring bonds are uncorrelated. At the time slice t there are t+1 possible endpoints to the 2^t paths emanating from (0,0). As discussed earlier, the zero temperature DPRM is simply a matter of global optimization which, for a given realization of the random energy landscape (i.e., collection of random bonds on the lattice), entails finding the path of overall least energy, where the total energy of a path is given by the sum of the random bonds visited along the way. Many essential features of the 1+1 dimensional DPRM were established early on; in particular, it is known that its geometric properties are controlled by transverse fluctuations off the central axis that scale as $x_{rms} \sim t^{\zeta=2/3}$, while sample to sample fluctuations in the energy of the globally optimal trajectory scale as $e_{rms} \sim t^{\omega=1/3}$, there being an index relation, $\omega = 2\zeta - 1$, connecting the energy and wandering exponents.

As stressed by Zhang [9], the globally optimal path through a given realization of the random energy landscape is, however, quite susceptible to small changes in that random environment, there being many neighboring paths whose energies are very close to that of the ground state, but whose configurations differ considerably. These concerns have great physical import, of course, since true physical systems often possess quenched disorders that are actually dynamical variables, albeit on rather long timescales and very small amplitudes. In his own numerical investigation into these issues, Zhang concentrated on the effects of an uncorrelated slow drift in the random energies of the bonds; that is, he considered adding to each random bond energy an uncorrelated perturbation drawn with uniform probability between 0 and δ, with $\delta << 1$. In the context of disordered two-dimensional magnets, this corresponds to uncorrelated RB perturbations upon a RB interface. In the DPRM global optimization problem, this procedure yields two realizations of the random energy landscape that are different, though produced from similar distributions and possessing substantial overlap. Because of the work of Shapir [10], one knows that there exists a crossover length scale $t^* \sim \delta^{-1/\varphi}$, where $\varphi_{RB} = 1/6$ for random bond perturbations and $\varphi_{RF} = 1/2$ for random field perturbations, beyond which the small differences between the

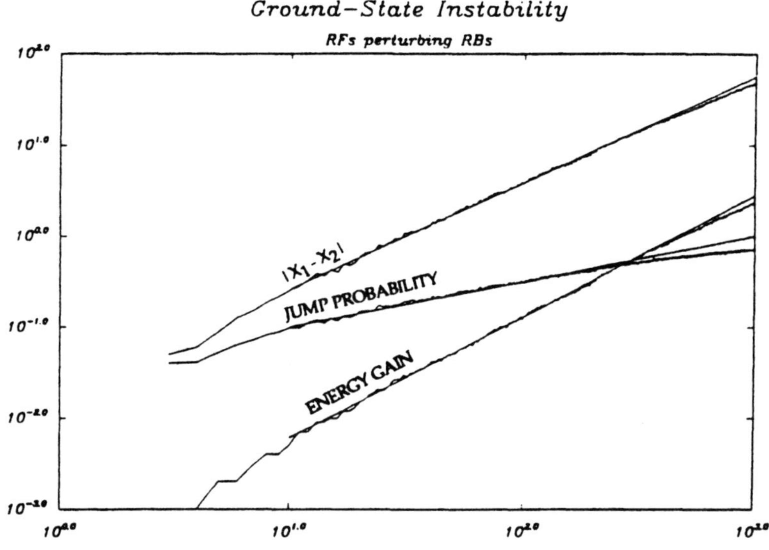

Figure 1. RF perturbations of the DPRM. From the left- Top curve: mean jump distance. Middle curve: jump probability. Bottom curve: energy advantage of new best path over old best path.

perturbed and unperturbed realizations of the random energy landscape manifest themselves in an asymptotic fashion. There are, however, a number of interesting scaling properties associated with ground-state instabilities of this disordered system that reveal themselves immediately. We focus our attention upon them first.

3. The Numerical Data

Consider, for example, the fact that the two globally optimal paths in the two different, but highly correlated random environments, are typically quite distinct. If x_1, x_2 denote the tranverse positions of these two best paths, then we find for RFs perturbing RB interfaces that the mean jump scales with path length as, see Fig. 1, $|x_1 - x_2| \sim t^{\alpha=1.16\pm0.02}$, entirely consistent with Shapir's prediction that

$$\alpha_{RF} = \varphi_{RF} + \zeta_{RB} = 1/2 + 2/3 = 7/6$$

Our own data for the case of RBs perturbing RBs, which Zhang considered in his original work and is presented here for sake of comparison, is shown in Fig.2, corroborating the mean jump exponent $\alpha_{RB} = \varphi_{RB} + \zeta_{RB} = 1/6 + 2/3 = 5/6$. In both instances, we have used $\delta = 0.1$ and performed disorder averages over 4000 realizations of the random energy landscape. Furthermore, it is apparent that in the case of RFs perturbing RBs, where the crossover length scale $t^*_{RF} \sim \delta^{-2} \sim 100$, the data begin to pull away from the straight line fit. By contrast, for RBs perturbing RBs, the data

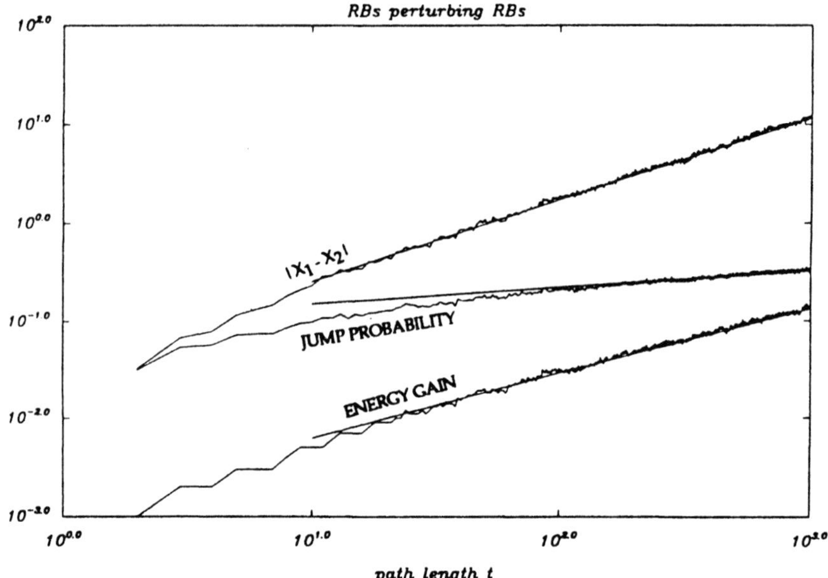

Figure 2. Same as previous figure, but for RB perturbations of the DPRM- the case considered by Zhang.

follows the line well beyond accessible system sizes since $t^{\star}_{RB} \sim \delta^{-6} \sim$ million steps in this case.

Note that this mean jump scaling index is quite large because of the ultrametric structure of the ensemble of locally optimal paths. The very best path in a given realization of randomness is stable with respect to its immediate family thanks to the substantial ancestry they have in common. Hence, there is an intrinsic resistance to change, which Zhang alludes to as a Hopfield memory effect. Nevertheless, because of the random perturbations upon the original disordered landscape, a distant relative of the original best path can accumulate enough energy gains to become the globally optimal trajectory in the perturbed landscape, incurring a large transverse jump in the process. Since the triumphant neighbor is rarely a local relative, the jumps make important contributions to the statistical averages.

In addition to the mean jump size, we have also studied the probability that a jump actually happens ($x_1 \neq x_2$) for the case of RF perturbations of the random energy landscape. A glance at Fig. 1 reveals that for paths shorter than the crossover lengthscale, the data are well fit by a line of slope 1/2. For RBs perturbing RBs, the data fall along a line of slope 1/6, as explained by Feigel'man and Vinokur [12]. Again, for the RB case, there is no indication, whatsoever, of incipient crossover phenomena. Note that, while there are geometric arguments [12-14] that predict these jump probability exponents, for both RFs and RBs, they coincide with the crossover index φ.

All this is easily understood on the basis of Shapir's scaling ansatz for the interfacial roughness $\mid x(t) \mid$,

$$\mid x_\delta(t) \mid = \mid x_o(t) \mid g(\delta t^\varphi) = \mid x_o(t) \mid \{1 + g'(0)\delta t^\varphi + ...\}$$

from which it follows that the displacement in the perturbed environment is given by

$$\Delta x(t) = \mid x_\delta(t) \mid - \mid x_o(t) \mid \sim \mid x_o(t) \mid g'(0)\delta t^\varphi \sim \delta t^{\varphi + \varsigma}$$

whence the mean jump exponent, while

$$P(jump) \sim \Delta x(t)/ \mid x_o(t) \mid \sim \delta t^\varphi$$

For reasons that are unclear, the energy-based derivation of Feigel'man and Vinokur[12] for the jump probability exponent appears not to carry through for RF perturbations of the RB landscape. Nonetheless, the linear dependence of both the mean jump distance and the jump probability on the strength of the perturbation were manifest in the RF simulations we performed for other values of δ [15]. Zhang had pointed out similar behavior in the RB case.

Given the quickly growing jump probability, it is natural to wonder whether the old best path in the original random energy landscape retains some honor by remaining a locally optimal path in the new disordered environment. In our numerical studies, we found that the energy change of the old best path in the different energy environments had an exponent of unity for RFs, which suggests that the new locally optimal path does indeed overlap substantially with the old best path. Finally, we investigated the energy advantage, in the new environment, that motivated the jump away from the old optimal path. See Figs. 1 and 2. It scales with an exponent $\omega'_{RF} = 4/3$ for RF perturbations, $\omega'_{RB} = 2/3$ for RB perturbations. The latter index had been noted of course, by Zhang. Nevertheless, our RF simulation provides additional support to the conjecture that the instability exponents obey a scaling relation, $\omega' = 2\alpha - 1$, analogous to that of the unperturbed problem.

Finally, in Fig. 3, we illustrate the long term implications of RF perturbations upon the RB landscape. Whereas the data for geometric and free energy fluctuations in the original unperturbed random energy landscape scale nicely and are consistent with the exponents $\zeta_{RB} = 2/3$ and $\omega_{RB} = 1/3$, in the new environment RF perturbations incur crossover to the stronger fluctuations characteristic of correlated roughening, the exponents $\zeta_{RF} = \omega_{RF} = 1$ in agreement with the those predicted by Imry-Ma type arguments [16] for the 2d RF Ising model. Note that, while deviations manifest themselves first for the sample to sample fluctuations of the energy, the march to asymptotic scaling is well under way for both quantities once the length scale $t^*_{RF} \sim 100$ is crossed.

It is clear from our numerical studies that correlated random perturbations can incur severe ground-state instabilities in uncorrelated disordered systems. In the case of the DPRM, a RF perturbation upon the RB landscape can have increasingly drastic consequences for the configuration of globally optimal paths, causing large jumps to distant relatives beyond the immediate family. Ultimately, of course, the scaling is

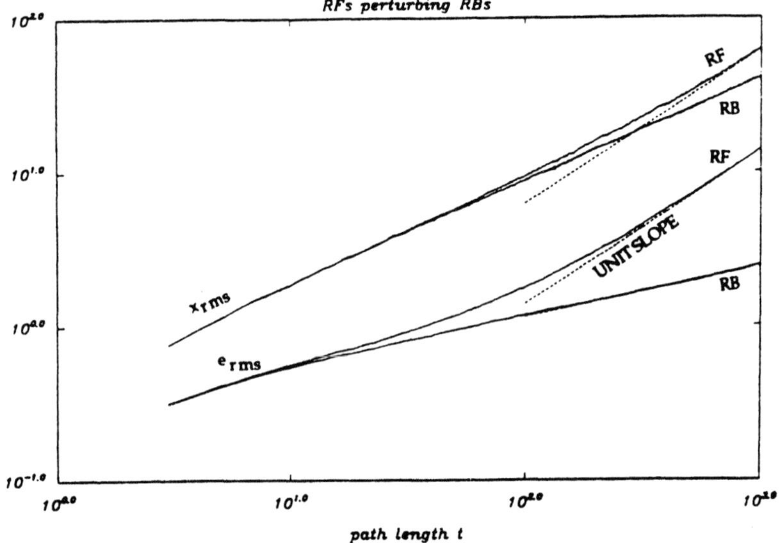

Figure 3. Position (top pair) and energy (bottom pair) fluctuations of the globally optimal path. Upper curves within each pair correspond to the best path in the disordered environment perturbed by random fields. For RFs perturbing RBs, both the position and energy fluctuations eventually scale with unit slope, in agreement with Imry-Ma predictions for the 2d RF Ising model. Lower curves, associated with the original unperturbed energy landscape, exhibit standard 1+1 DPRM exponents, $\zeta_{RB} = 2/3, \omega_{RB} = 1/3$.

controlled entirely by the RF fixed point, characterized by exponents quite different from those of the RB problem. These effects are presumably observable in 2d RB magnets that are subject to a very weak external magnetic field, giving rise to small perturbing RFs within the sample [17]. Thermally activated jumps of domain walls to minimal energy configurations might manifest themselves as large observable noises in measurements of the magnetization, susceptibility, etc. For the kinetic roughening of stochastically grown surfaces, the importance of a spatially correlated perturbation in the atomic beam would be dramatic, leading to radically different surface morphologies with substantially different scaling properties.

Acknowledgment

We tip our hat to Y.-C. Zhang for writing the nice PRL that inspired this work. THH was supported by grants from the Research Corporation and the Petroleum Research Fund, while DH was participating in the Undergraduate Research Program of the Pew Charitable Trust.

REFERENCES

[1] J. Krug and H. Spohn, *Solids far from Equilibrium*, (G. Godreche, ed.)Cambridge University Press, Cambridge 1991.

[2] F. Family and T. Vicsek, eds., *Dynamics of Fractal Surfaces*, World Scientific, Singapore 1991.

[3] M. Kardar, G. Parisi and Y.-C. Zhang, *Phys. Rev. Lett.* **56** 889 1986.

[4] T. Nattermann and R. Lipowsky, *Phys. Rev. Lett.* **61** 2508 1988.

[5] M. Kardar and Y.-C. Zhang, *Phys. Rev. Lett.* **58** 2087 1987.

[6] M. Kardar, *J. Appl. Phys.* **61** 3601 1987.

[7] M. Mezard, G. Parisi and M.A. Virasoro, *Spin glass theory and beyond*, World Scientific, Singapore 1987.

[8] M. Mezard and G. Parisi, *J. de Physique* I1 809 1991.

[9] Y.-C. Zhang, *Phys. Rev. Lett.* **59** 2125 1987.

[10] Y. Shapir, *Phys. Rev. Lett.* **66** 1473 1991.

[11] M. Mezard, *J. de Physique* **51** 1831 1990.

[12] M.V. Feigel'man and V.M. Vinokur, *Phys. Rev. Lett.* **61** 1139 1988.

[13] T. Nattermann, *Phys. Rev. Lett.* **60** 2701 1988.

[14] Y.-C. Zhang, *Phys. Rev. Lett.* **60** 2702 1988.

[15] T. Halpin-Healy and D. Herbert, *unpublished*

[16] T. Halpin-Healy, *Phys. Rev.* **A42** 711 1990.

[17] S. Fishman and A. Aharony, *J.Phys.* **C12** L729 1979.

DIRECTED POLYMERS IN A RANDOM POTENTIAL

Jin Min Kim

*Department of Theoretical Physics, The University, Manchester M13 9PL,
England*

Abstract. Directed polymers in a random potential are described by both extensive
simulations and the weak noise expansion. The energy fluctuation ΔE of the lowest
energy E grows as t^ω with $\omega \approx 1/(d+1)$. Our simulation shows that the probability
distribution $P(a)da$ of the normalized energy $a = \frac{E-\langle E \rangle}{\Delta E}$ is universal, i.e., it is independent of the model. In dimension $d = 2+1$, no evidence for a finite temperature phase
transition was found. Instead, the crossover length t^* is very large at high temperature
and has been found to vary as $\ln t^* \sim T^2$.

1. Introduction

Over recent years, there have been considerable efforts in studying directed polymers in a random potential [1]. Besides being one of the simplest problems involving
disorder, it is related through various mappings to other physical problems such as
the growth of an interface in the Eden model [2], ballistic deposition [3], the Kardar-
Parisi-Zhang (KPZ) equation [4], and the Burgers' equation [5]. Despite the simple
mathematical representation of the directed polymer problem, they are not understood
except in two dimensions.

The Hamiltonian of a directed polymer can be written as

$$H = \int dt[\gamma(\frac{d\mathbf{x}}{dt})^2 + \mu(\mathbf{x},t)] \tag{1}$$

where t is the polymer length perpendicular to the $d-1$ dimensional substrate, $\mathbf{x}$
is the transverse vector of the polymer in $d-1$ dimensions and $\mu(\mathbf{x},t)$ is a random
potential with $\langle\mu(\mathbf{x},t)\mu(\mathbf{x}',t')\rangle = 2D\delta^{d-1}(\mathbf{x}-\mathbf{x}')\delta(t-t')$. There are two competing
terms: one is a bending energy γ forcing the polymer to straighten and the other is

a quenched impurity $\mu(\mathbf{x}, t)$ which prefers that the polymer is deformed through the points of minimum potential.

We have already mentioned that the directed polymer in a random potential has strong connections with the KPZ equation [1]. The free energy $F(\mathbf{x}, t)$ of the directed polymer plays the same role as the continuous height variable $h(\mathbf{x}, t)$ of ballistic deposition. From the relation $h(\mathbf{x}, t) \sim \ln Z(\mathbf{x}, t)$ where the partition function $Z(\mathbf{x}, t)$ is the sum over all polymer configurations with Boltzmann weighting factor, $h(\mathbf{x}, t)$ satisfies the KPZ equation for a ballistic growth model [4],

$$\frac{\partial h(\mathbf{x}, t)}{\partial t} = \nu \nabla^2 h + \frac{\lambda}{2}(\nabla h)^2 + \eta(\mathbf{x}, t), \tag{2}$$

where $\eta(\mathbf{x}, t)$ is a random noise proportional to $\mu(\mathbf{x}, t)$.

In this work, we consider directed polymers on a discrete hypercubic lattice with periodic boundary conditions. The polymer path is restricted to $|\mathbf{x}(t) - \mathbf{x}(t + 1)| = 0$ or 1 with random potential $\mu(\mathbf{x}, t)$ assigned on the discrete site and a bending energy γ given to the case $|\mathbf{x}(t) - \mathbf{x}(t + 1)| = 1$.

The quantity of interest in the directed polymer problem is the free energy fluctuation ΔF which scales as [6]

$$\begin{aligned} \Delta F(t) = \langle (F - \langle F \rangle)^2 \rangle^{1/2} &\sim L^\chi f(t/L^z) \\ &\sim t^\omega \qquad t \ll L^z \\ &\sim L^\chi \qquad t \gg L^z \end{aligned} \tag{3}$$

where $\langle A \rangle$ is the average of A over the position $\mathbf{x}$ and L is the substrate size. There is a length scale $t^{1/z}$ describing the transverse fluctuation of directed polymers. In $d = 2$ (which we shall write as $d = 1 + 1$ to indicate that there is one parallel and one perpendicular direction), the exponents are known to be $\omega = 1/3$ and $1/z = 2/3$ from the stationary solution of the Burgers equation [7]. In higher dimensions, the values of the strong coupling exponents are not settled yet. Moreover, there is a controversy over the existence of a finite temperature phase transition in $d = 2 + 1$ [8,9].

Here, we study the directed polymer in a random potential by both numerical simulations and the weak noise expansion. We find that the probability distribution of a minimum energy is asymmetric with long tails due to the broken reflection symmetry($h \rightarrow -h$) of the KPZ Eq. In $d = 1 + 1$, we recover the known results $\omega = 1/3$ and we measure ω in $d = 2 + 1$ and $d = 3 + 1$, which satisfy the conjecture $\omega = 1/(d + 1)$ [10]. We find no evidence of a phase transition in $d = 2 + 1$. Instead, we find an extremely large crossover length t^*, viz. $\ln t^* \sim T^2$ for high temperature T.

2. Zero Temperature Directed Polymer in a Random Potential

At zero temperature, the entropy is ignored and the problem is much simplified by finding an optimal path and its energy. The optimal energy $E(\mathbf{x}, t)$ of the polymer ending at $(\mathbf{x}, t)$ is calculated recursively, for example in $d = 1 + 1$,

$$\begin{aligned} E(x, t) = Min[&E(x, t - 1) + \mu(x, t - 1), E(x + 1, t - 1) + \mu(x + 1, t - 1) + \gamma, \\ &E(x - 1, t - 1) + \mu(x - 1, t - 1) + \gamma] \end{aligned} \tag{4}$$

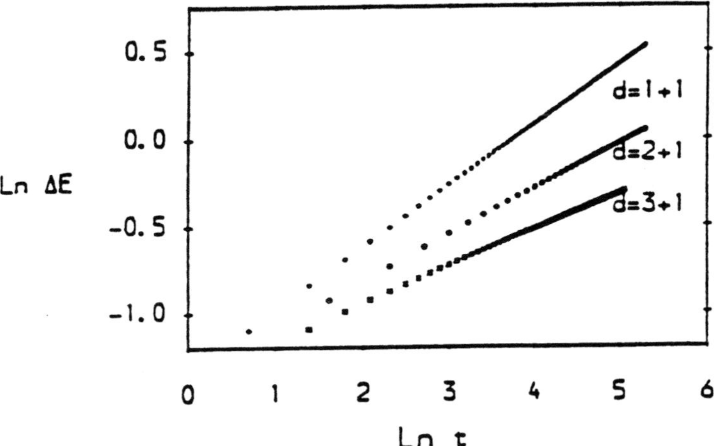

Figure 1. Energy fluctuation ΔE as a function of length t with Gaussian $\mu(\mathrm{x}, t)$ and $\gamma = 0.5$.

where Min takes the minimum value.

We start from a L^{d-1} substrate with periodic boundary conditions and as a random potential $\mu(x, t)$, we use Gaussian random numbers with $D = 1/24$. $\Delta E = \langle (E - \langle E \rangle)^2 \rangle^{1/2}$ is monitored as a function of t to determine ω, the exponent governing the rate of growth of the energy fluctuation. ΔE grows with time and obeys $\Delta E \sim t^\omega$ for $t \ll L^z$. The $ln (\Delta E) \sim ln\ t$ curves are shown in Fig. 1 for the largest system used, $L = 40,000$ $(d = 1 + 1)$, 1000 $(d = 2 + 1)$, and 100 $(d = 3 + 1)$. From the least square fitting of the data, we obtain

$$
\begin{aligned}
\omega &= 0.332 \pm 0.003 & d &= 1 + 1 \\
\omega &= 0.248 \pm 0.004 & d &= 2 + 1 \\
\omega &= 0.20 \pm 0.01 & d &= 3 + 1,
\end{aligned}
\tag{5}
$$

being consistent with the conjecture $\omega = \frac{1}{d+1}$ [10].

We calculate the probability $P(a, t)da$ at time t that the normalized energy $a = \frac{E - \langle E \rangle}{\Delta E}$ is between a and $a + da$. Fig. 2 shows very good data collapse for ten different times implying that $P(a, t)$ depends on the normalized energy a only. Since $P(a, t)$ is independent of time for $t \ll L^z$ as seen in Fig 2, the expectation value of higher moments $(m \geq 2)$ are

$$
M_m \equiv \langle (E - \langle E \rangle)^m \rangle \sim (\Delta E)^m \sim t^{m\omega}
\tag{6}
$$

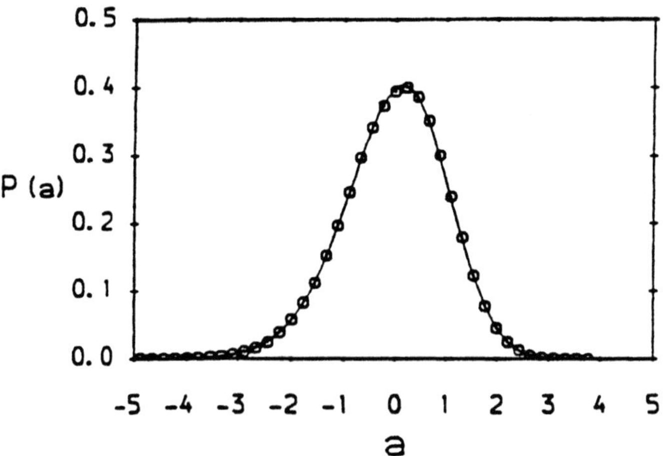

Figure 2. The probability distribution $P(a,t)$ of minimum energy E as a function of $a = \frac{E-\langle E\rangle}{\Delta E}$ at ten different times (t =30,60,... ,300 in $d = 1+1$; continuous lines) and the probability distribution $P_h(a,t)$ of height h as a function of $a = \frac{h-\langle h\rangle}{\Delta h}$ in a restricted solid-on-solid growth model for $d = 1+1$ ($t = 300, L = 40\,000$; circles).

and the nth cumulant is $C_n(t) \sim t^{n\omega}$ ($n \geq 2$). In $d = 1+1$, we calculate the cumulants and moments up to fourth order and they are consistent with $t^{n\omega}$ for nth order.

We find that $P(a)$ is not a gaussian distribution but a skewed distribution. Since there is a preferred growing direction which breaks the reflection symmetry of Eq. (2), the energy distribution should have the same symmetry with non-zero third cumulant. From a numerical simulation of the directed polymer, we get $\hat{C}_3 = -0.29 \pm 0.02$ in $d = 1+1$ where the normalized third cumulant $\hat{C}_3$ is $C_3(t)/(C_2(t))^{3/2}$. There is a trend that the asymmetry grows slowly with dimension.

Since there is a linear relation between the height in a growth model and the energy in the directed polymer model, we have studied the distribution of heights in the restricted solid on solid (RSOS) growth model [10] in $d = 1+1$. Defining $a = (h - \langle h\rangle)/\Delta h$, one sees from Fig. 2 that the scaling form of the height distribution $P_h(a)$ is identical to within numerical accuracy to $P(a)$ in our model of directed polymers. Even though the height is a discrete variable, the value of the normalized third cumulant in the model is around -0.29 which is the same as that of the directed polymer model. This result strongly implies the universality of the function $P(a)$ [11].

3. Finite Temperature Directed Polymer in a Random Potential

Consider a directed polymer on a discrete "hyper-pyramid" structure with random potential $\mu(\mathbf{x}, t)$ assigned to each site $(\mathbf{x}, t)$. The partition function $Z(\mathbf{x}, t)$ for the polymer ending at $(\mathbf{x}, t)$ can be obtained recursively. For example in $d = 1 + 1$

$$Z(x, t) = Z(x, t-1)\exp(-\frac{\mu(x, t-1)}{T}) + Z(x-1, t-1)\exp(-\frac{\mu(x-1, t-1) + \gamma}{T})$$
$$+ Z(x+1, t-1)\exp(-\frac{\mu(x+1, t-1) + \gamma}{T}),$$
$$(7)$$

where T is the temperature.

The whole partition function $Z(t)$ can be defined in terms of the sum over all paths starting at $(0, 0)$ as $Z(t) = \sum_{\mathbf{x}} Z(\mathbf{x}, t)$ and the free energy is given by $F(t) = -T \ln Z(t)$. In considering the thermodynamics of polymers, we take the symbol $\langle A \rangle_T$ to be the thermal average of the quantity A :

$$\langle A \rangle_T \equiv \frac{\sum_{\mathbf{x}} A Z(\mathbf{x}, t)}{\sum_{\mathbf{x}} Z(\mathbf{x}, t)} \tag{8}$$

and $\overline{A}$ as the sample average of A. The exponent z can also be obtained from the transverse fluctuation of the polymer $\overline{\langle x^2 \rangle_T} \sim t^{2/z}$. At infinite temperature, the random potential has no effect and the polymer follows a random walk with $z = 2$ and $\langle x \rangle_T = 0$ (high-temperature phase). However at low temperature, the walk is attracted by the local minimum potentials and it becomes super-diffusive (low-temperature phase). To study the transition between these two phases, we define a dimensionless quantity

$$g(T, t) \equiv \overline{\langle \mathbf{x} \rangle_T^2} \ / \ \overline{\langle \mathbf{x}^2 \rangle_T}, \tag{9}$$

such that g is unity at $T = 0$ and zero at $T = \infty$ [12].

At high temperature, after factoring out a term e^{3t} in Z and taking the continuum limit, Eq. (7) becomes

$$\frac{\partial Z(\mathbf{x}, t)}{\partial t} \approx \nabla^2 Z(\mathbf{x}, t) + \frac{\mu'(\mathbf{x}, t)}{T} Z(\mathbf{x}, t), \tag{10}$$

where $\mu'(\mathbf{x}, t)$ is a random potential with variance $2D'$. By a Cole-Hopf transformation, Eq. (10) becomes the KPZ equation. After a straightforward calculation, we obtain

$$g(T, t) \sim \frac{D'}{T^2} t^{\frac{3-d}{2}} \tag{11}$$

to first order in D'/T^2. Below the dimension $d = 2 + 1$, g increases with t showing no stable high temperature phase. For $d > 2 + 1$, since g approaches zero with t at high temperature, there exists a stable high temperature phase. At $d = 2 + 1$, the value of

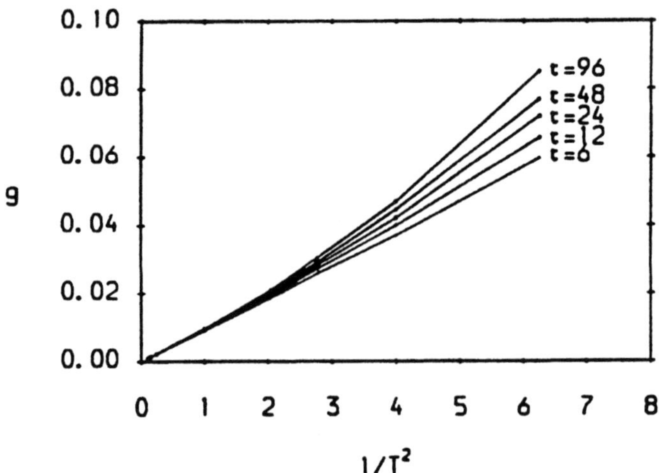

Figure 3. $g(T,t)$ as a function of $\frac{1}{T^2}$ for length $t = 6, 12, 24, 48$ and 96 in d=2+1.

g is independent of the length t in the first order calculation. If we add second order corrections, g becomes

$$g(T,t) \sim \frac{D'}{T^2}(1 + C\frac{D'\ln(t/t_0)}{T^2}), \qquad d = 2 + 1, \tag{12}$$

where C is a positive constant and t_0 is a cut-off length. Since g grows logarithmically with t, the infinite temperature fixed point is unstable and the cross over length t^* to the low temperature phase is very big; $\ln t^* \sim T^2$ for high temperature T. A polymer of length t follows a random walk in the transverse direction when $t \ll t^*$ and crosses over to the strong coupling regime (zero-temperature phase) when $t \gg t^*$. Therefore our weak noise expansion results indicate no evidence of a phase transition in $d = 2 + 1$ [12].

As a check on our results, we also calculated g numerically for systems of length 96 in $d = 2 + 1$ by the transfer matrix method as described in Eq. (7). A graph of $g(T,t)$ versus $1/T^2$ is shown in Fig. 3 for $t = 6, 12, 24, 48$ and 96. One may argue that there is a phase transition [8], because all the curves for the different lengths seem to meet each other near $T = 1$ and then stick together up to infinite temperature with $g(T = 1) \neq g(T \to \infty)$. However, our data show that $g(T = 1, t)$ increases very slowly with t, being consistent with Eq. (12) with a $(\ln t)/T^4$ correction to $1/T^2$, which itself indicates a flow to strong coupling at very large t. So, we conclude that there is no phase transition in $d = 2 + 1$.

4. Conclusion

At zero temperature, we have shown that the distribution $P(a)$ of a minimum energy is skewed as a function of normalized energy $a = \frac{E-\langle E \rangle}{\Delta E}$. This asymmetry indicates that the third cumulant of the energy is non-zero, this is due to the broken reflection symmetry of the KPZ Eq. In fact the $P(a)$ is identical to within our numerical accuracy to the equivalent distribution of heights in a RSOS model. We found that our results for ω are consistent with $\frac{1}{d+1}$. However, there exists a large scale simulation of a hypercube stacking discrete model, which gives slightly smaller value of ω in higher dimensions [9]. Although our model is a continuous version of the KPZ equation, the discrepancy remains unexplained. At finite temperature, we also find no evidence of a phase transition in $d = 2 + 1$.

Acknowledgements

The results presented here are a summary of work with A. J. Bray and M. A. Moore. I would like to thank them for their collaboration.

REFERENCES

[1] M. Kardar and Y. C. Zhang, *Phys. Rev. Lett.* **58** 2087 1987.

[2] M. Eden, *Proc. 4th Berkeley Symposium on Math. Stat. and Prob., Vol. 4* (ed. F. Neyman), Univ. of California Press, Berkeley 1961.

[3] P. Meakin, P. Ramanlal, L. M. Sander and R. C. Ball, *Phys. Rev. A* **34** 5091 1986.

[4] M. Kardar, G. Parisi and Y. C. Zhang, *Phys. Rev. Lett.* **56** 889 1986.

[5] J. M. Burgers, *The Nonlinear Diffusion Equation* Reidel, Boston 1974.

[6] F. Family and T. Vicsek, *J. Phys. A* **18** L75 1985.

[7] D. Forster, D. R. Nelson, and M. J. Stephen, *Phys. Rev. A* **16** 732 1977.

[8] B. Derrida and O. Golinelli, *Phys. Rev. A* **41** 4160 1990.

[9] B. M. Forrest and L. H. Tang, *Phys. Rev. Lett.* **64** 1405 1990.

[10] J. M. Kim and J. M. Kosterlitz, *Phys. Rev. Lett.* **62** 2289 1989.

[11] J. M. Kim, M. A. Moore and A. J. Bray, *Phys. Rev. A* **44** 2345 1991.

[12] J. M. Kim, A. J. Bray and M. A. Moore, *Phys. Rev. A* **44** R4782 1991.

FRACTAL TEAR LINES OF PAPER AND DIRECTED POLYMERS

János Kertész, Viktor K. Horváth and Ferenc Weber
Institute for Technical Physics, HAS, Budapest, P.O. Box 76, H-1325, Hungary

Abstract. We have studied the morphology of fracture lines in paper by tensile straining of sheets with constant velocity until rupture. The resulting quasi one-dimensional surfaces (tear lines) show scaling correlations over more than two orders of magnitudes. The roughness or Hurst exponent is 0.67 ± 0.05 which agrees well with the corresponding value of the directed polymers[1] in $1 + 1$ dimensions. Inspite of the sensitivity of the properties of paper to minor details of manufacturing, this exponent seems to be universal for a wide variety of papers.

1. Introduction

Fracture is one of the common processes leading to fractal patterns [1,2,3,4]. Most of the effort to understand the morphology of fracture have been concentrated so far on the concept of self-similarity although it have been pointed out that self-affinity should often be the adequate framework to interpret the scaling properties of the occuring structures [4,5,6]. The apparent global one-dimensionality of tear lines of paper sheets suggests that, if fractality occurs at all, it should be of self-affine chracter.

It is very difficult to find a precise definition of paper [7]. From our point of view, it should be considered as a very inhomogenous, anisotropic quasi two-dimensional material consisting mainly of plant fibres. Paper has extremely variable physical (mechanical, optical, electrical, thermal) properties depending on the ingredients, the preparation history, the temperature, on the humidity etc. According to paper scientists it is not very hopeful to hunt for universal behavior in such a variable kind material... However, this is just the purpose of the present contribution.

---

[1] Remark for paper scientists: Directed polymer is just a *terminus technicus* in statistical physics and it has not much to do with cellulose chains.

We have found that tear lines produced by tensile stress on different kinds of paper sheets are self-affine objects and the Hurst exponent characterizing their scaling properties is about 2/3. This value is remarkably close to the corresponding exponent of the so called 'directed polymer' problem which may be a model for the fracture geometry.

In the next Section we summarize some information about the mechanism of breaking of paper and collect arguments in support of the relation to the directed polymer model. In Section 3 the experiments are described and finally we give a short discussion.

2. Breaking mechanism and directed polymers

It has been a long debate in paper science whether the breaking of interfibrous bonds or the rupture of the fibres themselves is more important in explaining the failure mechanism. Zero span tension experiments show that the forces to break the fibres are larger than the interfibrous forces. The so called 'percentage adhesion', the ratio of the normal and the zero span strength varies from 10 to 70 % indicating a change in the rupture mechanism. On the other hand careful counting of broken fibres in tensile rupture lines with normal span shows that about 70 percent of the fibres does break during rupture. It is widely accepted that H-bonds (present in the fibres and constituting the interfibrous forces) play an essential role in the mechanical properties of paper. However, "there is now ample evidence that fracture in paper is largely governed by the geometry of the network"[8] and this is the point where we would like to start from.

The physical properties within a sheet of paper averaged over a cm^2 or so vary only by a few percents but they may vary by a few tens of percents on a mm scale (which is comparable to the length of the fibres). For example, it is easy to make visible the areal density fluctuations in a sheet of paper by transmitted light where the dark parts correspond to a higher areal density and light parts to a lower one. Usually a unique relationship^2 is assumed between the mechanical properties and the areal density [9,10].

1964 Tydeman and Hyron suggested a remarkable model for the tensile rupture of paper [9]. They supposed that flaws should be located at the low density regions. These flaws were not specified, they were just considered to be stress raisers. Photographs of light transmitted paper strips were analysed by microdensitometry in order to search for directed paths across the strip for which

$$E = \sum \rho_i = \min \tag{1}$$

where ρ_i was the local density at a 1mm×0.23mm scanning spot. These paths represent weak lines across the strip. The strips were brought to tensile rupture. The

---

^{2} This unique relationship is not essential from our point of view. It is important that we assume randomly distributed flaws which serve as stress raisers.

numerical method to solve (1) was rather rudimentary and inaccurate, however the results convincingly show that the calculated and observed lines are highly correlated.

The Tydeman-Hyron model is just the zero temperature directed polymer problem of statistical physics! It has been shown that the problem can be related to the Kardar-Parisi-Zhang equation of kinetic surface roughening [11] and even exact mapping between specific surface and directed polymer models can be established [12]. In this respect the light transmitted paper is just the usual two-dimensional graphical representation of the random potential in which the conformation of the directed polymer is looked for.

For us the most important result to be taken from statistical physics is that directed polymers are self-affine lines. Denoting the coordinates of the tear line by $y(x)$ where y is the direction of tension the squared width w^2 scales as follows:

$$w^2 = <y^2>_L - <y>_L^2 \sim L^{2\zeta} \tag{2}$$

where $<>_L$ means averaging over the x values in a window of length L and ζ is the roughness or Hurst exponent. Fortunately the problem is solved exactly in the physically relevant $1+1$ dimensions [11]; $\zeta = 2/3$ was found.

It was pointed out some time ago [13] that for two-dimensional rupture due to random distribution of flaws a directed path is needed. Recently Hansen et al. [14] carried out simulations on the electrical analog of the scalar version of an elastic-perfect plastic model and on a random fuse model. They found that the lines of breakdown could be characterized by an exponent $\zeta \approx 0.7$, a value close to 2/3. They saw in this finding support of the view that rupture lines could be related to directed polymers. In a recent paper a model experiment was reported where an exponent $\sim 0.73 \pm 0.07$ was found [15].

3. Experimental results

We have carried out rupture experiments with paper sheets on an Instron tensile testing machine. We used much larger samleps then the standards of the paper industry [6]: Usually 300×450mm sheets were taken with a span of 410mm in the machine direction. A notch of 1cm in the middle of one of the edges was cut into the paper in order to initiate the crack propagation. Except of one single case out of ~ 30 experiments, the rupture lines started indeed from the notches. Special grips were constructed in order to avoid slipping. The speed of tension was 2mm/min and a record of the force vs. strain function was taken. We tested five different kinds of papers provided by the Institute for Paper Research, Hungary (two kinds of writing paper, packing paper, board, glazed).

There are bifurcation or branching points along the rupture lines clearly indicating the direction of propagation. However, the separation of the two parts unambiguously determines the true rupture lines. These were digitized by using a 400 dpi scanner. The rupture lines showed an overall directed morphology; on a small ($\sim$ mm or less) scale overhangs were observable. This is similar to the effect of the intrinsic width in surface phenomena [16] and it is expected not to influence the general scaling behavior.

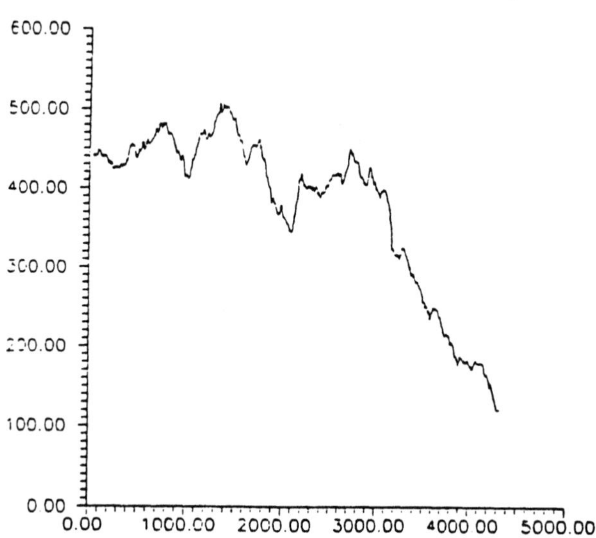

Figure 1. A digitized tear line. Note the different scales on the axes

During digitizing the SOS condition was imposed. Figure 1. shows the digitized picture of a typical rupture line of paper.

The digitized patterns were then analyzed according to (2). The log w vs. log L plots show broad straight parts while deviations occur for $L \sim$ mm values and for L close to the sample width, as expected. The roughness exponent ζ was calculated from the slope of a fitted line to the straight parts. We have carried out 2-10 experiments with the five different kinds of papers. Figure 2. shows some plots.

The obtained exponents were all in the range 0.63-0.72 while other properties like strength, density, thickness varied by factors of two or three. A more detailed presentation of the results will be given in a forthcoming publication [17].

4. Discussion

We have carried out tension experiments on five different kinds of papers and have found that the Hurst exponent characterizing the rupture lines does not depend much on the type of paper used in the experiment. This universality is remarkable in the light of the complex and variable structure of papers. The value of the exponent, $\zeta \approx 2/3$, supports the view that rupture lines in paper represent a phyisical realization of directed polymers.

The directed polymer picture is based on the assumption that the flaws already present before rupture determine the geometry of the tear line. In a disrodered system

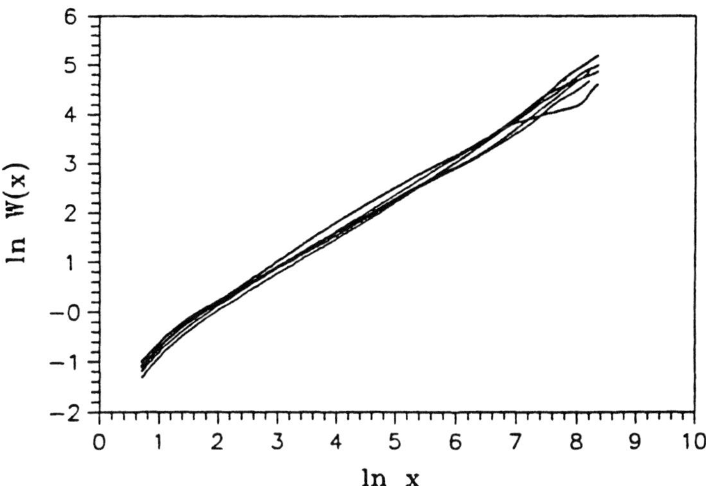

Figure 2. Scaling plots of tear lines in paper

where the random elements are ideally plastic this assumption seems to be justified [14]. Plastic behavior is definitely present in paper (as it can be seen from the stress strain diagram). Of course, it is difficult to identify the 'elements' in a network and their stress strain characteristics. The directed polymer picture can be valid if the flaws are strong enough to pin the stress even during propagation of the crack.

In recent publications [6] fracture surfaces of three dimensional ductile and brittle samples were found to be self-affine and the scaling exponent was $\zeta \approx 0.8$ irrespective of the rupture modes and the toughness values. It would be interesting to see whether the 3 dimensional analog of directed polymers (directed surfaces in random media) would be an adequate approach to explain this universality. The available data [18] seem to contradict this suggestion but no large scale simulation has been carried out in order to determine the corresponding exponent.

As far as the rupture lines of paper are concerned, there are many interesting questions to be answered about the effect of anisotropy, the spanning distance etc. Of course, we do not claim that all failure lines in paper caused by tensile stress must have the mentioned geometry. Obviously, tissue paper rather disintegrates to fibres instead of breaking. The viscoelastic behavior should certainly be taken into account when characterizing the response of paper to tension. It is possible that for fast enough straining the mechanism leading to directed-polymer-shaped rupture lines sets in. Here the time $\sim 1\text{mm}/$velocity of strain should be compared to the characteristic time scale given e.g. by the Maxwell time of viscoelasticity.

In principle the second exponent ω of the directed polymers problem could also be deterimed from experimental data. Densitometric studies *à la* Tydeman and Hyron [9] should be carried out and the scaling of $< E >_L$ from (1) should then be considered. The exponent ω is defined as $< E >_L \sim a_1 L + a_2 L^\omega$. The problem with such an investigation is – besides its difficulty – that the relation between local density, stress

273

and strength is not clear. Nevertheless, we believe that such a study would be very useful and an exponent $\omega \approx 1/3$ would reinforce the picture about the connection between directed polymers and rupture lines.

Acknowledgements

We are grateful to the Institute for Paper Research, Budapest and particularly to Attila Rab for their valuable help. Thanks are due to Alex Hansen and Tim Halpin-Healy for useful discussions. The reported research was supported by the Hungarian Science Foundation (OTKA-1218).

REFERENCES

[1] *Statistical Models for the Fracture of Disordered Media* (H.J.Herrmann and S. Roux eds.) North Holland 1990.

[2] *Disorder and Fracture* NATO ASI B 235 (J.C. Charmet et al. eds.), Plenum, New York 1990.

[3] T. Vicsek, *Fractal Growth Phenomena* (2nd edition) World Sci., Singapore 1992.

[4] R.H. Dauskardt, F. Haubensak and R.O. Ritchie, *Acta Metall.* **38** 143 1990.

[5] P. Meakin, in Ref. 1, p291 .

[6] E. Bouchard, G. Lapasset and J. Planes, *Europhys. Lett.* **13** 73 1990.
K. N. Maloy, A. Hansen, E. L. Henrichsen and S. Roux, *Phys. Rev. Lett.* **68** 213 1992

[7] *Handbook of Paper Science* Vol. 1,2 (H.F. Rance, ed.) Elsevier, Amsterdam 1982.

[8] *Handbook of Physical and Mechanical Testing of Paper and Paperboard* (R.E. Mark ed.) Vol.1., Marcel Dekker Inc., New York 1984.

[9] P.A. Tydeman and A.M. Hiron, *B.P. & B.I.R.A. Bulletin* **35** 9 1964.

[10] J.L. Thorpe, *Appita* **36** 1977 1982.

[11] M. Kardar, G. Parisi and Y.-C. Zhang, *Phys. Rev. Lett.* **56** 889 1986.

[12] L.H. Tang, J.Kertész and D.E. Wolf, *J. Phys.* A **24** L1193 1991.

[13] J. Kertész and T. Vicsek, *J. Phys.* C **13** L343 1980.

[14] A. Hansen, E.L. Hinrichsen and S. Roux, *Phys. Rev. Lett.* **66** 2476 1991.

[15] C. Potier, M. Ammi, D. Bideau and J.P. Troadec, *Phys. Rev. Lett.* **68** 216 1992.

[16] J. Kertész and D.E. Wolf, *J. Phys.* A **21** 747 1988.

[17] J. Kertész, V.K. Horváth and F. Weber (to be published) .

[18] T. Halpin-Healy, *Phys. Rev. Lett.* **62** 442 1990.

OTHER ASPECTS OF ROUGH SURFACES

THE SCALING OF ANOMALOUS SURFACE IMPEDANCE

R.C. Ball

Cavendish Laboratory, Madingley Road, Cambridge CB3 0HE, UK

Abstract. .Experiments with many rough electrode surfaces find an anomalous, frequency dependent, impedance Z_s scaling as frequency ω according to $Z_s \sim \omega^{-p}$. Assuming that this can be generalised to $Z_s \sim z_s^p$, where z_s is the bare surface impedance (i.e. Z_s for a flat surface), it is shown that $p = (\tau_2 + 2 - d)/(D + 2 - d)$ in d space dimensions for a surface of fractal dimension D , where τ_2 is the exponent governing the scaling of the second moment of its harmonic measure.
The exponent p is also calculated for a wide class of self-affine surfaces. It is found to depend not only on their roughness exponent, but also on the intermittency of self-affine features and whether narrower features are confined within broader ones.

1. Introduction

A wide range of experiments show rather clear evidence of power law dependence of surface impedance on frequency[1-10], and attempts have been made to relate this to scaling or fractal structure of the surface geometry[11-19]. It is widely assumed that the frequency dependence of the surface impedance is fundamentally capacitative in origin, coming from bare (flat) surface impedance z_s which is predominantly capacitative in series with a bulk medium of ohmic response. Then the distribution of distances to the surface gives a distribution of characteristic times which can lead to a power law impedance. This is the regime which will be pursued here.

For fractal surfaces in two dimensions Halsey and Leibig[21] have argued that the formula $p = \tau_2/D$ applies, where τ_2 is the scaling exponent for the second moment of the harmonic measure of the surface[23]. We reproduce this result below in its d-dimensional generalisation.

For self affine surfaces the disparity between physically different models with similar p exponents has always suggested that no simple relationship exists between the

277

frequency exponent p and the fractal scaling alone. An experimental studies by Bates
et al.[5,6] measured, for a range of surfaces, both the geometrical Holder exponent and
the impedance frequency exponent and found no simple correlation between them.
Further below we will show that the correct relationship is indeed more subtle.

Three classes of model will be considered. Fully fractal surfaces can certainly lead
to an anomalous exponent, though the appropriateness of such a description to typical
rough electrode surfaces is unproven. More likely is the self-affine case, where the mean
square amplitude of surface corrugation scales as

$$\langle \Delta z^2 \rangle \sim \Delta x^{2H}, \tag{1}$$

where Δx is the lengthscale of the measurement in the surface plane, and H is the
Holder (roughness) exponent. We will also consider the generalisation of this to a
surface whose rough features are intermittent, leading to $\langle \Delta z^n \rangle \sim \Delta x^{nH+c}, n > 0$.
There are cases where the intermittency is crucial to the existence of an anomalous
impedance exponent.

For a fully fractal surface, $H = 1$. For $H \neq 1$ there is a unique lengthscale x_c on
which $\Delta z \simeq \Delta x$. For $H < 1$, as typically measured for scratched and other simply
roughenned surfaces, we then have a 'shallow' regime for $x > x_c$, and a 'steep' regime
for $x < x_c$. Unfortunately it is only practical to measure H in the shallow regime,
whereas predictions of anomalous impedance from self affinity depend upon the steep
regime. There are also models based on distributions of pores of various widths and
depths, which relate rather directly to the steep self-affine regime[11-13,16,20]. Such
cases have been discussed through various levels of reduction to equivalent electrical
circuits.

2. General Problem

Assuming that we may use the laws of ohmic conduction, the governing equations for
the electrostatic potential ϕ reduce to

$$\nabla^2 \phi = 0 \text{ in the bulk,}$$

$$\phi/\zeta - \frac{\partial \phi}{\partial n} = 0 \text{ at the surface,} \tag{2}$$

where $\zeta = \sigma z_s$ is the bare surface impedance z_s scaled by the bulk conductance σ;
dimensionally ζ is a length. We set ϕ to be unity at a distant-counter electrode; the
other electrode is the rough surface of interest, which is earthed via the local surface
impedance z_s to give the boundary condition in (2).

The total impedance (scaled by σ) of the system, Z_0, is now given in terms of
the inward current by

$$\frac{1}{Z_0} = \int \frac{\partial \phi}{\partial n} dS \tag{3}$$

278

and we want to isolate the piece of Z_0 which depends on ζ (and hence z_s). This is achieved by looking at the derivative

$$\psi = \zeta \frac{\partial \phi}{\partial \zeta} \text{ which obeys } \nabla^2 \psi = 0 \text{ in the bulk, } \psi/\zeta - \frac{\partial \psi}{\partial n} = \frac{\partial \phi}{\partial n} \text{ at the surface,} \quad (4)$$

with $\psi = 0$ at the counter-electrode, so that the only source term in the ψ equations is the flux $\partial \phi / \partial n$ at the surface. In terms of the ψ field we have $d(1/Z_0)/d(\ln \zeta) = \int \partial \psi / \partial n \, dS$, which is simply the total inward current of the ψ field. These equations have a very direct physical interpretation in that the change in the current is controlled by the flux onto the surface which would have escaped with a change in the surface impedance, as discussed in terms of random walks by Halsey and Leibig[22].

The fields ϕ and ψ can be related by considering them as responses to source currents injected into the system with both electrodes earthed. The crucial point is that the Green function $G(\mathbf{r}, \mathbf{r}')$ for the voltage induced at $\mathbf{r}'$ in response to unit current input at $\mathbf{r}$ is symmetric, $G(\mathbf{r}, \mathbf{r}') = G(\mathbf{r}', \mathbf{r})$. We will make the simplifying (but unimportant) assumption that the distant counter-electrode is a sphere at radius R. Then at the rough elctrode ϕ (and hence $\partial \phi / \partial n$) is proportional to the field due to current Q injected isotropically just inside the counterelectrode. Conversely the current associated with the field ψ can be found from the response near the counterelectrode due to its current source $\partial \phi / \partial n$ at the rough electrode.

Detailed but essentially trivial manipulation then relates the dependence of the total impedance on the bare surface impedance to the second moment of the current density onto the rough electrode. In terms of the normalised flux of the original ϕ field, $E(\mathbf{x}) = Z_0 \partial \phi / \partial n$, whose surface integral is unity, the rather elegant result is that

$$\left(\frac{\partial Z_{tot}}{\partial z_s} \right)_\sigma = \frac{dZ_0}{d\zeta} = \int dS(\mathbf{x}) E(\mathbf{x})^2, \quad (5)$$

where Z_{tot} and z_s are the unscaled total and bare surface impedances.

3. Fractal Surface Case

The boundary condition of interest, (2), means that on a sufficiently fine scale near the rough electrode the field ϕ must have only weak variation. We can estimate the relevant cross-over scale for a (statistically isotropic) fractal surface by considering the lengthscale b beyond which the electrode appears to be a good absorber of the field. In a region of average potential ϕ a good absorber of size b has total flux of order $b^{d-2}\phi$, whereas a poor absorber would have total flux of order $S(b)\phi/\zeta$. Here $S(b)$ is the true surface area, given in terms of the fractal dimension D of the surface and the cutoff scale c below which it is smooth by $S(b) \simeq b^D c^{d-1-D}$. Equating the two fluxes then gives the scaling relation

$$b \simeq (\zeta c^{D+1-d})^{\frac{1}{3+D-d}}. \quad (6)$$

Beyond the scale of b the electrode does not differ significantly from one which perfectly absorbs the incoming flux, so that on these coarse scales E should correspond to the harmonic measure of the electrode strucure. In particular if the total (integrated) flux into a particular region of scale b is $q(b)$ then the corresponding local value of E should be $q(b)/S(b)$. This then leads to

$$dZ_0/d\ln\zeta = \zeta \int dS(x)E(x)^2 \simeq \zeta/S(b)\sum q(b)^2. \tag{7}$$

The sum is over the surface decomposed into regions of scale b, and in the notation of reference () scales as $(b/L)^{\tau_2}$, where L is the overall dimension of the electrode. Eliminating b in favour of the physical parameters ζ and c then gives the result

$$\frac{dZ_0}{d\ln\zeta} \simeq L^{d-2}(L/c)^{2-d-\tau_2}(\zeta/c)^p, \tag{8}$$

where the 'anomalous' exponent of the surface impedance is given by

$$p = \frac{\tau_2 + 2 - d}{D + 2 - d}. \tag{9}$$

If we assume that we can take this over to the case where z_s is purely capacitative so that ζ varies inversely with frequency ω, then p is truly the frequency exponent of the apparent surface impedance.

The analytic continuation from real poistive to pure imaginary z_s may be argued as follows. Suppose that the surface term in Z_{tot}, $Z_s \simeq z_s^{p'}$ for pure imaginary $z_s \simeq 1/i\omega$. Then we can compute the response to a voltage increasing (until some time t_1) with time t as $V(t) \simeq \exp kt$, where k is real and positive, from the Fourier decomposition of this signal. (The cutoff at t_1 cannot affect the earlier response.) This leads to $Z_s \simeq k^{-p'}$. On the other hand we can simply regard $V(t)$ as having a pure imaginary frequency $-ik$, for which z_s is real and positive leading to $Z_s \simeq k^{-p}$. The only resolution is $p = p'$.

In two dimensions the exponent relation (9) has been confirmed in simulations of Diffusion Limited Aggregates by Halsey and Leibig[21]. Such aggregates have also been grown electrochemically in both two and three dimensions and the exponent p could be checked by in situ measurements of the a.c. response of the electrical cells during growth.

One direct theoretical check on the result is presented by the hierarchically buckled fractal surface studied by Blunt and Ball in two dimensions, where to first order in a small parameter δ one has $D = 1 + \delta, \tau_2 = 1 - \delta[24]$ and $p = 1 - 2\delta[19]$, consistent with (9) to first order.

4. Self-Affine Surfaces

The moment formula (5) applies equally to self-affine surfaces, but these present a richer variety of cases of which only some lead to non-trivial impedance exponents. An important starting point is the observation that for a long channel (or pore) of width w, the laplace field decays exponentially into the channel with a penetration depth λ which varies as $\lambda \simeq w, w > \zeta; \lambda \simeq (w\zeta)^{1/2}, w < \zeta$. In the steep regime where the depth z scales as $z \simeq w^H x_c^{1-H}$, we must distinguish carefully between the penetrable case $z < \lambda$ and its converse. The comparison of z with λ is shown in Figure 1 below; a crucial difference from the self-similar case $H = 1$ is that on almost all scales the structure is either penetrable or not, never marginally so.

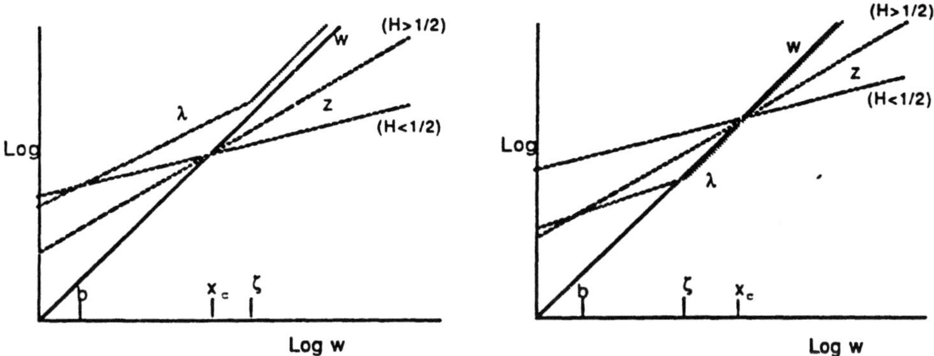

Figure 1. The height z and penetration depth λ for features vs their width w in the two cases (a) $\zeta > x_c$ and (b) $\zeta < x_c$. In case (a), all features are penetrable when $H > 1/2$ in the sense that $\lambda > w$, whereas for $H < 1/2$ there is a crossover to impenetrable behaviour at scales $w < b$, where the crossover length b depends on ζ. In case (b) there is always impenetrable behaviour below x_c but for $H > 1/2$ this crosses back to penetrable behaviour below a ζ-dependent length b.

We can also admit that the self-affine features be intermittent in the sense that only a fraction $f(w)$ of the projected surface area covered by features of scale w, and assume a scaling law $F(w) \simeq (w/y_c)^c$. This law does not at first sight appear consistent with self-similarity, but can be interpreted in terms of the self-affine structure covering a subset of a surface fractal (with dimension $d - 1 - c$) up to a scale y_c.

An important distinction lies in the accessibility of a feature, that is whether or not it lies within a larger one which could screen out the laplace field, notwithstanding the intrinsic penetrability of the smaller one. We will make a considerable further assumption that either a fraction of order unity of the smaller features lie outside the bottom of larger ones, or negligibly many do so. This may loosely be interpreted in terms of whether the surface presents fine detail on its mountains, or only in its valleys. Since for the self-affine case there is no marginal penetration, we have a simpler task in evaluating equation (5) as the flux will be relatively uniformly spread over the accessible surface, giving

$$dZ_0/d\ln\zeta = \zeta \int dS(x)E(x)^2 \simeq \zeta/T \qquad (10)$$

281

where all we need to find is T, the amount of accessible surface per unit of projected surface or 'relative area'. Note that T is at least unity, and that any calculated contribution smaller than this will be subdominant to simple flat surface contributions. Any case where T is independent of ζ gives the trivial exponent value $p = 1$.

We begin by considering the case $\zeta > x_c$, which is only interesting in the case $H < 1/2$ (otherwise the accessible surface area is independent of ζ - see Figure 1(a)). The crossover scale b below which the structure is not penetrable is given by

$$b/x_c = (\zeta/x_c)^{1/(2H-1)}, \tag{11}$$

so the contribution to the relative area due to the penetrable features is given by

$$T_1 = \int_b^{x_c} d(\ln w)(w/y_c)^c (w/x_c)^{H-1}$$

(where the last factor is the ratio of channel height to width). This is ζ-dependent only if it is dominated by its lower limit, giving $T_1 \simeq b^{c+H-1} y_c^{-c} x_c^{1-H}$ for $c + H < 1$. Thus provided y_c is not too much larger than x_c we have the non-trivial result

$$p = \frac{c - H}{1 - 2H}, \tag{12}$$

for the case $\zeta > x_c$, $H < 1/2$, $c + H < 1$ and fine detail only in the valleys. This contains the results of Blunt[20].

If there is fine detail on the mountains also, then we must consider the relative area T_2 due to all the impenentrable features below scale b. This is given by

$$T_2 = \int_{x_m}^b d(\ln w)(w/y_c)^c (a/w)^{1/2},$$

where x_m is the width of the smallest features and the last factor is the ratio of penetration depth to width. If $c > 1/2$ then T_2 is dominated by its upper limit and gives the same value as T_1 and hence the same result for p as before. If on the other hand $c < 1/2$ we obtain $T_2 \simeq (x_m/y_c)^c (\zeta/x_m)^{1/2}$ which dominates over T_1 giving the simple result

$$p = 1/2 \tag{13}$$

when $a > x_c$, $H < 1/2$, $c < 1/2$ and there are fine details on the mountains. Cases calculated by Sapoval [16] are included here. There is a further case where the calculated intercept $b < x_m$, but this is simply everywhere penetrated and gives $p = 1$.

Now we turn to the case $\zeta < x_c$ shown in Figure 1(b). Since the largest (steep) features will always be impenetrable interest lies only with the case where the mountains have fine detail. When $H < 1/2$ all features below the scale of x_c are impenetrable; then it turns out that for $c > 0$, a value of T greater than unity can only arise in the case $c < 1/2$, giving the same expression for T_2 as discussed in the $\zeta > x_c$ case and in turn the same result $p = 1/2$.

For $H > 1/2$ with $w < \zeta$ there is a crossover from impenetrable to penetrable below a crossover scale b given again by expression(11). The penetrable features now give relative area

$$T_1 = \int_{x_m}^{b} d(\ln w)(w/y_c)^c(w/x_c)^{H-1}$$

whilst the impenetrable features give

$$T_2 = \int_{b}^{\zeta} d(lnw)(w/y_c)^c(\zeta/w)^{1/2}.$$

If $c > 1/2$ so that T_2 is dominated by its upper limit then it also dominates over T_1, but the result is less than unity. Thus we consider only $c < 1/2$: for $c + H > 1$ this gives a total $T \simeq (b/y_c)^c(b/x_c)^{H-1}$ which is still les than unity, whereas for $c + H < 1$ T_1 could be large but the result is is independent of ζ. Thus we appear never to get a non trivial exponent $p \neq 1$ in the case $H > 1/2$.

The self-affine results can be summarised as follows in terms of the $c - H$ plane. There is a triangular region A delimited by $c > 1/2$, $H > 0$ and $H + c < 1$ in which p is given non-trivially by (12) for $\zeta > x_c$ but for $\zeta < x_c$, $p = 1$ applies. In the adjacent square region B delimited by $0 \leq c < 1/2$ and $0 < H < 1/2$ Sapoval's result $p = 1/2$ applies if small scale features are exposed outside larger ones; otherwise the behaviour of region A applies here also. Outside of regions A and B, at least for non-negative c and H the trivial case $p = 1$ applies.

5. Concluding Remarks

The anaomalous impedance exponent for a fractal surface (9) is of interest not just in its appplication, but also because it presents an experimental way of measuring a multifractal exponent in a physically based way. Whether higher moments of the hamonic measure are in principle accessible in related ways is an interesting and open question.

The self-affine surface results presented here are important because they relate previous disparate and highly specific results in a coherent way. For $H < 1/2$ the simple result $p = 1/2$ applies, unless either small features are shielded inside larger ones or they are reduced sufficiently in number by intermittency $(1/2 < c < 1 - H)$ leading to the new result (12). For $H > 1/2$ and also for $H + c > 1$ there is no anomaly in the surface impedance.

Acknowledgements

These calculations were stimulated by discussions with M.J. Blunt, to whom I am also indebted for critical comments including references up to 1988. I also acknowledge discussions of an early draft (1988) of the present manuscript with T. C. Halsey.

REFERENCES

[1] I. Wolf, *Phys. Rev.* **27** 755 1926.

[2] R. de Levie, *Electrochimica Acta* **10** 113 1965.

[3] P.H. Bottelberghs and G.H.J. Broers, *J. Electroanal. Chem.* **67** 155 1976.

[4] R.D. Armstrong and R.A. Burnham, *J. Electroanal. Chem.* **72** 257 1976.

[5] J.B. Bates, J.C. Wang and Y.T. Chu, *Solid State Ionics* **18&19** 1045 1986.

[6] J.B. Bates, Y.T. Chu and W.T. Stribling, *Phys. Rev. Lett.* **60** 627 1988.

[7] I.D. Raistrick, *Solid State Ionics* **18&19** 40 1986.

[8] W. Scheider, *J. Phys. Chem.* **79** 127 1975.

[9] G.J. Brug, A.L.G. Van Den Eeden, M Sluyters-Rehbach and J.H. Sluyters, *J. Electroanal. Chem.* **176** 275 1984.

[10] J. Bruinink, *J. Electroanal. Chem.* **51** 141 1974.

[11] S.H. Liu, *Phys. Rev. Lett.* **55** 529 1985.

[12] T. Kaplan and L.J. Gray, *Phys. Rev. B* **32** 7360 1985.

[13] S.H. Liu, T. Kaplan and L.J. Gray, *Solid State Ionics* **18&19** 65 1986.

[14] T.C. Halsey, *Phys. Rev. A* **35** 3512 1987.

[15] L. Nyikos and T. Pajkossy, *Electrochimica Acta* **30** 1533 1985.

[16] B. Sapoval, *Solid State Ionics* **23** 253 1987.

[17] J.P. Clerc, A.M.S. Tremblay, G. Albinet and C.D. Mitescu, *J. Physique Lett.* **19** 913 1984.

[18] M. Keddam and H. Tremblay, *C.R. Acad. Sci. Ser. 2* **302** 281 1986.

[19] R.C. Ball and M.J. Blunt, *J. Phys. A* **21** 197 1988.

[20] M.J. Blunt, *J. Phys. A* **22** 1179 1989.

[21] T.C. Halsey and M. Leibig, *Phys. Rev. A* **43** 7087 1991.

[22] T.C. Halsey and M. Leibig, *Europhys. Lett.* **14** 815 1991.

[23] T.C. Halsey, P. Meakin and I.Procaccia, *Phys. Rev. Lett.* **56** 854 1986.

[24] R.C. Ball and M.J. Blunt, *J. Phys. A* **20** 5961 1987.

ITERATION STUDY AND INFORMATION FRACTAL FOR d=2 ELECTRODES

B. Sapoval and R. Gutfraind

Laboratoire de Physique de la Matière Condensée[1], Ecole Polytechnique, 91128, Palaiseau, France

Abstract. It is shown that the iteration of a simple approximate equivalent circuit gives the impedance of a d=2 self-similar electrode. It permits to find geometrically the working set of the electrode or the information fractal.

1. Introduction

Transport across rough or porous surfaces is a basic problem in the study of the electrical response of electrodes in contact with electrolytes [1]. In the design of high current batteries it is natural to consider porous electrodes as means of increasing the net output current. More generally, any process that is limited by transport across a surface or interface can be enhanced in this way. This is probably one of the reasons why in nature, so many systems are found that have a ramified, or porous structure that can be considered as approximate examples of high surface area "fractal" structures [2]. Lung alveoli, plant roots, and villi in the human intestine are familiar examples. In the case of a membrane, neutral reacting species are brought to the surface by diffusion currents instead of electrical currents and the transport across the membrane plays the same role as the redox reaction on the electrode. The same type of problem may appear in the case of heterogeneous reactions in chemistry and hete rogeneous catalysis. In the last case molecules have to diffuse before reacting on irregular surfaces.

The problem is best stated in the electrochemical frame: A variety of experimental studies have indicated a relationship between constant phase angle (CPA) impedance

---

[1] Unité de Recherche Associée N° 041254 du Centre National de la Recherche Scientifique.

of the form

$$Z = R_o + k(j\omega)^{-\eta} \tag{1}$$

and surface electrode roughness in electrochemical cells [1,3]. In equation (1) R_o represents the electrolyte resistance, ω is the frequency of the applied voltage, j is $(-1)^{1/2}$, η is the CPA exponent and k is a constant. There have been essentially two approaches to understand this phenomenon. First, it can be attributed to a particular distribution of microscopic transport parameters on a planar surface [4]. Second it can be related to the electrode surface geometry. This is the subject of this paper. In the recent years, experimental and theoretical work has been focused on the relationship between fractal geometry and CPA electrode response [3]. Both self-similar and self-affine fractal surfaces have been explored with the objective of determining if the CPA exponent η of equation (1) is related to the dimensionality D or Hurst exponent H.

Two phenomena are involved in the (linear regime) response of a rough electrode: the effect of the finite resistivity r of the electrolyte and the finite rate of charge transfer occuring at the interface which is characterized by an admittance per unit surface $(r^{-1} + j\gamma\omega)$. The transport equations in the volume and at the surface are

$$J = E/\rho = -\nabla V/\rho \tag{2}$$

$$-J_s = V(r^{-1} + j\gamma\omega) \tag{3}$$

where E is the electric field and J the vector current in the electrolyte , J_s the outward current normal to the interface and V the local potential in the electrolyte at some point "very near" the interface. This equations must be solved together with the Laplace equation in the bulk of the electrolyte

$$\Delta V = 0. \tag{4}$$

Mathematically, the problem is to find the properties of the Laplacian field with the mixed boundary condition due to charge conservation described by (2) (3)

$$-\nabla V/V = (r^{-1} + j\gamma\omega)\rho. \tag{5}$$

A different but related problem is the study of bulk and membrane diffusion that could exists in biological fractal systems with the same geometry. It has been shown that the two problems are equivalent [5]. This equivalence has been used by Meakin and Sapoval [6] to show that for a variety of self-similar ramified fractals in d=2 the constant phase angle exponent is related to the fractal dimensionality of the electrode surface D by

$$\eta = 1/D. \tag{6}$$

This relationship between η and D is supported by a scaling argument based on the fact that the information dimension of the harmonic measure is 1 in d=2 [7]. Liebig and Halsey[8] proposed recently a slightly different relation. We show below that a very simple iteration in the complex plane permits to reach this result and to find an exact expression for the impedance. It also permits to define an "information fractal" of dimension $D = 1$ which permits a simple description of the "active" region in the fractal.

2. Iterated equivalent circuit in d=2

The purpose of this section is to show that an equivalent circuit approach permits a simple description of this complex problem in a manner that can hopefully be generalized to higher dimensions. We start with the smaller element of the 2d fractal electrode. It is supposed to be a linear piece of metal of length l and thickness b. It possesses a capacitance of impedance

$$Z_0 = (-j\gamma\omega lb)^{-1} \tag{7}$$

where γ is the specific capacitance of the interface. We consider a fractal generator made of a segment which is cut in P equal parts, of which N are kept and M identical segments are added (See Fig.1a.). The electrode is built by iteration. The fractal dimension is

$$D = ln(M + N)/lnP. \tag{8}$$

We build the electrochemical cell by iteration from the smaller element. At stage $n + 1$ the impedance Z_{n+1} of the larger unit can be represented in terms of the impedance Z_n of the cell at stage n using the equivalent circuit shown in Fig. 1b, which has an impedance given by

$$\frac{1}{Z_{n+1}} = \frac{N}{Z_n + R_n} + \cfrac{P - N}{1 + \cfrac{1}{2\beta/Z_n + \cfrac{1}{\beta' R_n + \cfrac{Z_n}{M/(P - N) - 2\beta}}}} \tag{9}$$

The numerical factors β and β' are to be discussed later. This formula is apparently complex but we will only use the low frequency and the high frequency approximations which are represented by the very simple circuits of Fig.1c and 1d. It is convenient to study this recursion in term of the reduced variable $z_n = Z_n/R_n$. The electrolyte resistances R_n are squares of side $l, Pl, P^2l, ...$ and thickness b. $R_0 = \rho l/lb = \rho/b$ and in $d = 2$ the resistances R_n does not change from one stage of iteration to the next : $R_n = R_{n-1} = ... = R_0$. Inserting the new variables z_n in (9) one obtains

$$\frac{1}{z_{n+1}} = \frac{N}{z_n + 1} + \cfrac{P - N}{1 + \cfrac{1}{2\beta/z_n + \cfrac{1}{\beta' + \cfrac{z_n}{M/(P - N) - 2\beta}}}} \tag{10}$$

It is interesting to note that the reduced variable z_0 has a modulus equal to $|z_0| = (\rho\gamma\omega l)^{-1} = \Lambda/l$ where $\Lambda = (\rho\gamma\omega)^{-1}$ is the only physical length in the problem [3]. We

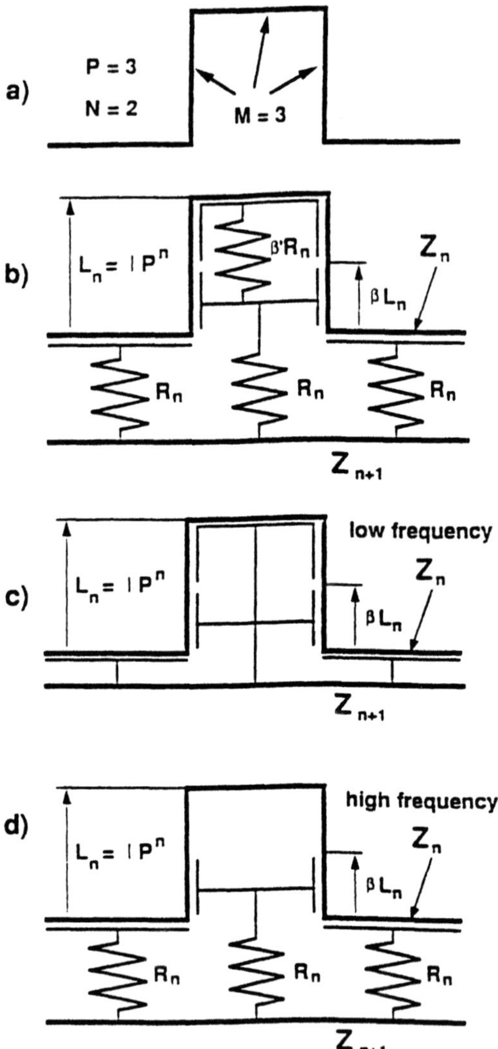

Figure 1. Fractal generator and the equivalent circuit used in the iteration. a) the fractal generator. b) the equivalent circuit. c) and d) the low and high frequency approximations of this circuit.

work at frequencies such that $\Lambda/l = |z_0| \ll 1$ and for the first stages of iteration (10) reduces to $z_{n+1} \approx z_n/(N+M)$. Whatever the values of β and β' the same dependence is obtained because in this regime the cell impedance is essentially that of the surface. The iteration can proceed this way up to a critical value $n = \nu(\omega)$ for which $z_{\nu(\omega)}$ is of the order of 1

$$z_{\nu(\omega)} \approx z_0(N + M)^{-\nu(\omega)} \approx 1. \tag{11}$$

The critical value $\nu(\omega)$ is

$$\nu(\omega) \approx Log|z_0|/Log(N + M) = Log(\Lambda/l)/Log(N + M). \tag{12}$$

After the stage $\nu(\omega)$ we reach a regime where $z_n < 1$ and the iteration procedure cannot be performed simply. In the case where $z_n < 1$ the impedance of the electrode becomes much smaller than the impedance of the electrolyte. This is a situation where the distribution of potential lines is determined by the usual harmonic potential distribution. We write

$$z_n = z_{n,yte} + z_{n,ode} \tag{13}$$

where $z_{n,yte}$ and $z_{n,ode}$ represents the electrolyte and the electrode contributions respectively. If we look for $z_{n,yte}$ for n large enough one can neglect the contribution of the electrode in the recursion. In this situation we know that the electrolyte resistance must be independent of the order of iteration (in d=2) because the electrolyte geometry keeps similar from one stage to the next. This is an important constraint which implies that (10) must approximate in this regime by

$$1/z_{n+1} \approx P/(1 + z_n). \tag{14}$$

This approximation implies that, in first approximation, β must be equal to 1/2 and that β' is large enough. The limit value for the resistance of the electrolyte is obtained by setting $z_{n+1,yte} = z_{n,yte}$ in the previous relation and one finds $z_{n=\infty,yte} \approx (P - 1)^{-1}$. But if β is equal to 1/2 and β' large, then relation (14) being a good approximation of (10) can be applied to the total impedance. The recursion relation for the electrode contribution is obtained by inserting $z_n = (P - 1)^{-1} + z_{n,ode}$ in (14)

$$z_{n+1,ode} = z_{n,ode}/P. \tag{15}$$

This result tells us that the electrode contribution is inversely proportional to the size of the electrode (which increases by a factor P at each iteration). For an electrode of size L the iteration must be performed $n(L)$ times such that the electrode size L is given by

$$L = l \cdot P^{n(L)} \quad \text{or} \quad n(L) = Log(L/l)/Log P. \tag{16}$$

From (15) the electrode impedance is given by $z_{n(L),ode} \approx P^{-n(L)+\nu(\omega)} z_{\nu(\omega),ode}$ and after substitution the admittance is found to be

$$Y = Lb\rho^{\eta-1}(\gamma\omega)^{\eta}l^{\eta-1} \quad \text{with} \quad \eta = 1/D. \tag{17}$$

Numerical simulations of the equivalent diffusion problem has shown that indeed relation (17) is verified with good precision[6,9].

3. Active zone and information fractal

If relation (17) is correct as suggested by the agreement with the numerical simulation then one should be able to construct directly the set on which the current is really absorbed (the red surfaces) using the iteration process itself. This is shown in Fig.2a. The region where the current arrives on the generator electrode is shown by the thin line. This thin line can be considered as the generator of the set on which the current arrives really on the electrode. It is the generator of a fractal object w hich can be called the "information fractal" of the corresponding electrode. The "information fractal" is shown in Fig. 2b. Its fractal dimension is exactly 1. As we have seen above the parts of the structure with a size such that $n(L) < n(w)$ are reache d uniformly. It is only for parts of large size that the screening effects which determine the structure of the information fractal play a role. At a given frequency the characteristic size $L(\omega)$ is given by the relation $n(L) = \nu(\omega)$ or $L(\omega) = l(\Lambda(\omega)/l)^{1/D}$. In Fig.2c the thick line represents for a given frequency the red zones which are effectively reached by the current at that frequency ω. In our simplified picture, the current reaches equally all the elements of the structure of size smaller than $L(\omega)$ but this is an approximation only. At a given frequency, the working region of the electrode is fractal with dimension D at small scale and fractal with a dimension 1 at large scale. In a pure diffusion problem that would be the working part of a membrane or of a catalyst. Note that this notion of active information set can be extended to arbitrary geometries.

We think that this interpretation can be applied also to random fractals like D.L.A., although in this case the interpretation cannot be so easily performed. This is supported by the results obtained by Meakin and Sapoval [6] using random walk simulations. It was in this case that the exponent $\eta = 1/D$ as well as the L dependence of the impedance admittance were first reported.

4. Comparison with a direct search for the information fractal

In the case of the usual Koch curve for which $D = ln4/ln3$ one can try to guess the generator of the information fractal and build this object by geometrical recursion. The geometry of the proposed information fractal is shown in Fig.3a. To compare this figure with the real current distribution we have calculated this distribution using a relaxation method to compute numerically the solution of the Laplace equations using the mixed boundary condition (5). This particular boundary condition is accounted for by using an additional surface impedance linking the surface sites to the sites in the bulk of the electrolyte. The calculation can be made in d.c. conditions thanks to linear response theory. The method includes a first order coarse mesh approximation as well as a fine mesh refinement near the fractal surfaces.

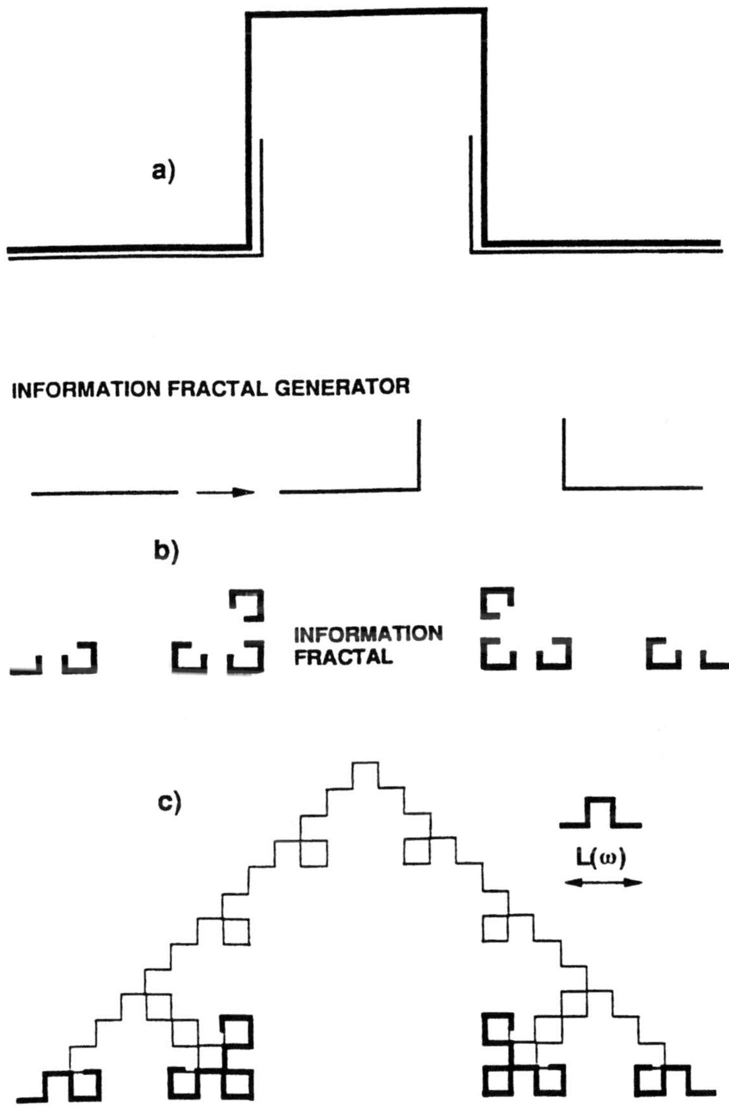

Figure 2. Definition of the "information fractal" and of the active (red) surfaces. a) the thin line represents approximately where the current arrives on the fractal generator (at infinite frequency). b) the information fractal after 3 iterations. c) the active (red) surfaces.

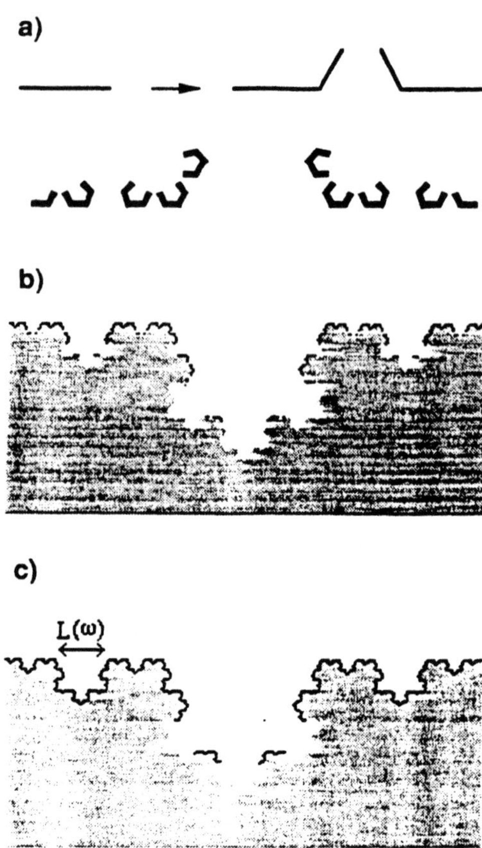

Fig. 3:"Information fractal" for the usual ($D = ln4/ln3$) Koch curve. a) an intuitive guess for the geometry of the information generator and the information fractal after 3 iterations. b) and c) Comparison with results of the numerical study of the current distribution by the relaxation method. b) corresponds to infinite freqency where $L(\omega)$ is in principle zero and c) corrseponds to the frequency for which $L(\omega)$ is shown.

Starting by the largest current density one can estimate the surface required to get a given percentage of the total current. In figure 3b and 3c the surface required for 91% of the total current is indicated. Notice that openings less or equal to $L(\omega)$ contribute almost with the total surface whereas those greater than $L(\omega)$ are penetrated almost half of the openings at each side of the triangular configuration. This confirms, in principle, the approximations made in the development of the theory. Details about the numerical method as well as the criteria used to select the percentage of the total current will be given elsewhere [10].

This research was supported by N.A.T.O. grant C.R.G.900483.

REFERENCES

[1] W. Scheider, *J. Phys. Chem.* **79** 127 1975.

[2] B. Mandelbrot, *The fractal geometry of nature* W.H.Freeman and Co, San Francisco 1982.

B. Sapoval, *Fractals* Aditech, Paris 1990.

[3] For a list of references up to 1990 see, e.g. B.Sapoval, in *Fractals and disordered systems* (Ed. A. Bunde and S. Havlin) Springer-Verlag, Heidelberg 1991 p.207.

[4] J. Ross Macdonald, *Electrochim. Acta* **35** 1483 1990and references therein.

[5] B. Sapoval, *Acta Stereologica* **6/III** 785 1987.

J. Electrochem. Soc. **137** 144C 1990.

Ext. Abst., Spring Meeting of the Electrochem. Soc. Montreal Canada, **90-1** 772 1990.

[6] P. Meakin and B. Sapoval, *Phys. Rev.A* **43** 2993 1991.

[7] P.Jones and T. Wolff, *Acta Math.* **161** 131 1988.

[8] M.Liebig and T.C. Halsey, *Europhys. Lett.* **14** 815 1991.

[9] B. Sapoval, R. Gutfraind, P. Meakin, M.Keddam and H. Takenouti,   to be published.

[10] R. Gutfraind and B. Sapoval,   to be published.

Surface Disordering:
Growth, Roughening and
Phase Transitions

EQUILIBRIUM SHAPE OF WETTING LAYERS ON ROUGH SURFACES

Peter Pfeifer and Kuang-Yu Liu

Department of Physics, University of Missouri, Columbia, MO 65211 USA

Abstract. The film-vapor interface of a liquid film, adsorbed by van der Waals forces on a surface with roughness exponent H and crossover length ~ 1 Å, is studied. It is shown that the interface is locally convex with radius equal to the Kelvin radius if the film is thick (capillary wetting), and locally concave with radius equal to the film thickness on a flat surface if the film is thin (van der Waals wetting). The adsorption isotherm is obtained in closed form from a monolayer upward. In the capillary regime, it yields the H dependence predicted by scaling arguments. In the van der Waals regime, it has an H dependence governed by three competing lengths (crossover length, pinning distance, length comparing the surface potential to surface tension). The transition between the two regimes is much slower than previously expected.

1. Introduction

When a liquid and its vapor are in contact with a solid and all three phases are in equilibrium, a question of fundamental interest is that of the position and shape of the liquid-vapor interface. Different contexts in which this question arises are:

(A) In wetting transitions on a flat surface, one studies how the interface changes, due to thermal fluctuations as the temperature is increased, from a partially wetting film to a completely wetting film (critical wetting). Or one may study how a thin completeley wetting film turns into a partially wetting droplet as the film grows with increasing vapor pressure.

(B) In films that completely wet the solid at all vapor pressures but where the surface is rough instead of flat, one would like to know how the interface changes, due to the underlying quenched disorder of the surface, as the film grows with increasing vapor pressure.

(C) For a porous solid immersed in a nonwetting liquid, one wishes to know how the interface changes as the liquid intrudes the solid under increased hydrostatic pressure.

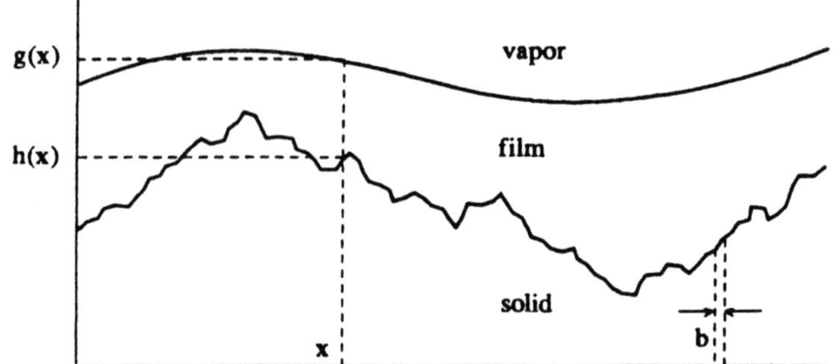

Fig. 1. Film-vapor equilibrium on a surface with roughness exponent H.

Situations B and C are of interest in applications where they are used to investigate the dramatic effect on various physical processes that disordered surfaces and confined geometries often exhibit. Adsorption experiments (B) are used to infer surface geometry, surface-adsorbate interactions, adsorbate-adsorbate interactions, and surface impurities. Intruded pore volumes (C) are the source of many experimental pore-size distributions.

Here we address Situation B, for films adsorbed by long-range van der Waals forces, which has received considerable attention recently [1-23]. Previous theoretical treatments can be classified according to whether the rough surface is modeled as random field [1, 2]; as weak perturbation of a flat surface [3, 4]; as self-similar surface [5-13]; as general self-affine surface with possible overhangs [11]; as self-affine surface without overhangs (roughness exponent H) [4, 14-17]; or as surface with nonscaling departures from planarity [4, 18, 19]. Experimental work relating to these scenarios is described in [8, 20-23]; see also [11, 12, 14-16].

The case we treat is a self-affine surface with roughness (Hölder) exponent H, $0 < H < 1$, and crossover length $b \sim 1$ Å as found in many vapor-deposited solids that are rough from atomic scales upward. Thus the surface height h(x) varies with lateral position x as

$$\sqrt{\langle (h(x + x') - h(x'))^2 \rangle_{x'}} \;=\; b \, (|x|/b)^H \tag{1}$$

for all distances $|x| \geq b$, where $\langle ... \rangle_{x'}$ is the average over all x' in the reference plane (Fig. 1). The regime $|x| \gg b$ for Eq. (1) is the global or shallow regime of the self-affine surface in which the surface scales with fractal dimension two (global dimension). The local or steep regime of (1), where $|x| \ll b$ and the surface would have fractal dimension $3 - H$ (local dimension), is physically not realized here. Film thicknesses are from a monolayer upward and hence all exceed b.[1] We assume that the roughness described by (1) extends to arbitrarily

---

[1]Earlier work [11, 16] considered also the possibility of b being substantially larger than a monolayer, in which case film thicknesses can range both in the steep and shallow regime. Experimental evidence suggests, however, that atomically rough surfaces without overhangs have b values for (1) of a few Å at most. We note that a value of $b \sim 50$ Å reported in [11, 16] for a rough Ag surface was estimated by a method different from (1) and is to be replaced by $b \sim 1$ Å [24] if b is defined by (1) as is done here.

large lengths |x|, with the understanding that any real material always possesses an upper limit for |x| (lateral correlation length) beyond which (1) no longer holds.

This leads to the following questions to be investigated in this paper. (i) Is the roughness as present in the shallow regime of (1) sufficient to alter the thermodynamics of the film in a significant (experimentally measurable) way from that on a flat surface? (ii) Does the quenched disorder desribed by H induce a transition similar to the wetting transition induced by thermal fluctuations in Situation A? The answer will depend on the value of H.

2. Pinning versus Depinning of the Interface

The starting point is the free energy of a film with arbitrary interface [3, 4, 11, 18]:

$$F = n \int d^2x \int_{h(x)}^{g(x)} dz\, U(x, z) + \sigma \int d^2x \sqrt{1 + |\nabla g(x)|^2} \ . \tag{2}$$

Here g(x) is the film height at position x; U(x, z) is the potential energy per particle in the film (at position x and height z) due to the substrate; n is the number density of the film; and σ is the film-vapor surface tension. The free energy F is for N particles relative to bulk liquid composed of N particles. The equilibrium configuration of the film is that g(x) which minimizes F subject to the condition $n \int d^2x\, [g(x) - h(x)] = N$. This yields the Euler equation

$$n\, [U(x, g(x)) - \mu] \ - \ \sigma \nabla \cdot \frac{\nabla g(x)}{\sqrt{1 + |\nabla g(x)|^2}} \ = \ 0 \tag{3}$$

where μ is the chemical potential of the vapor coexisting with the film, relative to the vapor coexisting with bulk liquid. For given parameters n, σ, μ and surface potential U, Eq. (3) completely determines the shape g(x) of the film. The shape h(x) of the surface enters through U. For explicit expressions for U when the surface is nonflat, see [3, 4, 8, 10, 11, 14, 18, 19].

When minimizing the free energy F (effective Hamiltonian) with respect to the order parameter g, we neglect thermal fluctuations which might give the interface a nonzero width (diffuse interface). In this sense, our treatment is a zero-temperature theory. This is well-justified in three dimensions where the thermal width of a free interface $g_0(x)$ only grows as

$$\sqrt{\langle\langle (g_0(x + x') - g_0(x'))^2 \rangle\rangle_{x'}} \ = \ \sqrt{\frac{kT}{\pi \sigma}}\, \ln\,(|x|/a) \tag{4}$$

with lateral distance |x|, i.e., has a thermal roughness exponent of zero [2]. $\langle\langle ... \rangle\rangle$ includes the thermal average, k is Boltzmann's constant, T is temperature, and a is the molecular diameter. In the potential U the width is even smaller. So, as is confirmed by molecular dynamics simulations, g(x) may be taken as sharp for all T away from the liquid-gas critical point.

To solve (3) on a self-affine surface, we first recall the flat-surface case. For a flat surface, $h(x) \equiv 0$, the substrate potential is $U(x, z) = - \alpha/z^3$ (non-retarded van der Waals interaction with α depending on the solid-liquid pair[2]) whence $g(x) \equiv d$ from (3) with

$$d \;=\; (-\alpha/\mu)^{1/3} \;=\; [-\alpha/(kT \ln (p/p_0))]^{1/3} \;. \tag{5}$$

This gives the film thickness d on a flat surface expressed in terms of the vapor pressure p and p_0 coexisting with the film and bulk liquid, respectively. The potential U pins the interface at distance d from the surface independently of σ because a flat film automatically minimizes the surface free energy of the film.

On a rough surface, the notion of film thickness becomes position-dependent [12] but d still sets a natural length scale for the film [8, 11], to which we refer as the "pinning distance." Indeed, U will still try to pin the interface at distance d so as to minimize the potential energy. On the other hand, the surface tension will try to depin the interface (move it farther away from the surface) so as to reduce the interface area. Thus σ acts as singular perturbation with respect to the planar film, making the interface highly sensitive to departures from planarity of the substrate. The singular nature manifests itself in that σ occurs as coefficient of the highest derivative in (3). It prevents $g(x)$ from having a simple expansion in powers of σ. In fact, the problem of separating $g(x)$ into a part due to U and a part due to σ is quite similar to the Born-Oppenheimer separation of electronic and nuclear motion into a fast electronic part and a slow nuclear part: both problems contain a small parameter multiplying the highest derivative, making ordinary perturbation theory inapplicable.

The depinning tendency of σ increases as d increases because U is weak far away from the surface. The competition between pinning and depinning generates a regime where $g(x)$ is dominated by U (van der Waals wetting at small d) and a regime dominated by σ (capillary wetting at large d) [11, 12]. What makes the surface with roughness exponent H particular is that, for d varying in the shallow regime, the competition may extend over a wide d range and yield only a weak or gradual transition from van der Waals wetting to capillary wetting.

3. The Oblique-Corner Construction

One expects that in large oblique valleys the interface is depinned so as to form a large convex meniscus (Fig. 1), which minimizes the interface area and maximizes the film-substrate attraction. At the mountain top between two valleys, the interface is pinned so as to form a small concave cap of radius ~ d, maximizing the film-substrate attraction at a small cost in energy for the cap area. We put this into a thermodynamically consistent geometric construction as fol-

---

[2]We neglect that the potential $U(x, z)$ has a repulsive part and an x dependence (atomic corrugation) at short distances z, and that it turns into a $1/z^4$ power law (retarded van der Waals interaction) at distances $z > 100\text{-}200$ Å.

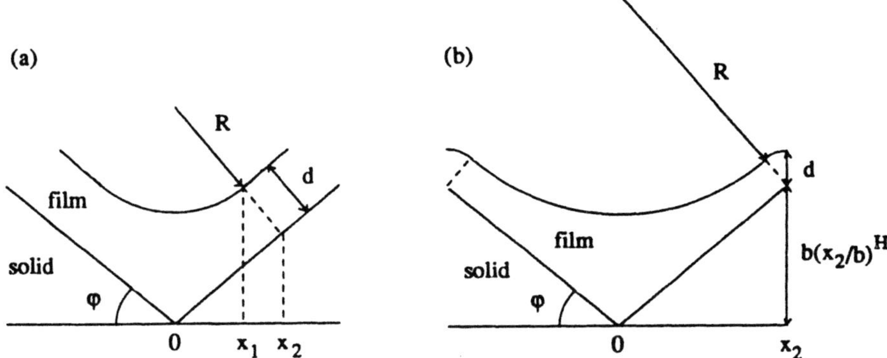

Fig. 2. (a) Film profile in a corner formed by two intersecting planes.
(b) Film profile in a valley of the self-affine surface.

lows. We first consider a single large valley, modeled as two planes intersecting at a corner with angle φ between the planes and the horizontal reference plane (Fig. 2a). For such an oblique corner, Cheng and Cole [18] have shown that the film profile is in very good approximation given by a central partial circle of radius R, joined smoothly (continuous first derivative) to two straight-line segments at distance d from the substrate. The relevant quantities for this construction, R and x_2 (Fig. 2a), are given by

$$R = d^3/d_0^2 \tag{6}$$
$$x_2 = (R + d)\sin\varphi \tag{7}$$
$$d_0 = \sqrt{\alpha n/\sigma} \tag{8}$$

[18]. From (5, 6, 8) it follows that $R = -\sigma/(\mu n)$, i.e., that R is independent of the substrate and is just the Kelvin radius at vapor chemical potential μ. The length d_0 measures the strength α of the substrate potential relative to surface tension σ. It is the nominal demarcation between van der Waals wetting and capillary wetting: for $d \ll d_0$, the Kelvin radius is much smaller than d and thus produces negligible depinning; for $d \gg d_0$, the Kelvin radius exceeds d and yields substantial depinning. At fixed d, the depinning interval $[-x_1, x_1]$ shrinks with decreasing angle φ as expected for decreasing nonplanarity of the substrate. The distance $2x_2$ is the minimum width a valley with angle φ must have to support a marginally pinned interface.

Now a self-affine surface has valleys with many different angles φ. The number of valleys varies from few for $\varphi \to 0$ (wide shallow valleys) to many for $\varphi \to \pi/2$ (narrow steep valleys). At any given vapor pressure, the film selects those valleys for filling by a convex meniscus of radius R that support a marginally pinned interface (Fig. 2b). This fixes the width $2x_2$ of the valleys by Eq. (7) and the angle φ by the condition that a horizontal excursion of x_2 on average yields a vertical excursion of $b(x_2/b)^H$, Eq. (1). The latter gives (Fig. 2b)

$$\tan\varphi = (x_2/b)^{H-1}. \tag{9}$$

The pinning at the left and right end of the valley produces a cap of radius d each.

If the surface extends across horizontal distance L (L >> max(d, R) $\geq$ b), the interface $g(x)$ on average consists of $L/(2x_2)$ segments each with profile as in Fig. 2b where R, x_2, φ are given in terms of d by Eqs. (6-9). It follows that $g(x)$ is dominated by concave caps of radius d when d << d_0, and by convex menisci of radius R when d >> d_0. From the film geometry and Eqs. (6-9) we obtain the film volume V as a function of d as

$$V = \frac{L^2 d}{2 \sin \varphi} \left[(\tan \varphi) \left(\frac{d^2}{d_0^2} + 1 \right) - \varphi \left(\frac{d^2}{d_0^2} - 1 \right) \right] \tag{10a}$$

$$\tan \varphi = \left[(\sin \varphi) \frac{d}{b} \left(\frac{d^2}{d_0^2} + 1 \right) \right]^{H-1}. \tag{10b}$$

Eq. (10) with (5) gives the film volume as function of vapor pressure p, as measured in adsorption experiments. Thus Eq. (10) provides a closed-form expression for the adsorption isotherm on a surface with roughness exponent H, crossover length b, and basal area L^2. By construction, it is uniformly valid for all $d \geq b$. All system specifics reside in the constant d_0. In the limit of low and high coverage, Eq. (10) yields the expansions

$$V/L^2 = \frac{b}{2} (d/b)^H + \pi d/4 + \dots \qquad (d \to 0, \text{ van der Waals wetting}) \tag{11}$$

$$= d + \frac{b}{6} [d^3/(bd_0^2)]^{H/(2-H)} + \dots \qquad (d \to \infty, \text{ capillary wetting}). \tag{12}$$

They show that adsorption in the shallow regime of a surface with roughness exponent H exhibits an H dependence both at low and high coverage. The transition from (11) to (12) may be interpreted as wetting transition driven by quenched disorder, similar to the transition from van der Waals wetting to capillary wetting on self-similar surfaces [11, 12, 16].

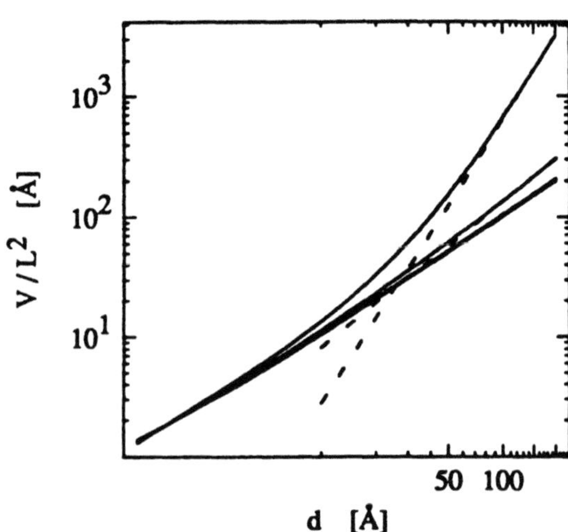

Fig. 3. Numerical calculation of the isotherm (10) for H = 0.9, 0.7, 0.5, 0.3 (full curves from top to bottom) and b = 1 Å, d_0 = 7 Å. The curves for H = 0.5 and H = 0.3 practically coincide at this resolution. The dashed straight lines are the asymptotic power law (12) for H = 0.9 and H = 0.7.

However, for $0 < H \le 1/2$, Eq. (12) predicts that the film volume for large d grows linearly with d, as it would on a flat surface, and has an H dependence only through correction to scaling. Thus, for $H \le 1/2$ the transition is a very weak one and the interface remains pinned at all pressures. For $1/2 < H < 1$, the surface roughness is sufficient to make the H-dependent term in (12) the leading one. It makes the film volume grow nonlinearly with d, corresponding to a depinned interface. This power law,

$$V(d) \propto d^{\max(1, 3H/(2-H))} \tag{13}$$

for $d \to \infty$, has been obtained previously from scaling arguments [14, 15], from mean interface positions [1, 2] and the assumption that the interface has the same roughness exponent H as the surface [16], and from renormalization-group calculations [17]. It does not support two other power laws for V(d) discussed in [16].

The first term in the low-coverage result (11) scales like the volume for van der Waals wetting in the *steep* regime [11, 16], which is unexpected because the construction underlying Eq. (10) is tailored to the shallow regime. The resulting H dependence of V for $d \ge b$ may be interpreted as "spillover" of the steep-regime behavior into the shallow regime, due to the competition of the two lengths d and b. This competition is seen in the first two terms of (11) which are comparable for $d \sim b$. In addition, there is a strong competition between d and the nominal beginning of the capillary regime, d_0. Indeed, typical values of d_0 range between 5 and 10 Å [18]. This makes the nominal van der Waals regime $b \le d \le d_0$ very short and generates an additional H dependence from the nearby capillary regime. The isotherm at low coverage is therefore governed by the competition of all three lengths b, d, d_0 and exhibits a more complicated H dependence than at high coverage.

We explore this dependence in Fig. 3 by numerical solution of (10). The value $d_0 = 7$ Å is for nitrogen on silver [12] and applies to the experiments in [8, 21]. The result is quite striking: if for $H > 1/2$ the van der Waals regime is defined operationally as the d range in which the asymptotic behavior (13) is not reached yet, it extends up to $d = 100$ Å for $H = 0.7$ and 0.9. This is very large in view of the smallness of d_0 and is accompanied by an accordingly strong H dependence of V(d) at $d < 100$ Å. [A detailed analysis of V(d) at $d < 100$ Å will be given elsewhere.] It leads to the conclusion that the transition to surface-tension dominated wetting for $H > 1/2$ is much slower than previously expected [14, 15] and that the pure power law (13) with $H > 1/2$ is unlikely to be experimentally observable (the current limits on d, dictated by vapor pressure control, are 50-100 Å).

The slowness of the transition reflects the fact that the surface is still close to the case where the interface remains pinned. For $H \to 1$, the surface geometry approaches that of a self-similar surface and the sharpness of the transition increases to that of the self-similar case at fractal dimension $D \to 2$. Indeed [11, 12], on a self-similar surface one has $V(d) \propto d^{3(3-D)}$ for $d \to \infty$ which at $D = 2$ agrees with (13) at $H = 1$. The singular limit $D \to 2$ is discussed in [12]. Thus, the wetting properties of a surface with roughness exponent H are intermediate in a well-defined sense between those of a flat surface and of a self-similar surface.

Acknowledgments

Acknowledgment is made to the Donors of The Petroleum Research Fund, administered by the American Chemical Society, for partial support of this work.

Note added in proof. There is an error in Fig. 3 in that the dashed straight line for H = 0.7 should pass through the points $(d, V/L^2) = (1\ \text{Å}, 10\ \text{Å})$ and $(d, V/L^2) = (200\ \text{Å}, 100\ \text{Å})$. This makes the conclusion in the text, that the power law (13) is experimentally not observable for H > 1/2, even stronger.

REFERENCES

[1] R. Lipowsky and M.E. Fisher, *Phys. Rev. Lett.* **56**, 472 (1986).

[2] M.E. Fisher, *J. Chem. Soc., Faraday Trans. 2* **82**, 1569 (1986).

[3] D. Andelman, J.F. Joanny and M.O. Robbins, *Europhys. Lett.* **7**, 731 (1988).

[4] M.O. Robbins, D. Andelman and J.F. Joanny, *Phys. Rev. A* **43**, 4344 (1991).

[5] P.G. de Gennes, in *Physics of Disordered Materials* (D. Adler et al. eds.), Plenum Press, New York, 1985, p. 227.

[6] S.M. Cohen, R.A. Guyer and J. Machta, *Phys. Rev. B* **34**, 6522 (1986).

[7] F. Brochard-Wyart, *C. R. Acad. Sc. Paris II* **304**, 785 (1987).

[8] P. Pfeifer, Y.J. Wu, M.W. Cole and J. Krim, *Phys. Rev. Lett.* **62**, 1997 (1989).

[9] P. Pfeifer, M. Obert and M.W. Cole, *Proc. Roy. Soc. Lond. A* **423**, 169 (1989).

[10] E. Cheng, M.W. Cole and A.L. Stella, *Europhys. Lett.* **8**, 537 (1989).

[11] P. Pfeifer and M.W. Cole, *New J. Chem.* **14**, 221 (1990).

[12] P. Pfeifer, J. Kenntner and M.W. Cole, in *Fundamentals of Adsorption* (A.B. Mersmann and S.E. Scholl eds.), Am. Inst. Chem. Eng., New York, 1991, p. 689.

[13] G. Giugliarelli and A.L. Stella, *Physica Scripta* **T35**, 34 (1991).

[14] M. Kardar and J.O. Indekeu, *Europhys. Lett.* **12**, 161 (1990).

[15] M. Kardar and J.O. Indekeu, *Phys. Rev. Lett.* **65**, 662 (1990).

[16] P. Pfeifer, M.W. Cole and J. Krim, *Phys. Rev. Lett.* **65**, 663 (1990).

[17] H. Li and M. Kardar, *Phys. Rev. B* **42**, 6546 (1990).

[18] E. Cheng and M.W. Cole, *Phys. Rev. B* **41**, 9650 (1990).

[19] E. Cheng, J.R. Banavar, M.W. Cole and F. Toigo, *Surf. Sci.* **261**, 389 (1992).

[20] C.L. Wang, J. Krim and M.F. Tony, *J. Vac. Sci. Technol. A* **7**, 2481 (1989).

[21] R. Chiarello, V. Panello, J. Krim and C. Thompson, *Phys. Rev. Lett.* **67**, 3408 (1991).

[22] I.M. Tidswell, T.A. Rabedeau, P.S. Pershan and S.D. Kosowsky, *Phys. Rev. Lett.* **66**, 2108 (1991).

[23] P. Pfeifer, G.P. Johnston, R. Deshpande, D.M. Smith and A.J. Hurd, *Langmuir* **7**, 2833 (1991).

[24] S.R. Lautenschlager and P. Pfeifer, Unpublished.

Surface Disordering:
Growth, Roughening and
Phase Transitions

DIFFUSION-LIMITED GROWTH AND MORPHOLOGICAL CHANGE IN BACTERIAL COLONY FORMATION

Mitsugu Matsushita, Masahiro Ohgiwari and Tohey Matsuyama*

Department of Physics, Chuo University, Kasuga, Bunkyo-ku, Tokyo 112, Japan
**Departmentof Bacteriology, Niigata University, School of Medicine, Asahimachidori, Niigata 951, Japan*

Abstract. Colonies of bacterial species called *Bacillus subtilis* have been found to grow two-dimensionally and self-similarly on thin agar plates through diffusion-limited processes in a nutrient concentration field. It is also found that the colony patterns change drastically when varying environmental conditions; concentrations of nutrient C_n and agar C_a in the agar medium. They are classified into five types; DLA-like, Eden-like, DBM-like, intermediate and homogeneously spreading. Physical implication of the pattern change is discussed.

1. Introduction

We are surrounded by various patterns and shapes, ranging from soot particle aggregates, snowflakes and mineral dendrites to clouds, mountains and coastlines. One wonders if there are any unifying principles underlying the formation of these diverse patterns and shapes found in nature. During the past decade or so there has been an upsurge of interest in the nonequilibrium growth phenomena, which has, above all, led to the development of a widely applicable framework of interfacial pattern formation in diffusion-limited growth in physical and chemical systems [1-3]. Here we consider the application to biological systems.

The pattern formation in biological systems is usually believed to be much more complicated than that in physical and chemical systems. Biological phenomena take place through complex intertwinement between inherently complex biological factors and environmental (physico-chemical) conditions. Setting morphogenesis of individual organisms aside, however, pattern formation of population of simple biological objects

303

may in some cases be dominated by purely physical conditions. Let us here pay attention to the colony formation of bacteria which are one of the simplest biological objects.

Usual bacteria such as *Escherichia (E.) coli* and *Salmonella (S.) typhimurium* are single cell organisms. Even very small number of bacteria (parent cells), once they are inoculated on the surface of an approapriate medium such as semi-solid agar with enough nutrient and incubated for a while, repeat the growth and cell division many times. Eventually the cell number of the progeny bacteria becomes huge, and they usually swarm on the medium to form a visible colony. The colony differs in size, form and color according to bacterial species [4]. It also changes its form sensitively with the variation of environmental conditions. The study of bacterial colonies seems to lead us into the zoo of growing random patterns. For instance, Matsuyama *et al.* [5] demonstrated that bacteria called *Serratia (S.) marcescens* exerts fractal morphogenesis in the process of spreading growth on an agar surface.

Here we discuss our recent experimental study [6-8] on a part of this rich-in-variety bacterial colony formation from the viewpoint of physics of random pattern formation. In order to study the morphological change due to environmental conditions in bacterial colony formation, we varied only two parameters; concentrations of nutrient (peptone) C_n and agar C_a in a thin agar plate as the incubation medium. Other parameters such as temperature (35°C) were kept constant. Also, throughout the present experiments we used one bacterial species called *Bacillus (B.) subtilis* only. Even the strain (OG-01) has been fixed throughout our experiments [6-8]. This strain is rod-shaped with flagella and motile in water by collectively rotating them. We also used immotile mutants (OG-01b) with no flagella in order to study the effect of bacterial cell movement on the surface of solid agar medium. Otherwise the experimental procedures are standard and were described elsewhere [6-8].

2. Diffusion-Limited Colony Growth

Present bacterial colonies grow two-dimensionally on an agar surface because the average pore size of the agar-gel network is smaller than the size of bacteria used. Bacterial cells cannot move into the agar medium.

The agar concentration C_a was fixed at some value in the range of $10 \sim 15$ g/l. Resultant agar plates are rather hard, and bacterial cells cannot move around on the surfasce of the plates. They only grow and perform cell division locally by feeding on the nutrient peptone.

We first observed the effect of nutrient on the growth of colonies [6]. They had been incubated for three weeks after inoculation at various initial nutrient concentrations C_n in the range of $0 \sim 1$ g/l. An increase of C_n in the range was found to enhance the colony growth. In particular, the growth does not occur without nutrient ($C_n = 0$). This implies that the bacterial growth and cell division at the interface of colonies (local growth processes) are governed primarily by the presence of nutrient where they live.

Now we fix the initial nutrient concentration at $C_n = 1$ g/l. Note that this concentration is much less than usually used, *e.g.*, 15 g/l. We then observed colony patterns with characteristically branched structure after 3 or 4 week incubation. They are clearly reminiscent of two-dimensional (2-d) DLA clusters. The DLA (diffusion-limited aggregation) model was first proposed by Witten and Sander [9] as a prototype model for random patterns grown through diffusion-limited processes [10]. Averaged over about 25 samples of colony patterns from the same strain, we obtained a fractal dimension of $D = 1.72 \pm 0.02$. This is in good agreement with that of 2-d DLA clusters.

In addition, we observed clear evidence for the existence of the screening effect that protruding exterior branches of a colony grow more and prevent interior ones from growing. We also observed repulsion behavior of two neighboring colonies inoculated at two points simultaneously. These behaviors are characteristic for the pattern formation in a Laplacian field.

Taking all these experimental results into account, it is now clear that the present bacterial colonies grow through DLA processes. There still remains one more problem of what makes the Laplacian field, or what diffuses. There are two main possibilities: (1) Nutrient diffuses in toward a colony, or (2) some waste material excreted by bacteria themselves diffuses out from the colony. Both cases are equally possible to produce the DLA-like pattern, although the effect of nutrient concentration described above strongly supports the first one. In order to determine which possibility takes place mainly, we put the nutrient at a corner of an incubation dish (elsewhere no nutrient initially), and inoculated bacteria at the center. It should be remembered that in the present case bacterial cells cannot move around on the surface of an agar plate. Nevertheless, the colony showed a clear tendency to grow toward where the nutrient was initially prepared. This result inevitably leads us to take the first possibility. We can, therefore, conclude that the present bacterial colonies grow through the *DLA* mechanism in the *nutrient* concentration field [7].

We have also confirmed the DLA-like colony morphology for common rod-shaped bacteria, *E. coli* and *S. typhimurium* [11]. This implies that the DLA growth is quite universal in the formation of bacterial colonies.

3. Morphological Change in Colony Formation

Let us next discuss the morphological change of colonies under the variation of environmental conditions. Colony patterns were found to change drastically when varying the concentrations of nutrient C_n and agar C_a. Figure 1 shows the phase diagram of various colony patterns observed in the present experiment. The abscissa indicates the inverse of agar concentration, which implies the softness of agar medium. Note also that the ordinate is represented by a logarithmic scale. Namely, agar plates become softer as the diagram is followed from left to right, while they become more nutrient-rich from bottom to top.

Here we classify colony patterns into five types, each of which was observed in the region labeled as A-E in Fig. 1, respectively. Typical patterns observed in the regions are also inserted in the figure.

In the region A, *i.e.*, at low C_n (poor nutrient medium) and high C_a (hard agar plates), we obtained DLA-like colony patterns, as described in the previous section.

Moving from the region A to B, *i.e.*, increasing C_n with C_a more or less fixed, colony patterns showed gradual crossover from DLA-like to Eden-like at high C_n (rich nutrient medium). The branch thickness of the colony increased gradually as C_n was increased.

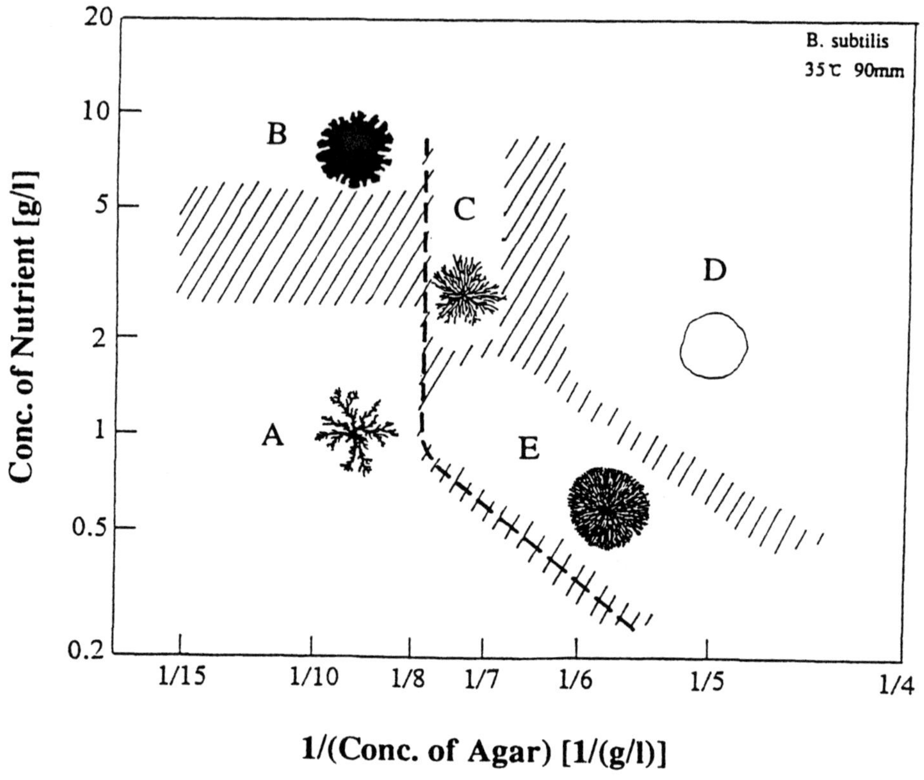

Fig. 1. Phase diagram of morphological change in colonies of *B. subtilis*.

Eventually branches fused together to form a compact pattern. However, an outwardly growing interface looks still rough. Moreover, two colonies inoculated simultaneously at two points grew and came close together in the region B, in contrast to the region A [6]. This implies that the diffusion length ℓ of the nutrient is very small. These properties are characteristic to the Eden growth. In order to confirm real Eden growth, however, one has to show that the growing interface is self-affine and obeys characteristic dynamic scaling [12].

In the wide region D with low C_a (soft agar medium) colonies spread confluently with smooth growing interface and no branching at all. They grew very fast and looked very transparent. The growth mechanism will be discussed from the population dynamics viewpoint in the final section.

In the narrow region E between A and D where the nutrient is poor and the medium softness is intermediate, the colony morphology took patterns clearly reminiscent of the so-called dense-branching morphology (DBM) [13,14]. Colonies in this region branch densely, while the advancing envelope looks characteristically smooth, compared with DLA-like colonies. In this sense the colony patterns seen in the region D seem as if dense branches fused together to form homogeneously spreading patterns. Up to the present any models that describe the DBM formation unambiguously have not been proposed yet, although DBM-like patterns have been observed in various systems [3,13,14]. We found, however, that two DBM-like colonies inoculated at two points simultaneously grow, come close together and then repel each other with almost constant gap distance approximately equal to the average distance of neighboring branches [6]. This implies that DBM-like patterns seen in the region E cannot be described by the difusion field of nutrient alone. We must look for an additional mechanism that smooths out the envelope of a growing branched pattern or makes the diffusion length ℓ constant. The candidates may be thin water layer and/or the variation of agar surface tension along the envelope due to some surfactant excreted by bacteria.

In the region C the morphology took something intermediate. Colonies grew with tip-splitting and narrow branching just as DBM. However, the advancing envelope was rather rough as DLA-like patterns.

4. Effect of Bacterial Cell Movement

We notice in Fig. 1 that the morphological change is conspicuous from the region A (DLA-like patterns) to E (DBM-like), compared with gradual change from A to B (Eden-like). The growth rate in the region E is also much higher than A. It took a colony about a month in A and a week in B to grow to the size of about 5 cm (half a diameter of a petri dish containing agar medium), while it took only a day in C and E and half a day in D. This suggests the essential difference in the way of colony growth. In fact, by microscope observation of growing zones (tips of outwardly growing branches or growing fronts) of colonies, we recognized two distinct types of growing processes. One is accompanied with no active movement of individual bacterial cells when the growth is taking place on the colony interface at high C_a (hard agar; regions A and B). The growth of this type is relatively "static": Only cells composing the outermost part of a colony feed on nutrient and increase their population mass by cell division. It takes more than a few days to develop a specific pattern. Cells in the inner part change to spores and enter into rest phase or hybernation. They never grow afterwards unless nutrient condition is improved again.

The other type is accompanied with the active movement of individual cells for intermediate and low C_a(soft agar; regions C, D and E). The colony growth of this type is driven by the cell movement and remarkably "dynamic": The outermost growing front of a colony is enveloped by a thin layer (thickness $\simeq 5\mu$m) of bacterial cells whose movement is very dull, while cells inside the layer move around actively and erratically. Sometimes they collide with the layer, break through it and rush out. However, the rushed-out cells become immediately dull and a part of the layer. As a consequence, the layer or frontline expands a little. Although the expansion of the frontline is not so speedy as the movement of cells inside, the growth rate of a colony as a whole is still much higher than that of the first type. Deep inside the growth front the cell movement is again inactive for the colonies grown in the regions C and E.

A thick broken line in Fig. 1 indicates the boundary beyond which the bacterial cell movement described above was observed. It should be noted that the line coincides

with almost vertical phase boundaries of morphologies between regions A and E, and
B and C in the figure. The verticality is understandable because the cell movement is
mainly influenced by the softness of agar plates. The bending from the verticality in the
lowermost part (boundary between A and E) may be attributed to the inactivity of cells
due to starvation coming from poor nutrient.

Since bacterial cells are known to be able to move around only by using their flagella
and the movement seems to induce the colony morphology change, we carried out the
same experiment described so far except by using immotile mutant with no flagella
(OG-01b). We then obtained a surprisingly simple phase diagram consisting of only the
regions A and B. Namely, the morphological change observed is only the gradual crossover
from DLA-like to Eden-like colony patterns in all the range of the agar concentration
examined. We have not observed DBM-like nor homogeneously spreading patterns. In
other words, we found that the regions A and B in Fig. 1 expand laterally to the entire
region studied in the present experiment, and the regions C, D, and E disappear. We
can, therefore, conclude that the bacterial cell movement seen for the wild type (OG-01)
really triggers the morphological change.

5. Population Dynamics Approach

Let us here discuss the growth mechanism of homogeneously spreading patterns seen in
the region D from another viewpoint. We noticed by microscope observations that the
population density of bacterial cells inside a colony in D is low and the growing front
is obscure microscopically, compared with those in C and E. This is the reason why
colonies in D looked transparent and hard to be observed. We assume, therefore, that
the bacterial cell movement can be described in terms of diffusion, although according
to close observations their motion does not look completely Brownian but somewhat collective. The spatio-temporal variation of population density of bacterial cells $b(\mathbf{r}, t)$ and
concentration of nutrient $n(\mathbf{r}, t)$ is then represented by reaction-diffusion-type equations;

$$\frac{\partial b}{\partial t} = \nabla \cdot (D_b \nabla b) + f(b, n), \tag{1}$$

$$\frac{\partial n}{\partial t} = D_n \nabla^2 n - \nu b n, \tag{2}$$

where D_b and D_n are the diffusion coefficients of bacterial cells and nutirent, respectively,
ν is the consumption rate of nutrient by bacteria and $f(b, n)$ denotes the reaction term
due to local bacterial growth. D_b is in general dependent on b and n. In the following,
however, we regard D_b as a constant. Moreover, we assume that $f(b, n)$ can be described
by the following logistic-like form;

$$f(b, n) = (\varepsilon n - \mu b)b, \tag{3}$$

where the term εn represents the growth rate of individual cells and μ is the coefficient of
competition among cells. This form is plausible because too many bacterial cells restrain
themselves from increasing their population.

Let us now consider the limiting case in which nutrient is so rich that n can be put constant, *i.e.*, $\epsilon n = \epsilon_0$ (linear growth rate of individual bacterial cells). Then eq. (1) becomes

$$\frac{\partial b}{\partial t} = D_b \nabla^2 b + (\epsilon_0 - \mu b)b. \qquad (4)$$

This is known as the Fisher equation [15]. This equation and its travelling wave solutions have been widely studied. In particular, this equation asymptotically yields isotropically spreading, homogeneous solutions with stable travelling wavefronts of constant speed in two dimensions. This may explain the simple growth pattern observed in the region D in Fig. 1. It is an interesting future problem whether the set of eqs. (1) and (2) gives rise to the interfacial instability of travelling wavefronts and eventually to DBM-like patterns.

Acknowledgments

One of the authors (M.M.) would like to thank Prof. N. Shigesada and Prof. H. Kawasaki for helpful discussions about the population dynamics approach.

REFERENCES

[1] T. Vicsek, *Fractal Growth Phenomena*, World Scientific, Singapore 1989.

[2] D. Avnir (ed.), *The Fractal Approach to Heterogeneous Chemistry*, Wiley, Chichester 1989.

[3] E. Ben-Jacob and P. Garik, *Nature*, **343** 523 1990.

[4] P. Singleton and D. Sainsbury, *Introduction to Bacteria*, Wiley, New York 1981.

[5] T. Matsuyama, M. Sogawa and Y. Nakagawa, *FEMS Microbiology Letters*, **61** 243 1989.

[6] H. Fujikawa and M. Matsushita, *Journal of the Physical Society of Japan*, **58** 3875 1989; *ibid.*, **60** 88 1991.

[7] M. Matsushita and H. Fujikawa, *Physica A*, **168** 498 1990.

[8] M. Ohgiwari, M. Matsushita and T. Matsuyama, *Journal of the Physical Society of Japan*, **61** 816 1992.

[9] T. A. Witten and L. M. Sander, *Physical Review Letters*, **47** 1400 1981.

[10] P. Meakin, in *Phase Transitions and Critical Phenomena*, Vol. 12 (ed. by C. Domb and J. L. Lebowitz, Academic Press, New York 1988) p. 335.

[11] T. Matsuyama and M. Matsushita, *Applied and Environmental Microbiology*, **58** 1992 (to be published).

[12] F. Family and T. Vicsek (ed.), *Dynamics of Fractal Surfaces*, World Scientific, Singapore 1991.

[13] Y. Sawada, A. Dougherty and J. P. Gollub, *Physical Review Letters*, **56** 1260 1986.

[14] D. Grier, E. Ben-Jacob, R. Clarke and L. M. Sander, *Physical Review Letters*, **56** 1264 1986.

[15] J. D. Murray, *Mathematical Biology*, Springer-Verlag, Berlin 1989.

CONTRIBUTED TALKS

1) H. van Beijeren: On the free energy of surfaces (invited)
2) J. Chevrier: Epitaxial growth of iron on silicon
3) M. Siegert: Instability in surface growth with diffusion
4) S.M. Bhattacharjee: Directed polymers and vertex models
5) T. Hwa: Nonequilibrium dynamics of driven lines
6) B. Sapoval: Vibrations of fractal drums
7) V.K. Horváth: Output noise analysis of growing surfaces
8) J.-P. Bouchaud: The roughening transition in the presence of inpurities
9) M. Paczuski: Dynamics of 1d Toom interfaces
10) M.C. Desjonquéres: Energetics of an adatom on stepped transition metal surfaces
11) J. Villain: Experiments vs. theory of MBE
12) A. Pavlovska: Disordering of metal surfaces studied by in situ UHV electron microscopy (with video)
13) A. Hansen: Roughness of brittle cracks
14) T. Hwa: Exact distribution of rare events for 1+1 dimensional directed polymers in random media
15) J. Willemsen: Kolmogorov-scaling of gravity waves
16) S. Buldyrev: Anomalous interface roughening in porous media: Experiment and model
17) Y.-C. Zhang: Complexity measure and its applications
18) A.-L. Barabási: The 3d Toom model and anizotropic KPZ

LIST OF PARTICIPANTS

PARTICIPANT	ADDRESS	E-MAIL ADDRESS
R. AHMADI	2-240 Knudsen Hall, UCLA 485 Hillgard Ave, LA, CA 90024, USA	ahmadi@ucla.physic.edu
S. BALIBAR	Lab. Physique Stat., ENS, 24 rue Lhomond, 75231 Paris Cedex 05, France	balibar@frulm63
R. BALL	Cavendish Lab, Madinglay Rd. Cambridge, C83 0HZ, UK	—
A.-L. BARABÁSI	Center for Polymer Studies Boston Univ, MA 02215, USA	alb@buphy.bu.edu
E. BAUER	Physikalisches Institut TU Clausthal, Germany	—
J. BECHHOEFER	Physics Dept., Simon Fraser Univ. Burnaby, BC, V5A 196, Canada	bech@chaos.phys.sfu.ca
S.M. BHATTACHARJEE	Institute of Physics Bhubaneswar-751005, India	sb% iopb@shakti.ernet.in
J.-P. BOUCHAUD	Lab. Physique Stat., ENS, 24 rue Lhomond, 75231 Paris Cedex 05, France	bouchaud@frulm62
R. BRUINSMA	Phys. Dept., UCLA, Los Angeles CA, 90024, USA	ilk6bru@uclamvs
S. BULDYREV	Center for Polymer Studies Boston Univ, MA 02215, USA	sergey@buphy.bu.edu
J. CHEVRIER	CRMC2 Marseille Case 913 Campus de Luminy, F-13288 Cedex 9, France	derrien@frmop11
M. CIEPLAK	Inst. of Phys., P.A.S. 02-681 Warsawa, Poland	cieplak@planifg1
G. COMSA	IGV-KFA, P.O.Box 1913 W-5170 Jülich, Germany	—
B. DERRIDA	SPT, CE Saclay, F-91191 Gif-sur-Yvette, France	derrida@poseidon.saclay.cea.fr
M.C. DESJONQUERES	DRECAM/SRSIM, Bat. 462, CE Saclay, F-91191 Gif-sur-Yvette, France	—
D. EAGLESHAM	AT&T Bell Labs, 600 Mountain Ave Murray Hill, NJ 07974, USA	dave@physics.att.com
H.-J. ERNST	DRECAM/SRSIM, Bat. 462, CE Saclay, F-91191 Gif-sur-Yvette, France	—
F. FAMILY	Physics Dept., Emory Univ. Atlanta, GA 30322, USA	phyff@emory.edu
G. FORGÁCS	Dept. Physics, Clarkson Univ. Potsdam, NY 13699, USA	forgacs@craft.camp.clarkson.edu
B. FOURCADE	I.L.L. B.P. 156X 38042 Grenoble, Cedex, France	fourcade@frill
F. GALLET	Lab. Physique Stat., ENS, 24 rue Lhomond, 75231 Paris Cedex 05, France	—

A. GEORGES	ENS (LIPTENS) 24 rue Lhomond F-75231 Paris Cedex 05, France	georges@frulm62
M. GIESEN	IGV-KFA, P.O.Box 1913 W-5170 Jülich, Germany	—
M. GRANT	Physics Dept., 3600 rue University McGill Univ., Montreal, H3A 278, Canada	grant@physics. mcgill.ca
T. HALPIN-HEALY	Physics Dept., Columbia University 3009 Broadway, New York, NY 10027, USA	healy@cuphyd.bitnet
A. HANSEN	GMCM, Bat. 11G, Universite des Rennes I F-35042 Rennes Cedex, France	hansen@servix.cicb.fr
V.K. HORVÁTH	Eötvös University, Dept. Atomic Phys. Múzeum krt 6-8, Budapest, H-1088, Hungary	h836hor@ella.uucp
T. HWA	Dept. Phys. Harvard U. Cambridge, MA 02143, USA	hwa@cmt.harvard.edu
D. IZZO	Theor. Polymerphys. Rheinstr. 12 Freiburg i. Br. W-7800, Germany	—
R. JULLIEN	Physique des Solides, Univ. Paris XI 91405 Orsay, France	jullien@frsol11
M. KARDAR	Room 12-109, Physics Dept., MIT Cambridge, MA 02139, USA	kardar@cht7.mit.edu
J. KERTÉSZ	Inst. for Techn. Phys., POB 76 Budapest, H-1325, Hungary	h2230ker@ella.uucp
M.J. KIM	Dept. of Theor Phys. Univ. Manchester M13 9PL, England	mbctpjk@cms. mcc.ac.uk
J. KRIM	Physics Dept., Northeastern Univ. Boston, MA 02115, USA	surface@nuhub.bitnet
J. KRUG	IBM Watson Research Yorktown Heights, NY 10598, USA	krug@yktvmz.bitnet
J. LAPUJOULADE	DRECAM/SRSIM, Bat. 462, CE Saclay, F-91191 Gif-sur-Yvette, France	—
R. LENORMAND	IFP BP 311 F-92506 Rueil, France	lenormand@ct.ifp.fr
M. MATSUSHITA	Dept. of Physics, Chuo Univ., Kasuga Bukyo-ku, Tokyo 112, Japan	FAX: 81-3-3817-1792
P. MEAKIN	Central Res. & Dev. Dept., du Pont & Co., Exper. Station, Wilmington DE 19898, USA	—
R. MESSIER	Pennsylvania State Univ.,Nat. Res. Lab University Park, PA 16802, USA	—
S.H. MEYER	Theor. Physics Univ. Kaiserslautern, Germany	meyer@dhdihep1
T. NATTERMANN	Theor. Physics, Univ. Köln D-5000 Köln 41, Germany	natter@thp. uni-koeln.de
M. PACZUSKI	Brookhaven National Lab. Bdng 510A Upton NY 11973, USA	maya@cmt1. phy.bnl.gov

G. PARISI	Dip. Fisica, II Università Via E. Carnevale Roma 00173, Italy	parisi@vaxtov.infn.it
A. PAVLOVSKA	Phys. Inst. Techn. Univ. W-3392 Clausthal, Germany	—
Y.-P. PELLEGRINI	DHA/SYS CESTA BP2 F-33114 Lebard, France	—
P. PFEIFER	Dept. of Physics, Univ. Missouri Columbia, MO 65211, USA	physpepf@umcvmb
A. PIMPINELLI	DRFMC/MDN, CENG 85X, F-38041 Grenoble Cedex, France	pimpinelli@gis1.cea. ceng.fr
B. POELSEMA	IGV KFA Jülich Pf. 1913 W-5170 Jülich, Germany	—
Z. RÁCZ	Phys. Dept., Clarkson Univ. Potsdam, N.Y. 13676, USA	racz@craft.camp. clarkson.edu
L.M. SANDER	Physics Dept., University of Michigan Ann Arbor, MI 48109, USA	lsander@um.cc.umich.edu
B. SAPOVAL	LPMC Ecole Politechnique, F-91128 Palaiseau, France	—
D. SAVAGE	Univ. of Wisc. 1500 Johnson dr., Madison, Wi 53706, USA	savage@cae.wisc.edu
M. SIEGERT	Physics Dept., Simon Fraser Univ. Burnaby, BC, V5A 196, CANADA	siegert@sfu.ca
H. SPOHN	Theor. Phys., Theresienstr. 37 W-8000 Mnchen, Germany	spohn@stat.physik. uni-muenchen.dbp.de
H.E. STANLEY	Center for Polymer Studies Boston Univ, MA 02215, USA	hes@buphy.bu.edu
L.-H. TANG	Inst. Theor. Phys. University W-5000 Kln 41, Germany	tang@thp.uni-koeln.de
H. VAN BEIJEREN	Inst. Theor. Phys. P.O.B. 800061 3508 TA, Utrecht, The Netherlands	—
T. VICSEK	Dept. Atomic Physics, Eötvös Univ. Puskin u. 5-7, Budapest, H-1088 Hungary	h845vic@ella.uucp
J. VILLAIN	DRFMC/MDN, CENG 85X, F-38041 Grenoble Cedex, France	—
J. WILLEMSEN	Univ. of Miami, RSMAS-4MP Miami, FL-33149, USA	willemsen@rsmas. miami.edu
E. WILLIAMS	Dept. Phys. U. of Maryland Colllege Park, MD 20742-4111, USA	williams@surface. umd.edu
D.E. WOLF	Fachbereich 10, Universität Duisburg Pf. 10 15 03, D-4100 Duisburg 1, Germany	hm345wo@duc220. uni-duisburg.de
Y.C. ZHANG	Institut de Phys. Theor., Univ. Fribourg, CH-1700, Switzerland	zhang@cfruni52.bitnet

Subject Index